版式设计指南

网格应用的基本原则

[美] 贝丝·托恩德罗（Beth Tondreau）○ 著
林子佳 ○ 译

LAYOUT ESSENTIALS

REVISED AND UPDATED

100 Design Principles for Using Grids

广西师范大学出版社
·桂林·

First published in 2009; this edition published in 2019 by Rockport Publishers,
an imprint of The Quarto Group,
100 Cummings Center, Suite 265-D, Beverly, MA 01915, USA. T(978) 282-9590 F (978) 283-2742
QuartoKnows.com

著作权合同登记号桂图登字：20−2019−190

图书在版编目（CIP）数据

版式设计指南：网格应用的基本原则 /（美）贝丝 · 托恩德罗著；林子佳译. —桂林：广西师范大学出版社，2020.6

书名原文：*Layout Essentials Revised and Updated:100 Design Principles for Using Grids*

ISBN 978−7−5598−2745−6

Ⅰ. ①版… Ⅱ. ①贝… ②林… Ⅲ. ①版式−设计 Ⅳ. ①TS881

中国版本图书馆 CIP 数据核字 (2020) 第 051371 号

策划编辑：高　巍
责任编辑：季　慧
助理编辑：马竹音
装帧设计：六　元

广西师范大学出版社出版发行
（广西桂林市五里店路 9 号　　邮政编码：541004
网址：http://www.bbtpress.com）
出版人：黄轩庄
全国新华书店经销
销售热线：021−65200318　021−31260822−898
广州市番禺艺彩印刷联合有限公司印刷
（广州市番禺区石基镇小龙村　邮政编码：511400）
开本：889mm × 1 194mm　　1/16
印张：13　　字数：153 千字
2020 年 6 月第 1 版　　2020 年 6 月第 1 次印刷
定价：88.00 元

如发现印装质量问题，影响阅读，请与出版社发行部门联系调换。

献给

帕特里克·詹姆斯·奥尼尔

（Patrick James O'Neill）

他善良、有爱心、风趣、有教养、擅长运动、爱挖苦人、有耐心、厨艺高超……至少还有92个其他优点。

导 论

“网格是页面布局中最容易被误解和误用的元素。只有根据素材量身打造的网格才能发挥作用。”

——德里克·伯索尔（Derek Birdsall）
《书籍设计的技巧》（*Notes on Book Design*）

网格系统能组织空间，并能容纳多种交流所需的一系列材料，它规定并维持秩序，但往往并不明显。网格是规划，而不是限制。

网格已经被使用了几个世纪，许多平面设计师会将网格与瑞士设计师联系在一起。瑞士设计师在20世纪40年代因对秩序的狂热创造了这种系统化的形式，它几乎可以将所有的东西可视化，但直到20世纪末，网格还被认为是单调乏味的。现在，随着大量数据和图片在多个平台快速更新，网格再次被视为必不可少的工具，无论是新手还是经验丰富的从业者都会依赖网格做设计。

这本书是一本入门读物，简单介绍了在设计中如何使用网格。这些原则中的每一条都价值非凡，有助于构建页面布局系统或交流系统，每一条都配有世界各地的设计师或设计公司在不同媒体上发布的项目图片。

这些原则通常不是单独使用的，一个项目中会包含多个原则。因此，本书举例说明了这些原则在相同项目的不同部分或完全不同的项目中是如何应用的，其中一个项目展示了交流系统的组成部分是如何遵照多个原则的。与第一版相比，本书包含了更多适用于印刷品、个人电脑端、平板电脑端等场景的设计应用的例子。

这本书中包含了一些颇有天赋的设计师的作品，这些作品富有启发性，内容扎实，令人愉悦，并且紧扣主题。我希望这本书中的项目能指导读者，激发他们的灵感，同时引导读者学到最基本的网格原则，确保作品反映出设计师或作者想要传达的思想。

目 录

“正如自然秩序系统控制着生命体和非生命体的生长和结构一样，从很早的时候起，人类活动本身就以追求秩序为特征。”

——约瑟夫·穆勒－布罗克曼
（Josef Muller-Brockmann）

“只有在所有的文字问题都解决之后，设计师才能超越网格结构所隐含的统一性，并用它创建一种动态的视觉表现，从而保持所有页面的趣味性，这才是真正的成功。”

——蒂莫西·萨马拉（Timothy Samara）

《塑造和突破网格》（*Making and Breaking the Grid*）

GETTING STARTED

迈出第一步

● 1.了解构成要素

网格的主要构成要素是分栏、模块、边距、空间区、流线和标记。开始一个新项目可能很难，从内容开始，然后设置边距和分栏，之后会根据需要调整。让我们从头说起。

分栏

分栏是垂直的容器，用于放置文字和图片。页面或屏幕上分栏的宽度与数量可以根据内容调整。

模块

模块是由连续空间分隔而成的一个个独立单元，这些单元一起形成一个重复、有序的网格。组合模块可以创建不同大小的行和列。

边距

边距是缓冲区。它是剪裁区域（包括边槽）与页面内容之间的空间。边距还可以容纳二级信息，如注释和说明文字。

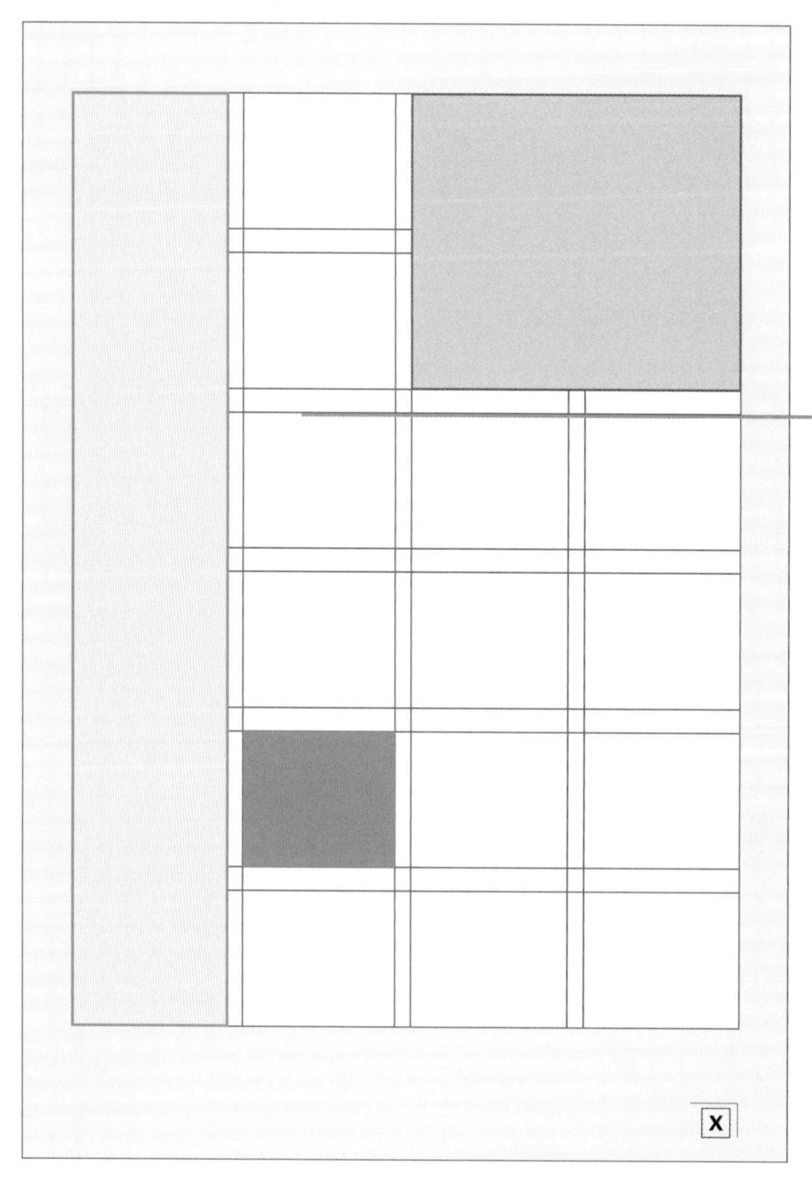

空间区

空间区是为了展示文字、广告、图片或其他信息形成的一组特定的模块或分栏。

流线

流线是将空间分隔成水平条带的横线。流线不是实际的线，而是一种利用空间和元素引导读者浏览页面的方法。

标记 X

标记帮助读者浏览文档，表示出现在每个页面相同位置的素材。标记包括页码、页首标题、脚注（页眉和页脚）以及图标。

2.了解基础结构

尽管下面的图显示了常见的网格结构，但是在基本结构上还可以有其他变化。报纸及网站的多栏网格已经从三栏扩展到五栏，甚至更多。

单栏网格

通常用于连续的文本，如论文、报告或书稿。文本框是单页面、对开页面或设备屏幕上的主要部分。

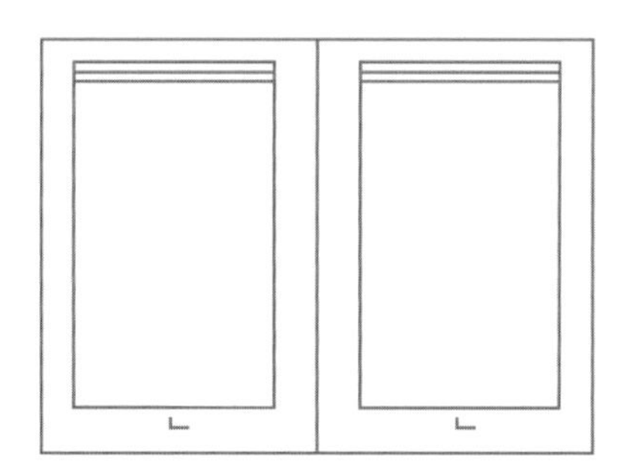

模块化网格

适用于处理报纸、日历和图表中的复杂信息。它结合了横栏和竖栏，将网格结构分隔成更小的空间块。

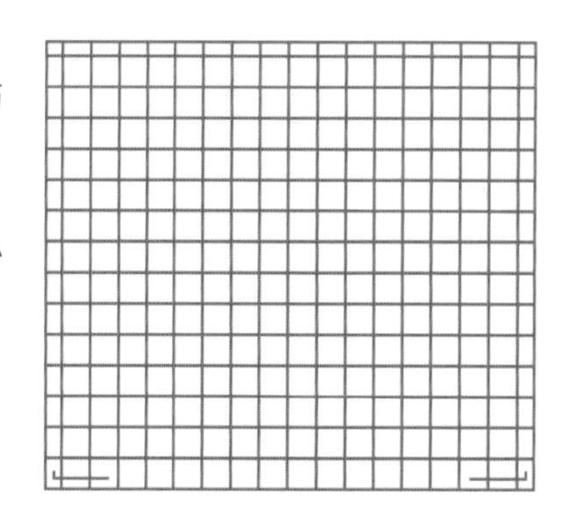

双栏网格

用于处理大量文本，或在单独的列中显示不同类型的信息。双栏网格的栏宽可以相等，也可以不等。当一栏比另一栏宽时，理想的比例是宽栏的宽度是窄栏的两倍。

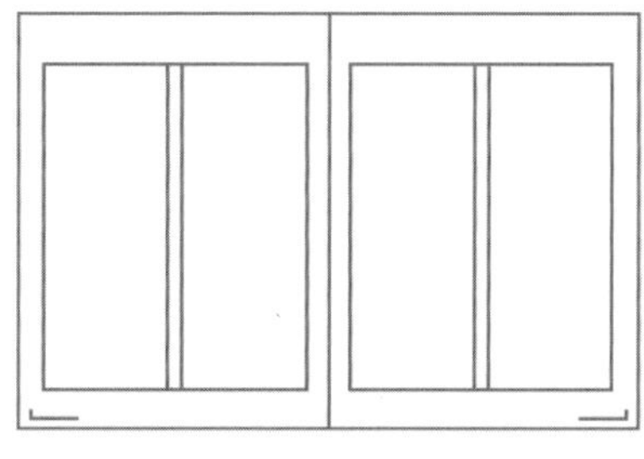

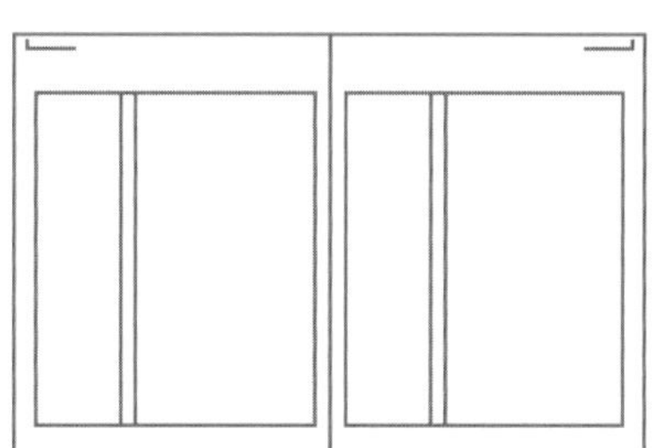

分级网格

将页面分成不同区域。为了方便读取和提高阅读效率，许多分级网格由横栏组成，有些杂志的内容就是水平排列的。

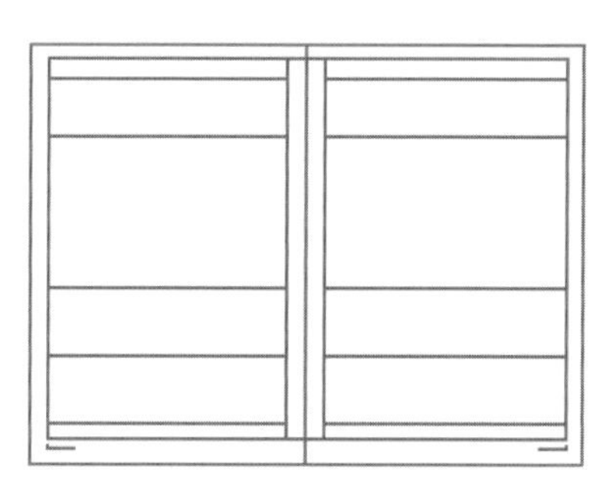

多栏网格

比单栏或双栏网格具有更大的灵活性，可以组合不同宽度的栏目，非常适用于杂志和网站。

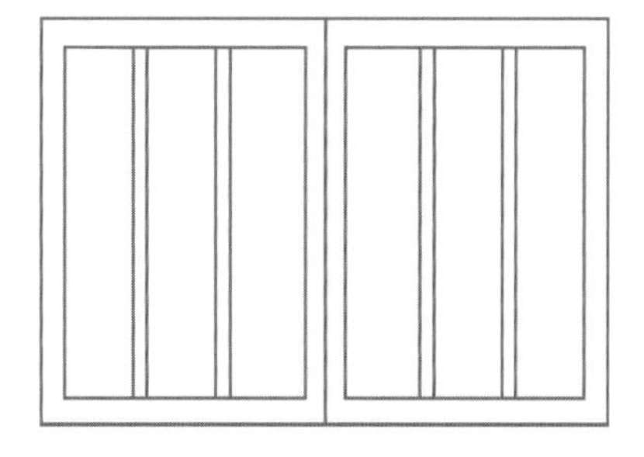

从提问开始

- 素材是什么？是否复杂？
- 目前已有哪些素材？
- 目标是什么？
- 读者（浏览者、用户）是谁？

● 3.评估素材

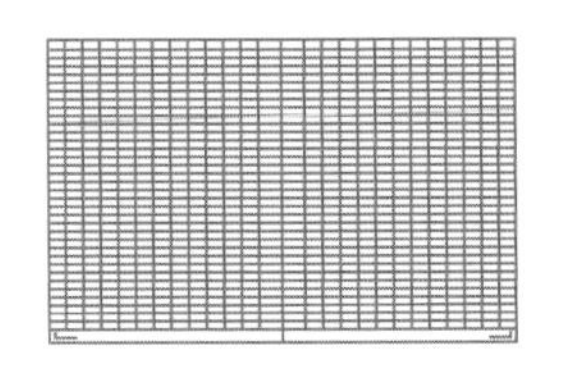

内容、边距、图片的数量、所需的页数、屏幕和面板都是决定如何设置网格的要素。首先，内容决定了网格的结构。使用哪种网格取决于特定的设计问题，但有一些普遍的准则：

在处理连续文本（如一篇文章或一本书稿）时，可使用单栏网格。与多栏网格相比，单栏网格的版式不那么复杂，更加大方，比较适合艺术图书或商品名录。

对于复杂的素材，双栏网格或多栏网格更具有灵活性。能够进一步分列的栏目可以产生最多的变体。多栏网格用于网站时，可以管理大量的信息，包括故事、视频和广告。

模块化网格适用于信息量大的素材，如日历或日程表，有助于将众多信息单元整理为可管理的模块。模块化网格也可以应用于信息区域较多的报纸。

分级网格可以水平地划分页面或屏幕，通常用于简单的网站。在这些网站中，信息模块是有序排列的，以便向下滚动页面时，更方便读者阅读。

所有网格都要创建秩序，并且要进行规划和计算。无论设计师是以像素、派卡还是毫米为单位来设计，确保所有数字都合理是创建网格秩序的关键。

项目
《好》（*Good*）杂志

客户
《好》杂志有限责任公司（Good Magazine, LLC）

设计公司
Open

设计总监
司各特•斯托厄尔（Scott Stowell）

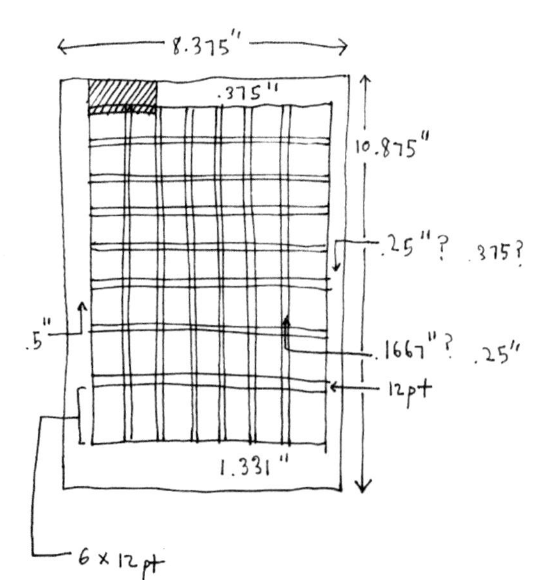

设计大师的草稿展示了网格是如何一步步成形的

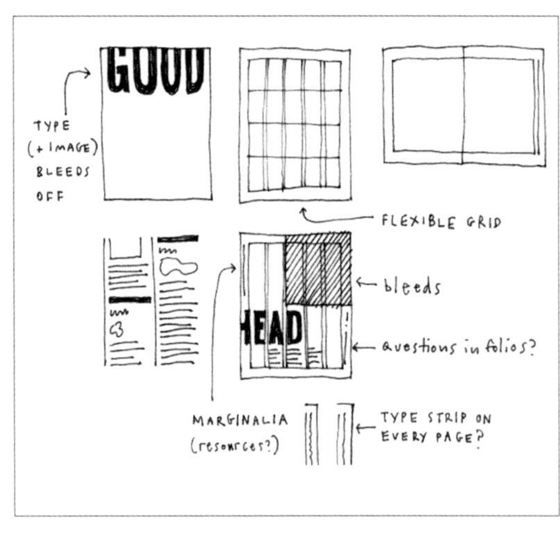

设计中的草稿显示了一本杂志可能最终形成的版式网格

● 4.计算一下

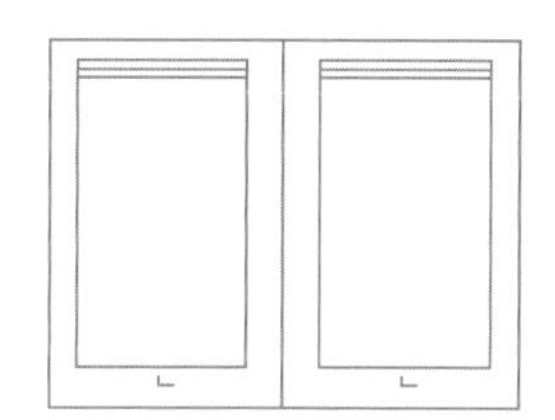

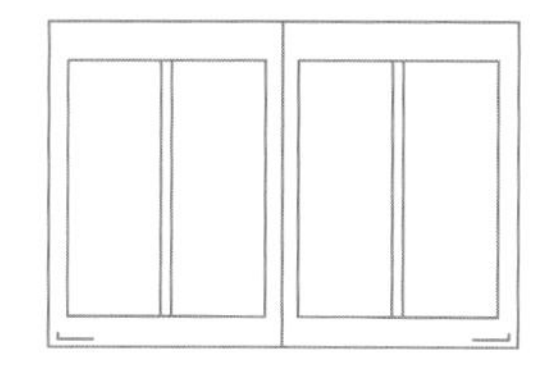

首先考虑主文本，然后分析项目的复杂性。大多数项目都有限制条件，如尺寸、页数、色彩。在关注内容的同时，也要考虑可能存在的项目标准。

一旦知道了页面或屏幕的大小和基础文本，就可以计算出各元素如何在页面上排列。如果只是文本处理，可以将文本根据分配好的页数排版。如果还需要包含图片、标题、框或图表，那么首先要确定文本所需的空间大小，之后将剩余的空间用于摆放图片、图表和其他信息。通常，要同时计算好所有元素的数量。

当确定了材料的基本处理方法及其所需空间之后，就可以深入研究标题和层次结构的细节了（参见下一个原则）。

The Stars of NGC 300

Like grains of sand on a cosmic beach, individual stars of large spiral galaxy NGC 300 are resolved in this sharp image from the Hubble Space Telescope's Advanced Camera for Surveys. The inner region of the galaxy is pictured, spanning about 7,500 light-years. At its center is the bright, densely packed galactic core, surrounded by a loose array of dark dust lanes mixed with the stars in the galactic plane. NGC 300 lies 6.5 million light-years away and is part of a group of galaxies named for the southern constellation Sculptor. Hubble's unique ability to distinguish so many stars in NGC 300 can be used to hone techniques for making distance measurements on extragalactic scales.

August 13

这本天文学图集的单栏文本呼应了外太空的概念

118 • SYMBOLS OF POWER

22. *Paumier*

JEAN-RENÉ FLAMAND, UPHOLSTERER

1809

Gilded wood, 15/16 green

H. 94 cm; W. 105 cm; D. 67 cm

Label: *Fontainebleau / Appartement de Sa Majesté / ayant son entrée sur la galerie / Cabinet intérieur* (Fontainebleau / His Majesty's apartment / with entrance to the gallery / Inner office)

Label: *Une méridienne en toille fine / carreaux plume et crain couverte d'une / ousse de 15-16 vert gros grain / Flamand* (A méridienne in fine check cloth / feather and horsehair with / 15–16 green petersham cover / Flamand)

Printed label by Chatard: *Garde Meuble de la Couronne / impérial / Aout 1809 / Service de l'Empereur a Fontainebleau / . . .* (Imperial / Crown Furniture Store / August 1809 / Emperor's Service at Fontainebleau / . . .)

Musée National du Château, Fontainebleau (inv. F 3914)

The *paumier*, also known as a *méridienne*, is a three-backed rectangular sofa, with a lower back on one side. This offers a variety of comfortable positions, either sitting normally, or reclining with the legs over the lower back, or the neck resting on it. The paumier appeared at the end of the eighteenth century and became more widespread under the Empire. Napoleon was particularly fond of resting on one.

Described as a *pommier* in the château inventories, or as a *méridienne* by the upholsterer Flamand, this seat was created by Flamand in 1809 for Napoleon's office in the château at Fontainebleau, which became a small bedroom in 1811. It is completely upholstered and stands on four very small circular, rounded feet that are decorated with wreaths of waterleafs. • Y.C.

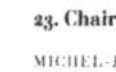

23. Chair

MICHEL-JACQUES BOULARD

1806

Mahogany, cotton, and *bazin*

H. 90 cm; W. 46 cm; D. 39 cm

Musée National du Château, Fontainebleau (inv. F 664)

Quite often an ordinary chair was used in the bathroom, as in this case. A seat specially designed for the bathroom would have a swiveling back and base to make the hairdresser's or barber's work easier, and be fitted out with cane, leather, or cloth.

This seat is one of a set of four made in 1806 for Napoleon's bathroom in the Château de Fontainebleau. Its striking feature is the extreme simplicity of its lines, accentuated by the close-fitting cotton and metal-thread-cloth cover, which is designed to protect the seat from splashes. The slightly trapezoid base consists of a simple, flat surround with no visible connecting pins. The rear feet and rectangular, slightly arched seat back form an elegant interplay of curve and counter-curve that suggests a concern for ergonomics. The rear feet are saber-shaped, while the front feet are baluster-shaped and have the seat's only decorative features, an interplay of layered moldings. • Y.C.

包含大量文本的商品名录使用双栏网格来放置文本和图片

项目
《天文学》（*Astronomy*）和《权力的象征》（*Symbols of Power*）

客户
哈里·N.艾布拉姆斯公司（Harry N. Abrams, Inc.）

设计公司
BTDNYC

设计总监
马克·拉里维埃（Mark LaRivière）

设计师
贝丝·托恩德罗（Beth Tondreau），苏珊娜·戴尔·奥尔托（Suzanne Dell' Orto），司各特·安布罗西诺（Scott Ambrosino）(仅限《天文学》)

使用单栏还是双栏网格取决于文本的内容和范围

排版技巧

文本的整体风格取决于文本的尺寸、间距、宽度和分行。连续的文稿使用同样的颜色比较便于阅读。如果文本很长，字号必须足够大，并且有足够的行距，才能保证轻松的阅读体验。如果栏宽很窄，为避免部分文字字距过大，可以把字号调小，或者让文字左对齐。因为不同的字体设置不同，字号的大小没有完美的标准。例如，10磅的Helvetica字体看上去比10磅的Garamond字体要大。字体非常值得研究。

5.对读者友好一点儿

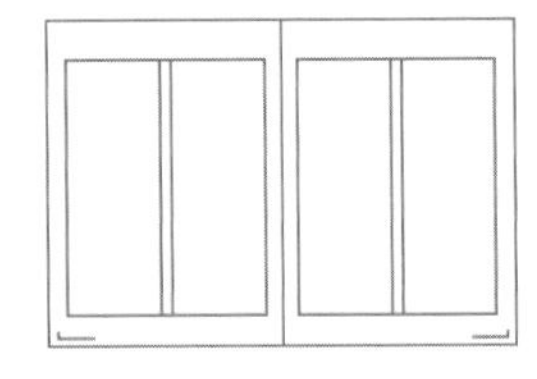

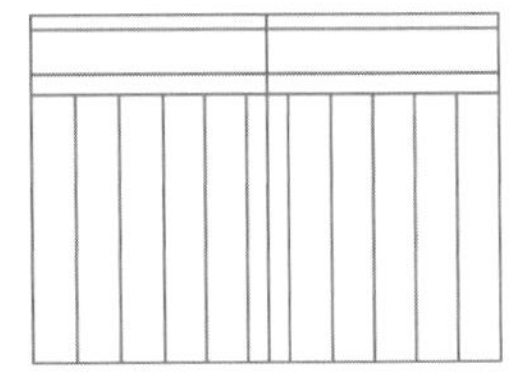

素材有标题吗？有副标题吗？有列表吗？有项目符号吗？如果没有，需要加上这些吗？把最重要的文本放大、加粗，以区别于次要的文本。不同的字体、字号和不同粗细的文字可以用于区分不同类型的素材，但版面要简洁。如果采取了多种文本样式，应保证每种样式的用途都一目了然，否则会使版面显得很混乱。

尽管字体的大小很重要，但间距也同样重要，标题的位置和它与周围元素的间距也能体现它的重要性。要使许多截然不同的材料易于理解，可以将其分成几个部分，以便阅读。使用侧边栏和方框可以将信息分割成易于浏览的信息块，这种形式的排版可以帮助读者快速理解内容。

项目（左图）
《权力的象征》

客户
哈里•N.艾布拉姆斯公司

设计公司
BTDNYC

设计总监
马克•拉里维埃

使用Bodoni字体表现拿破仑时期的经典版式

项目（右图）
《蓝图》（*Blueprint*）

客户
玛莎•斯图尔特生活全媒体公司（Martha Stewart Living Omnimedia）

设计总监
黛布•毕晓普（Deb Bishop）

现代的排版简洁、信息量大、干净利落

126 • SYMBOLS OF POWER

35. Woman's Formal Dress

ANONYMOUS
c. 1805
Embroidered cotton plain weave (mull)
L. 112 cm (center front); 211 cm (center back)
Museum of Fine Arts, Boston; Gift in memory of Helen Kingsford Preston (1989.347)

This white cotton gown reveals some of the subtle changes that took place in women's fashions during the Empire period. The antique-inspired columnar silhouette, popular at the turn of the century, is still evident but has been modified by the gently flaring skirt (formed by triangular fabric inserts, known as *godets*, at the side seams), the squared neckline, the back tie closure, and the small puffed drawstring sleeves. Empire-period embroidery often achieved the grandeur of Ancien Régime embellishment; here, however, the delicate embroidered motifs that adorn the dress show restraint. A small vine pattern of alternating leaves and berries forms a discreet *I* filled with embroidered dots on the bodice, while the skirt bears a slightly different vine of alternating flower blossoms and leaves, again filled with dots, in a pattern that cascades down the center front of the skirt to form a scallop at the hem. The overall effect is one of delicacy and simplicity, underscored by the transparency of the finely woven, white cotton mull fabric imported from India.

The dress was probably purchased in Paris in 1805 by Sarah Bowdoin (1761–1826), the wife of the Hon. James Bowdoin III (1752–1811), who served under President Jefferson as U. S. Minister to Spain and Associate Minister to France. The Bowdoins lived in Paris from 1805 to 1808, and Sarah's diary, written between 1806 and 1808, offers an interesting glimpse into her daily life there.[1] A keen observer of the customs and fashions of the period, Sarah recorded her excitement at walking in the Bois de Boulogne and seeing "almost *all* the fashionables returning; therefore, we had the pleasure of facing them."[2] During one of her favorite walks in the gardens at the Tuileries, Sarah observed the Empress Josephine leaving her Palace, and made careful notes in her diary detailing Josephine's fashionable attire: "She was dressed in a flesh coloured sattin [sic] with long sleeves, short dress with a trimming of very wide lace festooned at the bottom, a small purple velvet bonnet with three small white feathers at the bottom, a small purple camel's hair shawl over her shoulders."[3]

• L. W.

NOTES

1. Sarah's busy schedule included nightly entertainment of dinner guests, daily walks, shopping for dress fabrics and trimmings with her niece, visiting exhibitions at the Louvre, and doing needlework at home. (Sarah Bowdoin Diary, 1806–08, Bowdoin Family Collection, George J. Mitchell Department of Special Collections and Archives, Bowdoin College, Brunswick, Maine.)
2. Ibid., February 15, 1807.
3. Ibid., February 13, 1807.

如果只使用一种字体，可以通过结合使用罗马字母的大小写和斜体建立层次结构。对于更复杂的信息，可以使用不同的字体和字号设置文本块

START HERE

***PLANNING**

Mortgages, inspections, APRs—oh, my! If the idea of buying your own place ties you up in knots, just follow this road map and we'll walk you through every step of the process. Repeat after us: There's no place like home...

TEXT BY ESTHER HAYNES

Buying vs. renting: What makes more sense economically? In a nutshell: If you plan on living in the same city for at least five years, it's usually worth it to buy. **Advantages to owning:** Over the long term, real estate is rarely a bad investment. The recent market downturns in some areas just give you more choice and additional negotiating room. You'll build equity (read: grow your assets) and establish a good credit history, and you won't have to deal with a landlord (bye-bye, linoleum floors). **Disadvantages:** You'll have to take care of all your own maintenance and improvements (hello, hardwood floors) and pay property taxes. **Situations in which you should hesitate to buy:** If your current rent is unusually low; if you're considering a co-op or condo apartment that has high maintenance costs that aren't tax deductible (these will be noted in the listing); if you think you might get laid off or be relocated in the near future; if your credit is so bad that you can qualify only for loans with very high interest rates. **To help you make the right decision:** Check out the "Buy vs. Rent Calculator" at ginniemae.gov, where you can plug in all the numbers and see a comparison. The results can be shocking. For example, if you're paying $1,500 a month in rent and thinking of buying a place for $200,000 (with a 10% down payment), buying now could save you $110,000 over the next 10 years.

How much can you afford? A mortgage calculator (like the one at mortgage-calc.com) is the simplest way to get a ballpark idea. You punch in the loan amount, interest rate, and length, and it spits out your estimated monthly payment. (Just keep in mind that insurance, property taxes, and maintenance will add to your costs.) **If you want to find out for certain what your upper allowable limit is:** Get preapproved (not prequalified) for a loan. Find a certified mortgage broker (see page 6), who will have you fill out a loan application. Based on your income, debt, and credit score, you will receive a letter stating the maximum amount a bank is willing to lend you. **Bonus:** Preapproval will give you an advantage over other buyers because the seller will know you definitely have the money available to make a purchase. **Beyond the banks:** Calculators and lenders don't take your lifestyle into consideration. Do your own detailed budget that includes all nonessential expenses, from movies to mocha lattes, and talk to a tax pro to see how home ownership can add up for you. Then ask yourself how much you can comfortably spend on your mortgage each month. For peace of mind, be sure you stick with that figure, even if a bank will give you a bigger loan with a larger payment.

Down payments: Putting 20% down is great if you can swing it. You'll start with more equity, pay less total interest, and avoid some fees. But if you have high credit card debt, consider putting down less and using the rest of the money to pay off your cards—mortgage interest is tax deductible, but credit card interest isn't.

If you're broke: Don't panic! You are not a lost cause. Check to see if you qualify for a Housing Finance Agency program in your state that helps first-time buyers (ncsha.org). Another option is to ask a family member or friend to lend you some money interest free, or to cosign the mortgage. You can also save by negotiating for the seller to pay closing costs, which can add up to 3% to 6% of the house price.

CHECKLIST

Have these things handy when just starting out:

- **This guide**
- **Financial information** to begin loan preapproval: recent bank statements, tax forms, and pay stubs, plus employment history
- **Notebook and pen**
- **Digital camera** to document house visits
- **Map or GPS** to find and remember locations
- **The Department of Housing and Urban Development's checklist** of things to note, such as proximity to good restaurants (*hud.gov/buying/checklist.pdf*)

Mortgage banker: A lender, bank, or company that provides loans.

Mortgage broker: A person or firm that arranges loans from multiple lenders.

Preapproval: A lender's promise (subject to final guarantee) to supply you with a specific loan amount for a mortgage.

Prequalification: A lender's informal estimate of the maximum you'd be allowed to borrow.

Credit bureau score: A number showing how likely you are to default on a loan, based on your credit history (see yours at annualcreditreport.com). Check it and correct any mistakes, which can foil your mortgage plans.

Down payment: The amount of a home's price you pay in cash; usually from 0% to 20%.

85% OF HOME buyers used a REAL-ESTATE AGENT to help them during their SEARCH FOR A HOUSE.

变换字体和字号，以及将素材放在框里，都是漂亮地展示大量信息的方法

● 6.决定顺序

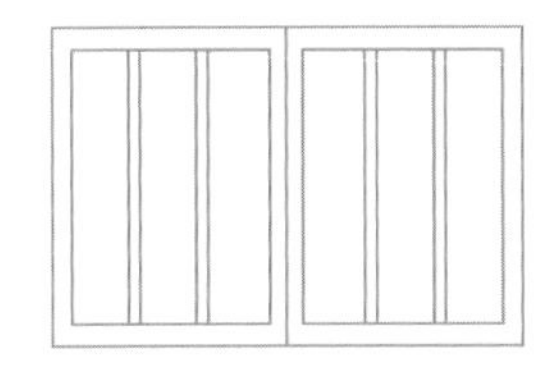

在一个页面中，很少会把所有图片都设计成同样的大小。就像文本传达信息一样，图片的大小表明了事件或主题的重要程度。有的设计公司先将图片按大小排序，然后再排版；有些公司则让设计师决定图片顺序，从而让作品有更醒目的效果。当然，有时候将一些复杂的图片放大只是为了更便于读者阅读。图片的大小变化除了能体现图片的不同功能和版式的灵活性外，还能吸引读者。

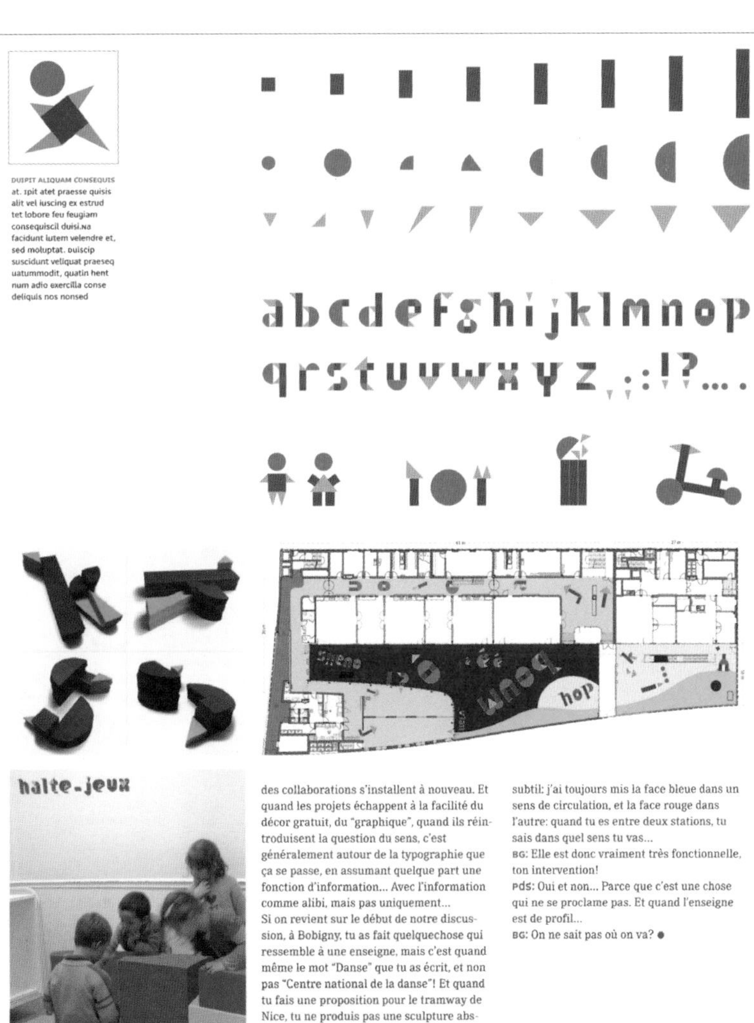

DUIPIT ALIQUAM CONSEQUIS
at. Ipit atet praesse quisis alit vel iuscing ex estrud tet lobore feu feugiam consequiscil duisi.Na facidunt lutem velendre et, sed moluptat. Duiscip suscidunt veliquat praeseq uatummodit, quatin hent num adio exercilla conse deliquis nos nonsed

des collaborations s'installent à nouveau. Et quand les projets échappent à la facilité du décor gratuit, du "graphique", quand ils réintroduisent la question du sens, c'est généralement autour de la typographie que ça se passe, en assumant quelque part une fonction d'information... Avec l'information comme alibi, mais pas uniquement...
Si on revient sur le début de notre discussion, à Bobigny, tu as fait quelquechose qui ressemble à une enseigne, mais c'est quand même le mot "Danse" que tu as écrit, et non pas "Centre national de la danse"! Et quand tu fais une proposition pour le tramway de Nice, tu ne produis pas une sculpture abstraite: c'est quand même d'un "T"!
PdS: Oui, mais il y a des gens qui vont passer devant pendant des années sans décrypter ce "T"... Il y a même quelque chose d'un peu plus subtil: j'ai toujours mis la face bleue dans un sens de circulation, et la face rouge dans l'autre: quand tu es entre deux stations, tu sais dans quel sens tu vas...
BG: Elle est donc vraiment très fonctionnelle, ton intervention!
PdS: Oui et non... Parce que c'est une chose qui ne se proclame pas. Et quand l'enseigne est de profil...
BG: On ne sait pas où on va? ●

56 : 2.2008

图片的宽度可以是半栏、一栏或两栏。偶尔打破网格会达到意想不到的效果，并引起读者对某一张图片的注意。也可以通过图片所占空间的大小显示图片的重要程度

项目
《阶段》杂志（*étapes: magazine*）

客户
金字塔/《阶段》杂志（Pyramyd /*étapes: magazine*）

设计师
安娜•图尼克（Anna Tunick）

不同大小的图片会在视觉上建立一种层级关系

● 7.综合考虑所有元素

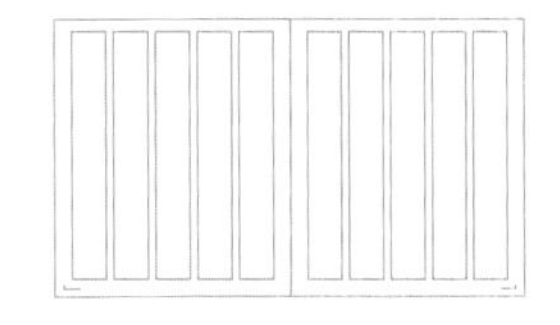

根据媒介或项目的不同，网格可以在一栏或一个区域中显示文字，在另一栏或另一个区域中显示图片，以此分隔元素。但大多数网格都是将文字和图片混合排列，并要确保无论图片还是文字都能传达给读者明确的信息。

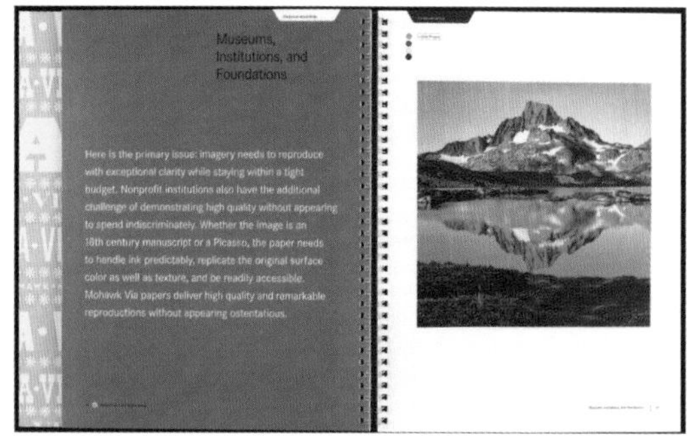

上图：强调文本。这张图中，文本在一个页面上，图片在另一个页面上

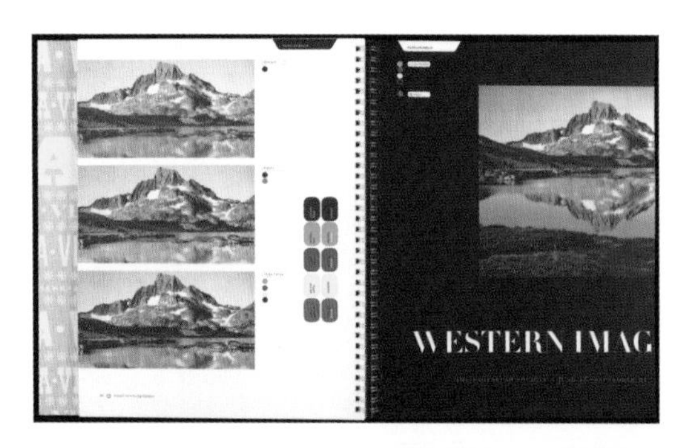

左图和下图：图片可以跨越不同的栏，以水平的方式排列，图片说明可以加在下面。也可以垂直堆放图片，图片说明放在图片的任意一边

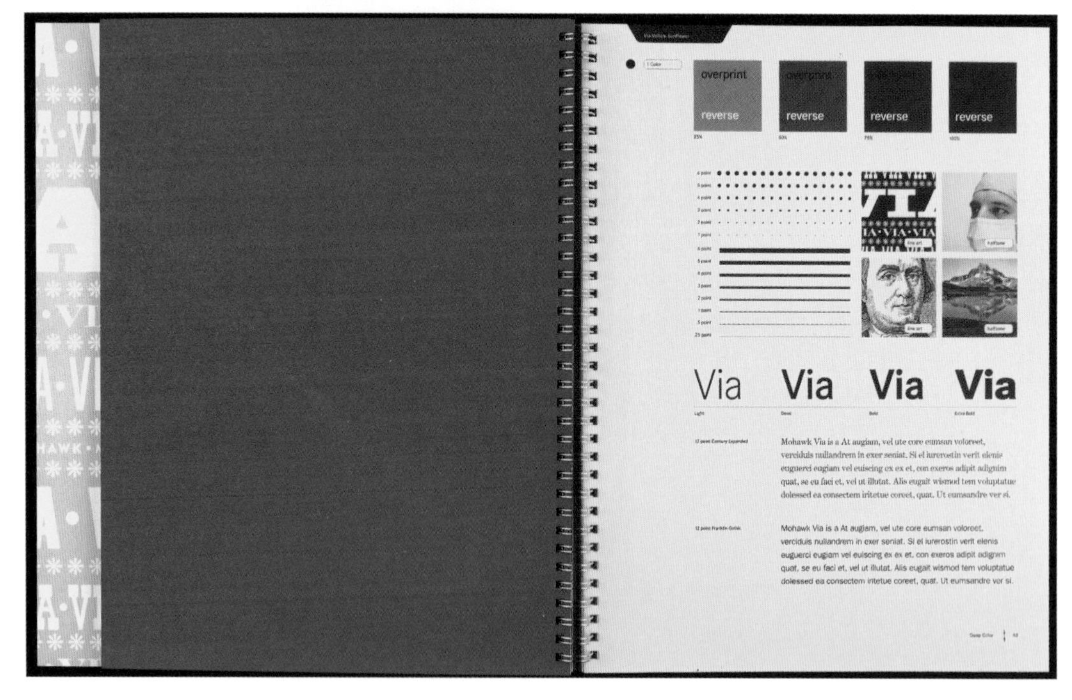

项目
《莫霍克大手册》（*MOHAWK VIA THE BIG HANDBOOK*）

客户
美国莫霍克精品纸业（Mohawk Fine Papers Inc.）

设计公司
亚当斯•莫瑞卡公司（Adams Morioka, Inc.）

设计师
肖恩•亚当斯（Sean Adams），克里斯•泰伦（Chris Taillon）

这张纸质促销单通过网格展示了不同大小的图片

8.用色彩划分区域

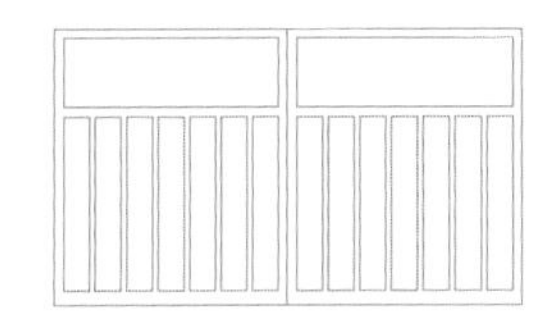

色彩是一种凸显模块或段落的手段。色彩可以定义空间，也有助于组织空间中的元素。色彩还能使页面生动，并为需要传达的某种信息提供标记。在设置颜色时，要考虑受众。饱和的色彩能吸引注意力，而不饱和的色彩则使素材的表达更低调。太多的色彩会让一块区域显得很拥挤，并让这部分内容难以阅读。

屏幕上的色彩 VS. 纸面上的色彩

我们生活在一个充满色彩的世界里，无论客户还是设计师，都需要在屏幕上查看很多带有色彩的内容。屏幕上的色彩明亮、饱和、艳丽，并且是RGB模式的，然而，屏幕上的色彩和纸上的色彩有很大的区别。使用传统的四色印刷时，需要仔细选择纸张，并用色轮校正颜色，使印刷品上的色彩与屏幕上的色彩更接近。

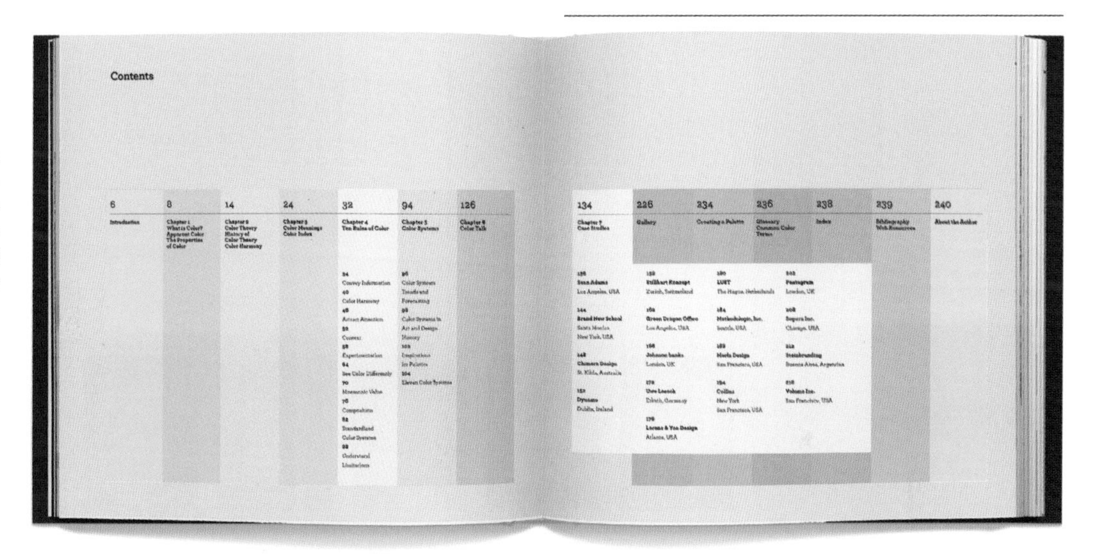

色彩可以用来显示不同的、零碎的信息

无论是在模块、文本框还是区块中使用，色彩都可以用来区分信息。模块可以是半装饰性的——将彩色框置于文本框之下，也可以是功能性的，用以区分不同类型的文本框

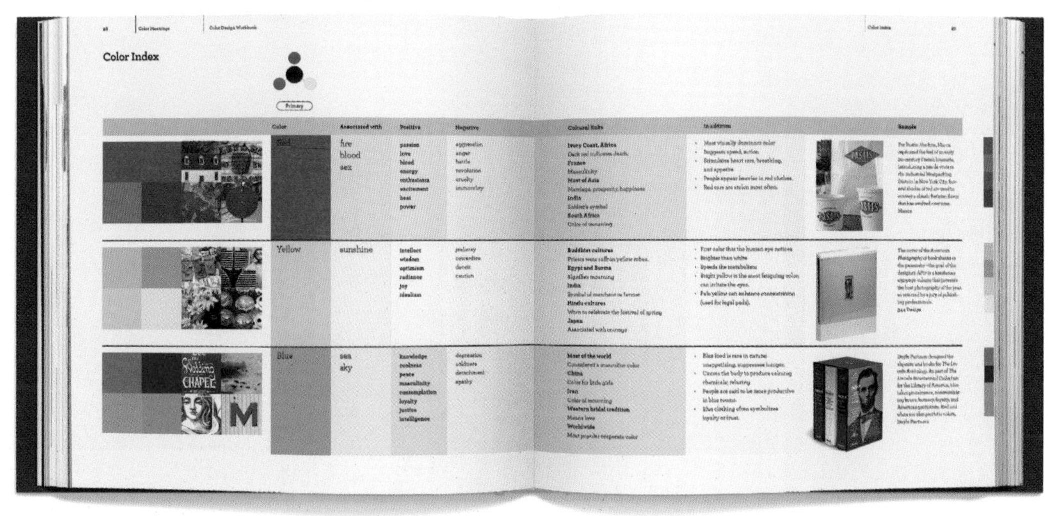

项目
《色彩设计手册》（*Color Design Workbook*）

客户
洛克波特出版社（Rockport Publishers）

设计师
肖恩•亚当斯

该书的对开页面显示了色彩的重要作用，以及色彩是如何让一个页面变得更明亮的

9.将留白作为平面元素

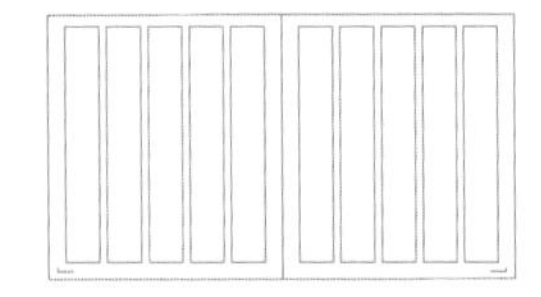

留白有助于充分传递信息。虽然网格必须足够清晰并能容纳大量信息，但没有必要填满每个部分。留白可以把信息分开，为阅读和理解文本提供适当的空间。经过设计，大量的留白可以创造出戏剧性的效果和阅读的焦点。留白可以凸显信息的重要性，页面上的空白区域可以创造一种独有的美感。

另见166～167页

Self-Portrait (5 Part), 2001
Five daguerreotypes, each 8 1/2 x 6 1/2 in.
(21.6 x 16.5 cm)

留白是一种有意为之的设计，目的是让读者有一个短暂的停顿

项目
《查克·克洛斯作品集》（*Chuck Close | Work*）

客户
普雷斯特出版社（Prestel Publishing）

设计师
马克·迈尔尼克（Mark Melnick）

与设计一样，艺术也与留白息息相关

10.通过节奏保持读者的兴趣

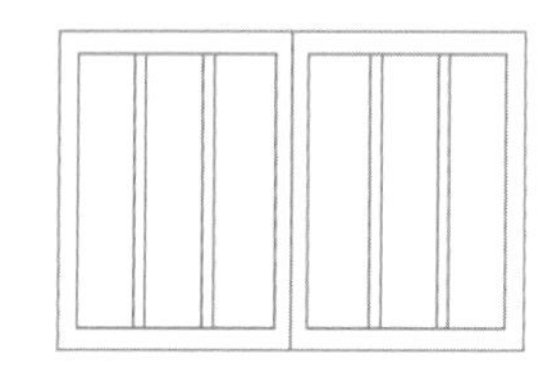

有些网格通过设立清晰、重复或连续的图片栏或信息栏尽可能多地编录素材，然而，大多数网格中的内容也可以形成“流动”的节奏感，从一个信息块到下一个信息块，从一个对页到下一个对页，或者从一个屏幕到另一个屏幕。页面上的素材表现出的节奏可以在不同程度上引起和保持读者的兴趣。节奏可以通过图片及字体的大小和位置，以及每张图片周围的间距来实现。

博比•马丁（Bobby Martin）和他所在的OCD的团队设计了这本具有重大影响力的出版物，里面包括了很多有深度的文章和与历史有关的图片。当时，他们将每一个对页都贴在墙上，以便检查和调整位置，为页面增加戏剧感和节奏感

项目
《金》（*King*，纪念马丁•路德•金遇刺50周年的特刊）

客户
《大西洋月刊》杂志社（*The Atlantic*）

设计公司
OCD | 原创设计冠军（Original Champions of Design）

创意总监
保罗•思培拉（Paul Spella）

艺术总监
戴维•萨默维尔（David Somerville）

设计师
博比•马丁，詹妮弗•奇洛恩（Jennifer Kinon）

流畅的布局仿佛讲述了一个十分清晰的故事

GRIDS AT WORK

在实践中使用网格

11.给你的主题设定一种风格

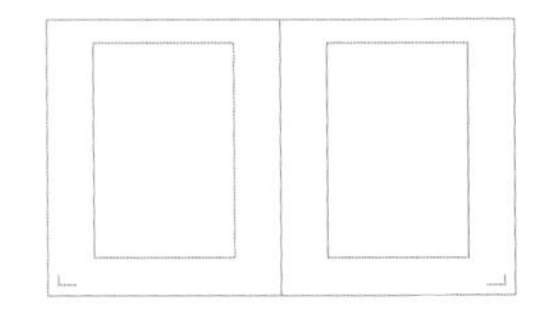

当为单页或对页的单栏网格挑选合适的字体时，应该考虑页面的主题。有的风格是经典的、中性的，可以与大多数素材搭配，而有的风格与主题一样，表达了一种观点。字体可以表达态度，也可能在不知不觉中削弱态度。通过使用具有历史感的字体，可以在版式中营造出年代感。相反，挑选非传统的字体则可以制造新意。字体宽度、字号大小、字体间距和行距都会影响页面整体的观感。

另见17页

对这个项目来说，Geometric这种无衬线字体避免了粉色封面给人的俗气感，而有了一种朋克的风格

PINK
THE HISTORY OF A
PUNK, PRETTY, POWERFUL
COLOR
EDITED BY
VALERIE STEELE
Thames & Hudson

there were at least "two pinks" in the 1950s: a feminine pink and
"young, daring—and omnisexual" pink.[69]

THE NAVY BLUE OF INDIA

Pink has long been a very popular color in India for both men's
clothing and adornment. In Rajasthan, for example, it is still commo
men wearing hot pink turbans. India's polychromatic sensibility h
many Westerners. Already in 1956, Norman Parkinson did an infl

必须仔细考虑行距，留出足够的空间，避免版式看上去太密集、压抑。反之，间距太宽又会让文字看上去像装饰，而不是可读的文本

左边的例子展示了《粉色》一书再版时的字号和字体——10 / 15.75磅的Geometric大写字体与10 / 15.75磅的Meridian字体。可见，无衬线字体比衬线字体在视觉上显得更大

项目
《粉色》（*Pink*）

客户
泰晤士与哈得孙出版社（Thames and Hudson）

设计公司
BTDNYC

干净的页面得益于一个简洁、不花哨的无衬线字体

FEMININE DESIRE AND FRAGILITY: PINK IN EIGHTEENTH-CENTURY PORTRAITURE

A. Cassandra Albinson

A SURVEY OF OBJECTS MADE IN FRANCE AND BRITAIN IN THE EIGHTEENTH CENTURY REVEALS A REMARKABLE NUMBER OF PINK ITEMS: coverings for furniture; colored prints; porcelain; paint for interiors; paint for works of art; suits for men; and especially dresses and ribbons for women. Aside from its plentitude, what did the color pink represent in women's clothing? Portraiture provides us with a fruitful entryway because we often have indications about the sitter's life and circumstances that can provide further understanding of the choices they made in terms of costume and adornment.[1] In comparable portraits of Jeanne-Antoinette Poisson, Marquise de Pompadour, (1721–1764) and Frances Abington (1737–1815), each woman is depicted close to the picture plane and from the waist up. Both Pompadour and Abington were exceedingly famous at the time they were painted, and their fame and fortune rested in large measure on their physical beauty and prowess. And each woman had the means to be painted by the most famous artist of her day: François Boucher for Pompadour, Joshua Reynolds for Abington.[2] Each woman gestures toward herself and suggests a touch, either in the immediate future—in the case of Pompadour, who holds a rouge brush as if poised to apply color to her cheeks—or concurrently with being painted in the case of Abington in the role of Miss Prue. Hands in each portrait are as important as faces and are stressed by the inclusion of bracelets at the wrist. Both portraits also feature a second figure around the sitter's midriff. In the case of Pompadour we see a miniature portrait of her lover, the French king, Louis XV, while a fluffy white dog sits on

83. François Boucher, *Jeanne-Antoinette Poisson, Marquise de Pompadour*, 1750, with later additions.
Harvard Art Museum.

82. Joshua Reynolds, *Mrs. Abington as Miss Prue in "Love for Love" by William Congreve*, 1771.
Oil on canvas. Yale Center for British Art, Paul Mellon Collection.

without one." A few years later, it was back, "beautifully refreshed" in a variety of styles—"pink evening shirts, pink-shirt dresses, even a pink swimming shirt," not to mention "one of 1953's prettiest little-evening blouses."[67]

"Across the US, a pink peak in male clothing has been reached as manufacturers have saturated more and more of their output with the pretty pastel," reported *Life* magazine in 1955. "Sole responsibility lies with New York's Brooks Brothers," whose pink shirt "was publicized for college girls and caught on for men too." Gradually, pink neckties, dinner jackets, golf jackets, trousers, and other garments also became increasingly visible. "Like most male fashions, including the Ivy League Look, this pink hue and cry has taken some time to develop." But by 1955, the "traditionally feminine color" had become "a staple for [the] male."[68]

Elvis Presley not only wore pink suits, jackets, and trousers, he also drove a pink car and slept in a pink bedroom. Was he influenced by African-American style? His fans wore lipstick in Heartbreak Hotel Pink, and rock and roll extolled the color with songs like "Pink Pedal Pushers" (1958), "Pink Shoe Laces" (1959), and "A White Sport Coat (and a Pink Carnation)" (1957). Meanwhile, the warm carotenoid pink of flamingos was increasingly associated with newly affordable, warm-weather vacations in places like Florida and the Caribbean. So perhaps there were at least "two pinks" in the 1950s: a feminine pink and an emerging "young, daring—and omnisexual" pink.[69]

THE NAVY BLUE OF INDIA

Pink has long been a very popular color in India for both men's and women's clothing and adornment. In Rajasthan, for example, it is still common to see many men wearing hot pink turbans. India's polychromatic sensibility has influenced many Westerners. Already in 1956, Norman Parkinson did an influential photo shoot in India for British *Vogue*. One of his striking images juxtaposed a Western model in the latest fashion with an Indian girl in a hot pink sari. Another, shot in Jaipur, the "Pink City," posed the model in pale pink with a group of Indian men in bright pink coats and turbans. Diana Vreeland, then editor of *Harper's Bazaar*, saw the images and allegedly said, "How clever of you, Mr. Parkinson, also to know that pink is the navy blue of India."[70]

Traditionally, there were many rules about color in clothing related to age, region, caste, occasion, complexion, and time of day. More recently, the individual's personal taste has played an increasingly important role. In their book

43. Unknown artist, *Portrait of an Indian Prince Wearing a Wedding Sehra* [headgear], ca. 1920–40, Rajputana Photo Art Studio.
Gelatin silver print and watercolor. 14¼ × 11 in. (36.5 × 28 cm). The Alkazi Collection of Photography.

● 12.决定页边距

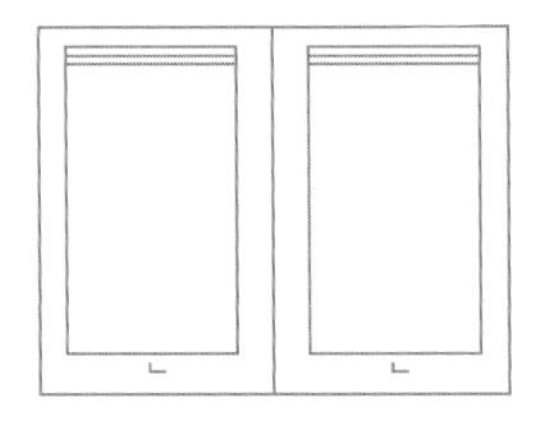

如果一个需要印刷的文件页数很多，最好是留出足够大的页边距，以防止文本在装订时被挡住。例如，一本书在屏幕上或打印时看起来比例协调，但在印刷并装订后可能发生很大变化。为装订线预留的空间大小取决于装订方法，无论这份文件是无线胶装、线装，还是骑马钉装订，都要确保没有任何内容被遮挡。有的装订方式与其他装订方式相比更容易遮挡文本。线装或磨脊胶装的书要比无线胶装的书更容易平摊开。如果使用无线胶装，文字可能会被遮挡在装订线内，而读者在把书摊开时可能不愿意用力翻开装订线的位置。如果采用螺旋装订的方式，需要在装订部位为螺旋孔留出足够的空间。

另见14～15页

项目
《酱料》（*Sauces*）

客户
约翰•威利父子出版社
（John Wiley and Sons）

设计公司
BTDnyc

这本八百多页的烹饪指南需要大面积的装订空间，设计师对症下药，给出了方案

图片选自约翰•威利父子出版社的《酱料》，版权属詹姆斯•彼得森（James Peterson）所有，经约翰•威利父子出版社授权再版

GOLD-PLATED CHICKEN WITH GINGER, SAFFRON, AND ALMONDS

This modern adaptation is not based on any particular recipe but is taken from several recipes in Taillevent's Viandier (fourteenth- and fifteenth-century manuscripts). Ginger, saffron, and mint are the principal flavorings: ginger and saffron were the spices most often called for in medieval recipes, and mint was one of the most commonly used herbs. The sauce is bound with almond butter, a typical medieval liaison (bread can also be used). Green-colored marzipan almonds and pomegranate seeds are used as the garniture. The almonds are a reference to the medieval cook's tendency to fashion one food from another to surprise and titillate the diner. They are sweet (and surprisingly good with the sauce), recalling the inclination to juxtapose the savory with the sweet in the medieval meal. The gold plating is extravagant and can be eliminated (or silver leaf can be substituted), but it is taken from an authentic recipe. Gold and silver foil are still used in Indian cooking to decorate desserts. Medieval diners were fond of bright colors, hence the gold, the pomegranate seeds, the saffron, and the colored almonds.

The chicken is prepared like a fricassée, but the recipe could be adapted to a sauté model as well.

YIELD: 4 SERVINGS

chicken, quartered (1 chicken)	3 pounds	1.4 kilograms
salt and pepper	To taste	To taste
butter or lard	4 tablespoons	60 grams
onion, chopped	1 medium	1 medium
white chicken stock	2 cups	500 milliliters
almond paste	2 ounces	50 grams
green food coloring or chlorophyll	Several drops	Several drops
pomegranate	1	1
saffron threads	1 pinch	1 pinch
hot water	1 tablespoon	15 milliliters
finely grated fresh ginger root	2 teaspoons	10 grams
mint leaves	1 small bunch	1 small bunch
almond butter (*see* Chapter 17, "Purees and Puree-Thickened Sauces," page 431)	2 tablespoons	30 grams
egg yolk	1	1
gold or silver leaf	4 sheets	4 sheets

1. Season the chicken pieces with salt and pepper. In a 4-quart straight-sided sauté pan, gently cook the seasoned chicken pieces, skin side down, in the butter or lard. After about 10 minutes, turn and cook the flesh side. Avoid browning the chicken or burning the butter. Remove the chicken.

8 ● SAUCES

2. Add the chopped onions to the butter in the pan, and sweat, without browning, until they are translucent.
3. Add the chicken stock to the pan. Arrange the chicken pieces in the liquid and cover.
4. Cook the chicken in a 350°F (175°C) oven or over low heat on the stove for 15 to 20 minutes.
5. While the chicken is cooking, work the almond paste with the food coloring or chlorophyll until it is bright green. Shape the colored paste into 12 almonds and set aside.
6. Remove and reserve the seeds from the pomegranate. Discard the flesh.
7. Soak the saffron threads in the hot water for at least 20 minutes.
8. Transfer the chicken to a plate and keep it warm. Add the grated ginger to the liquid in the pan and let it infuse for 5 minutes.
9. Strain the sauce into a 2-quart saucepan and reduce it to ¾ cup (200 milliliters). Skim carefully.
10. Gradually add the saffron, tasting so that its flavor becomes apparent but does not overpower the flavor of the ginger. Add the mint.
11. Whisk in the almond butter until the sauce has the desired consistency. Add salt and pepper to taste.
12. Beat the egg yolk with a large pinch of salt to make an egg wash.
13. Brush the top of the chicken pieces with the egg wash.
14. Apply the gold or silver leaf by holding the sheet about ⅛ inch from the surface of the chicken and systematically blowing on the back of the gold leaf with a 5-inch-long (13 cm) plastic straw.
15. Serve the chicken surrounded with the sauce, the pomegranate seeds, and the green almonds.

RENAISSANCE COOKING: THE SIXTEENTH CENTURY

Surprisingly little has been written about cooking in the sixteenth century. In France one important book was published, a translation of Bartolomeo Platina's *De Honeste Voluptate*. Whereas most other books were based on earlier works and were medieval in character, Platina gives us a deeper understanding of both the cooking and the priorities of Renaissance Italy and France. During the Renaissance and for several centuries thereafter, culinary methods were closely linked to health and medicine. Much of Platina's writing was influenced by medieval medicine, which itself was based on Greek medicine with its elaborate system of humors and emphasis on the use of diet to balance the basic "personalities": sanguine, phlegmatic, choleric, and melancholic. The ingredient that appears in greater quantities in sixteenth-century recipes is sugar. Although by no means inexpensive, refining methods made it more accessible than it had been during

A SHORT HISTORY OF SAUCE MAKING ● 9

充足的页边距确保了重要的菜谱信息简单易读，文字不会落入装订线内。另外，宽边距也有利于子标题和图表这些可能出现在文本块外的元素的排版。宽边距下的栏外脚注和页码也可以给人一种平静和休闲的感觉

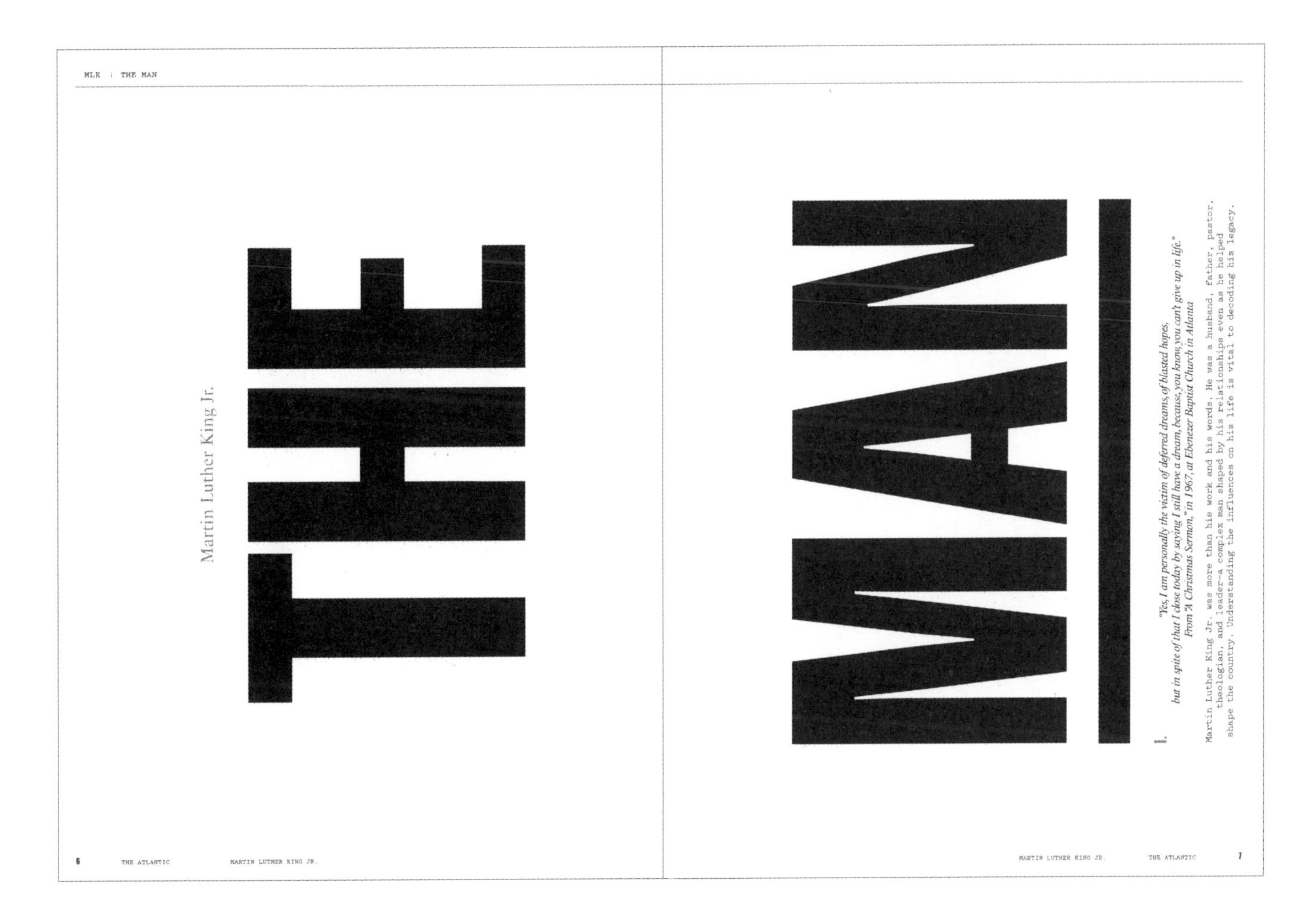

MLK | THE MAN
THE
Martin Luther King Jr.
MAN
"Yes, I am personally the victim of deferred dreams, of blasted hopes, but in spite of that I close today by saying I still have a dream, because, you know, you can't give up in life."
From "A Christmas Sermon," in 1967, at Ebenezer Baptist Church in Atlanta
Martin Luther King Jr. was more than his work and his words. He was a husband, father, pastor, theologian, and leader—a complex man shaped by his relationships even as he helped shape the country. Understanding the influences on his life is vital to decoding his legacy.
6 THE ATLANTIC MARTIN LUTHER KING JR.
MARTIN LUTHER KING JR. THE ATLANTIC 7

项目
《金》

客户
《大西洋月刊》杂志社

设计公司
OCD | 原创设计冠军

创意总监
保罗•思培拉

艺术总监
戴维•萨默维尔

设计师
博比•马丁，詹妮弗•奇洛恩

本页元素采用了不同的大小和间距，标注的行距也很紧凑，增加了页面的动感和张力

窄边距可以渲染出紧凑感和历史感。在本页案例中，紧贴地脚的页码和章节信息与页面的留白形成鲜明对比，强调了内容的重要性。

尽管这本杂志从严格意义上讲，采用的是单栏网格，但整本杂志都使用了较小的页边距。该杂志的实际尺寸为195毫米×264毫米，外边距为5毫米，顶端到页首标题的距离和底端到页码和脚注的距离为5毫米，突破了印刷与剪裁的极限。但这个布局取得了比较好的效果

经验法则

常常有人问：“一般的印刷页边距设置多少合适？”这个问题并没有标准答案，不过我建议先试试12.5毫米，然后在此基础上增减。在页边距之外再留出不超过6毫米的距离，这在印刷上被称为出血。最终采用的页边距取决于素材在页面中的占比，如果是印刷品，还取决于印刷设备。通常，设计师在设计印刷品时常犯的错误是文字过多而边距过小。在网页或平板设备上，页边距十分重要，如果不留出足够的边距，一部分内容可能会被遮挡。

● 13.按比例设计

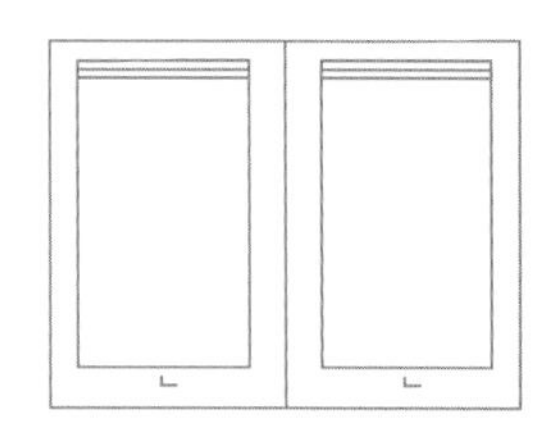

将比例牢记在心，地脚也不要漏掉。即使是看起来很简单的页面，也需要合理利用空间，这样才能在印刷页面和屏幕上合理地显示内容。

THE PLAGUE OF

DOVES

LOUISE ERDRICH

HarperCollins*Publishers*

隐藏的、模板化的设计巧妙地烘托了标题字体，字体加粗可以凸显质感，用小号字体也会显得比较克制

The Plague of Doves

IN THE YEAR 1896, my great uncle, one of the first Catholic priests of aboriginal blood, put the call out to his parishioners that they should gather at Saint Joseph's wearing scapulars and holding missals From that place they would proceed to walk the fields in a long sweeping row, and with each step loudly pray away the doves His human flock had taken up the plow and farmed among German and Norwegian settlers Those people, unlike the French who mingled with my ancestors, took little interest in the women native to the land and did not intermarry In fact, the Norwegians disregarded everybody but themselves and were quite clannish But the doves ate their crops the same When the birds descended, both Indians and whites set up great bonfires and tried driving them into nets The doves ate the wheat seedlings and the rye and started on the corn They ate the sprouts of new flowers and the buds of apples and the tough leaves of oak trees and even last year's chaff The doves were plump, and delicious smoked, but one could wring the necks of hundreds or thousands and effect no visible diminishment of their number The pole and mud houses of the mixed bloods and the bark huts of the blanket Indians were crushed by the weight of the birds They were roasted, burnt, baked up in pies, stewed, salted down in barrels or clubbed dead with sticks and left to rot But the dead only fed the living and each morning when the people woke it was to the scraping and beating of wings, the murmurous sussuration, the awful cooing babble, and the sight, to those who still possessed intact windows, of the curious and gentle faces of those creatures

5

居中的页码是经典设计风格的标志

项目

《鸽灾》（*The Plague of Doves*）

客户

哈珀•柯林斯（Harper Collins）

设计师

弗里茨•梅兹（Fritz Metsch）

这是一个“水晶高脚杯”式设计的例子，简洁的页面令一个文学天才的作品熠熠生辉。版式设计师和学者比特利斯•沃德（Beatrice Warde）在《“水晶高脚杯”：关于排版的十六篇文章》（*The Crystal Goblet:Sixteen Essays on Typography*）中写道：“设计应该是隐形的。”她提出，无声的设计好比一只水晶高脚杯，“它的一切都是为了揭示而不是隐藏它包含的美好的东西”

The Plague of Doves

cousin John kidnapped his own wife and used the ransom to keep his mistress in Fargo Despondent over a woman, my father's uncle, Octave Harp, managed to drown himself in two feet of water And so on As with my father, these tales of extravagant encounter contrasted with the modesty of the subsequent marriages and occupations of my relatives We are a tribe of office workers, bank tellers, book readers, and bureaucrats The wildest of us (Whitey) is a short order cook, and the most heroic of us (my father) teaches Yet this current of drama holds together the generations, I think, and my brother and I listened to Mooshum not only from suspense but for instructions on how to behave when our moment of recognition, or perhaps our romantic trial, should arrive.

The Million Names

IN TRUTH, I thought mine probably had occurred early, for even as I sat there listening to Mooshum my fingers obsessively wrote the name of my beloved up and down my arm or in my hand or on my knee If I wrote his name a million times on my body, I believed he would kiss me I knew he loved me, and he was safe in the knowledge that I loved him, but we attended a Roman Catholic grade school in the early 1960's and boys and girls known to be in love hardly talked to one another and never touched We played softball and kickball together, and acted and spoke through other children eager to deliver messages I had copied a series of these second hand love statements into my tiny leopard print diary with the golden lock The key was hidden in the hollow knob of my bedstead Also I had written the name of my beloved, in blood from a scratched mosquito bite, along the inner wall of my closet His name held for me the sacred resonance of those Old Testament words written in fire by an invisible hand Mene, mene, teckel, upharsin I could not say his name aloud I could only write it on my skin with my fingers without cease until my mother feared I'd gotten lice and coated my hair with mayonnaise, covered my head with a shower cap, and told me to sit in the bathtub adding water as hot as I could stand.

The bathroom, the tub, the apparatus of plumbing was all new Because my father and mother worked for the school and in the tribal offices, we were hooked up to the agency water system I locked the bathroom door,

9

黑体、字间距大的栏外标题（页眉）和页码使整页的版式具有质感。宽裕的页边距和充足的行距使阅读变得更容易

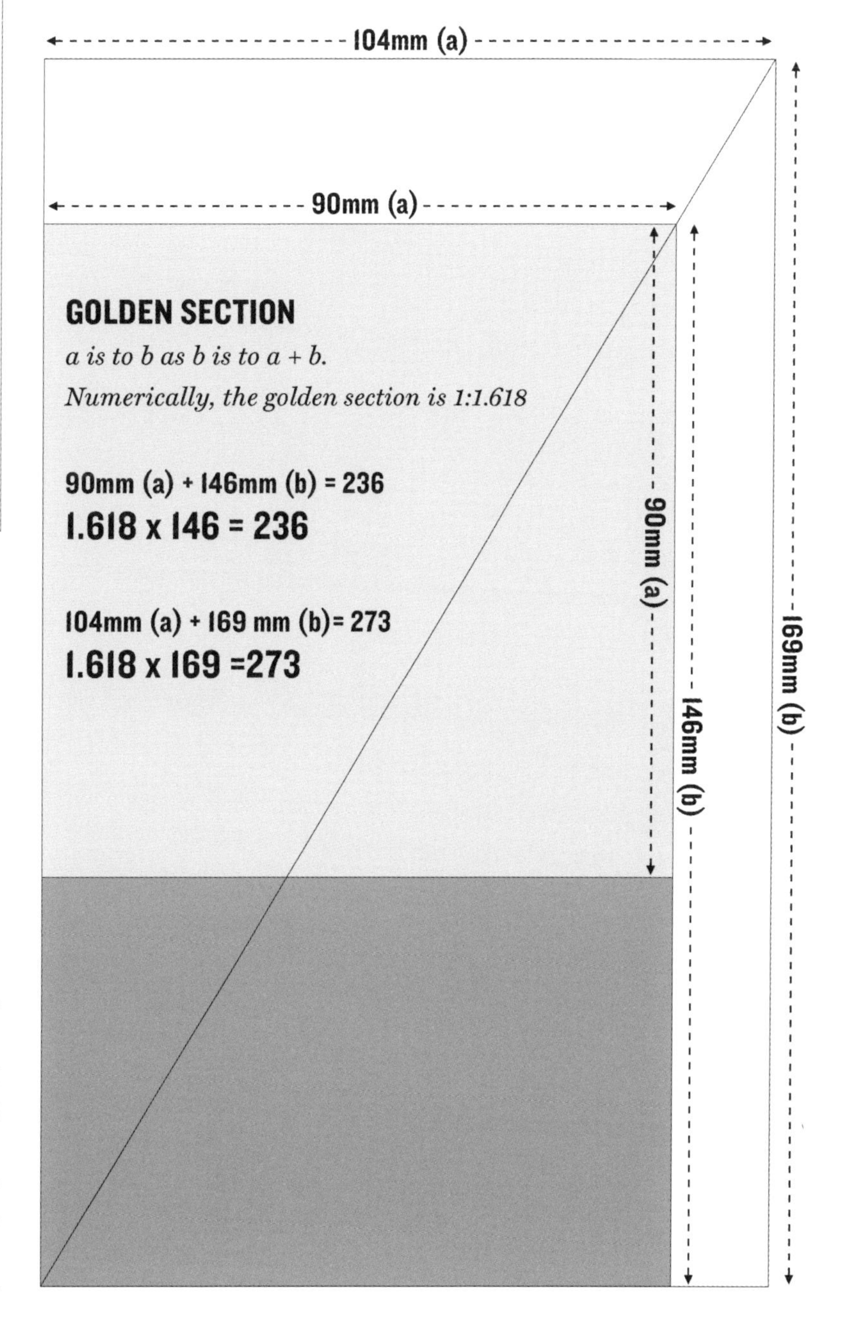

黄金比例

黄金比例在艺术和建筑领域已经使用了数千年。黄金比例也被称为黄金分割，是指事物各部分间的一定的数学比例关系，这个比例大约是1：1.618。换句话说，较小的部分（如宽度a）与较大的部分（如高度b）的比值等于较大的部分与二者之和的比值。

● 14.均分网格

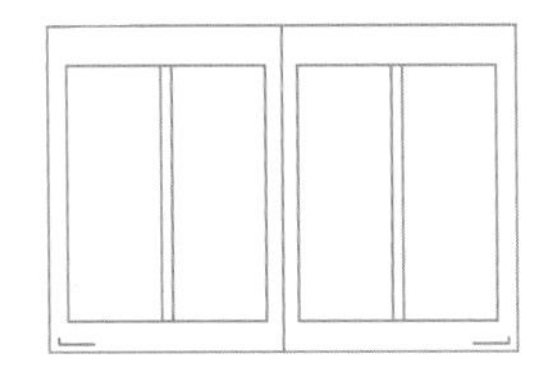

有两个均分栏的网格可以在一个页面上包含大量内容。对称的两栏给人一种秩序井然的感觉，可以容纳不同大小的图片和各种留白。

有两个均分栏的网格更适合国际化的出版物，可以同时用两种不同的语言呈现相同的信息。

EVGENY CHUBAROV
ЕВГЕНИЙ ЧУБАРОВ

传统的、对称的分栏方式为编辑和读者提供了一种秩序感和舒适感

[illegible]

14

In 1959, he moved to the town of Zagorsk to be closer to Moscow, to artists and art. It was at that time that he met Dmitry Tsaplin whose guidance was just as productive for Chubarow as the association with Vladimir Favorsky and Robert Falk was for his fellow-artists, the students of various Moscow art schools. Tsaplin was a unique figure in Russian art. A brilliant sculptor with an impressive record of exhibitions in the West, he returned to the Soviet Union and found himself unemployed, unappreciated and pushed back to the margins of official art. He had much to share with a budding artist, including the skill of art appreciation (incidentally, he was very much a self-taught artist himself and assimilated the art of form-building on par with the level of modern art of his times) and the strategy of artistic behavior, independent and critical towards officialdom. In practical terms, Tsaplin gave Chubarow two main impulses: first, the powerful energy of archaism, typical of Tsaplin's own form-building; and second, rejection of everything plot-driven and literary imposed on form-building from outside. Tsaplin was intolerant to questions about the concrete meaning of his works. He made no difference among animal painting, portrait or any other – for him only one distinction existed: a sculpture or not a sculpture. However,Chubarow came to this freedom from "the theme" only somewhat later.

Chubarow was invited to exhibit at "alternative art" exhibitions rather early in his career. (It was too early at that time to use the term "unofficial art" because 1962, when he exhibited his work together with some Moscow artists at the cinema houses Illusion and Udarnik, was a borderline year when illusions were still alive regarding the good will of the authorities and their willingness to abandon the totalitarian, one-dimensional ideology in favor of peaceful co-existence among different trends in art.) Chubarow was probably unconcerned with these general considerations; he was not the type of artist who wished to be a leader or ideologist of an art movement. It was important for him to learn his trade as best he could, to get some commissions if he was lucky, and try to earn his living by his trade. However there was one factor, which interfered with this reasonable, worldly behavior: an all-consuming passion for self-expression that he developed early in his life. Self-expression is an old-fashioned word, it belongs to the language of romanticism and means spontaneous, pre-rational art practice when a person is aware of some bursting inner energy and tries to organize it somehow, to express himself with no heed to strategies and positioning.

But neither was Chubarow a naive artist, a so-called artist of the "seventh day of creation" (he wouldn't even dream of positioning himself in this role like his naively cunning contemporary "Lyonya Purygin from Nara"). Many people at that time were beguiled by the outward pattern of his life and behavior: a naturally gifted self-educated artist from an under-privileged family, a marginal loner, a "clean sheet". Many well-wishers probably tried to teach him life's wisdom and culture. Meanwhile he already considered himself as a professional in art and moreover, in modern art, and probably felt more assured of himself than many of his contemporaries, who were torn apart between the skills learnt at school, the norms and rules imposed by the official art, and their own inner feeling of freedom sustained by the few extant relics of the culture of the 1920's. Chubarow knew none of those inner discords. No formal art education stood in his way because he had never had any while Tsaplin probably relieved him of any inferiority complexes on this account. Moreover, it must have been the same Tsaplin with his profound knowledge of contemporary European sculpture and the example of his own life, who infected Chubarow with a purely demiurgic, modernist complex of mistrust of any formalized structures imposed from outside. He also infected him with a faith in the artist's creative will – the world was shaped here and now irrespective of established norms. Thus Chubarow had chosen the best possible, in his situation, way of self-identification: he articulated only what derived from his "self-autonomy", to quote Pushkin, what was natural, organic, rooted in the soil. Several decades later, in an interview with Vitaly Patsukov, an indefatigable collector and keeper of the history of Russian underground art, Chubarow spoke at length about his roots: pre-Christian Bashkirian and Russian Old Believers'. I must mention at once that his root-seeking could have led him into a blind alley, which was the case with many others, if it remained at the level of declarative news and the plot, because what is rooted in the soil often turns into things artificial, ideologically biased and nationalistically crude. This way is only fruitful, in my opinion, when the vector of form – building is directed from within (from one's heart), no more no less. While the form building itself is interpreted in the context of modern art.

15

项目
《回归抽象》（*Return to the Abstract*）

客户
俄罗斯国家博物馆（The Russian State Museums）

设计
安东·金兹伯格（Anton Ginzburg），RADIA工作室（Studio RADIA）

双栏网格以俄语和英语双语的形式展示信息

оказывается совершенно неспособным проявлять свои функции прозрачности и ясности. Разбросанные в пространстве произведения отвердевшие «знаки-образы художника», чем-то напоминающие «флаговые структуры», устойчивую эмблематику социальных систем, перефразированные элементы поп-арта, откровенно обнажают идеологию артдиверсий Евгения Чубарова, его открытость к языкам массовой культуры. Но в художественных измерениях картинного пространства они транслируются, скорее, как узоры, как внезапные описки, спотыкания о некое «что-то не так», превращаясь или возвращаясь в комическое сатори. Эти «случайные» ошибки дарят нам шок соприкосновения с неведомым при контакте с, казалось бы, заведомо освоенным. Они опровергают диктат идеи однородной абстракции над витальностью знакотворчества как выпад против здравого смысла, как выпадение в целительное безумие. В такой стратегии их образы утверждают новый принцип абстракции, освобожденный от власти личного монолога художника, но реализующий себя в контексте нового смыслового поля, «запаковывая» непосредственное чувство в интеллектуальную рефлексию. Она естественно возникает в своих сгущениях и пустотах, наплывах и разрывах как горизонтальная модель нового художественного сознания, прорывая гипноз знаковой поверхности через жест своеобразной «деконструкции».

Сама технология живописи Евгения Чубарова, ее способность комментировать и описывать саму себя порождает эффект картины как некоего живописного объекта, где сама живопись раскрывается как чистая ностальгия по живописи, как воспоминание о картине, где в гуще информационного шума спрятан в коконе бывший «абстрактный шедевр», «нетленка», по выражению Ильи Кабакова. Ее «почерк», ее многослойный ландшафт, блестяще выстроенный со всеми своими ассоциативными рядами, где подлинные слои художественной реальности просвечивают сквозь профанные, подбрасывая загадки – все это свидетельствует о новых глубинных ориентациях в искусстве абстракции. Они говорят о ветшании и абстрактных авангардных моделей и о рождении ее абсолютно новой телесности, отрефлексированной и генетически преображенной. Ее новые формы манипулируют следами и обломками ее исторического прошлого и последствиями собственного личного внутреннего опыта художника. В них проступают сознательные цитаты мирового культурнаследия, включающие целые

Замкнутые и разомкнутые кривые Джексона Поллока присутствуют в современной культуре как трансляторы, передающие архаический мир и мир авангарда 10-20 годов ушедшего века. Они никогда не исчезают из визуальности нашей мифологии. Знаковая символика в пластике Евгения Чубарова корреспондирует с этими архетипами «шаманистического» текста, пульсируя в его композициях, раскрывая их еще не высказанные смыслы.

Джексон Поллок, «Страницы из блокнота», 1938

Closed and open-ended curves of Jackson Pollock exist in contemporary culture as translators, conveying the archaic world and that of the avant-garde art of the 1910s and 20. They never disappear from view in our mythology. The sign symbolism in the art of Evgeny Chubarov corresponds to the archetypes of Shamanist texts, pulsing in his compositions and disclosing the meaning of their yet unexpressed messages.

Jackson Pollock, "Pages from Sketch-book", 1938

Поиски первичных образов начала человеческой истории, его архетипов открывают парадоксальный диалог, необусловленный никакими внешними влияниями. В нем открывают себя английский мистик, поэт и художник XIX века Вильям Блэйк и московский художник нашего тысячелетия, родившийся в башкирской деревне, в «культуре» шаманизма – Евгений Чуберов. Тот же пейзаж, та же пластика, те же предельные психоделические состояния – словно они работали в одном каноне или видели одно и то же. Об этих же соответствиях заявляет режиссер независимого американского кино Джармуш в фильме «Dead man», где встречается белый человек, носящий имя английского поэта Блэйка и индеец, живущий архаической мифологией.

Вильям Блэйк, «Тело Авеля, обнаруженное Адамом и Евой», 1826

A search for original images related to the birth of human history, its archetypes inspired a paradoxical dialogue, unconditioned by any external influences. The English mystic, poet and artist William Blake and Evgeny Chubarov, a Moscow artist of our times, born in a Bashkirian village amidst Shamanist culture, both find a place in this dialogue. The same landscape, the same artistic style, the same extreme psychedelic states, it's as if they followed the same canon or had the same things before their eyes. These similarities have also been proclaimed by the independent American film director Jarmush in his film Dead Man, in which two men are confronted: a white man named Blake, as the famous English poet, and an Indian, who is immersed in his archaic mythology.

William Blake, "The Body of Abel Found by Adam and Eve," 1826

abstraction, one that is free from the pressure of the artist's monologue and one that realizes itself in the context of a new field of meaning, packaging spontaneous feelings into intellectual reflection. It emerges naturally as densities and empty spots, inflows and gaps, as a horizontal model of a new artistic consciousness, breaking through the hypnosis of the sign surface by way of a deconstructing gesture.

The technology of Chubarov's art, its capacity for self-commentary and self-description, creates paintings that have the effect of being objects of pure nostalgia for painting. A recollection of a painting where in the thick of the information noise, as in a cocoon, a former masterpiece of abstract art is concealed, "Imperishables" to quote Ilya Kabakov. Its style and complex landscapes, brilliantly structured with all their associations, where different layers of artistic reality show through the profane, suggesting all sorts of riddles – all this testifies to the new, deep-going orientations in abstract art. They demonstrate the withering of the abstract avant-garde models and the emergence of a new corporeality, carefully thought out and genetically transformed. These new forms manipulate with the traces and debris of history and the consequences of the artist's personal experience. Intentional quotation from the world cultural heritage is evident in this art, including whole movements and trends, skillfully woven into a new cultural context. Moreover, you find in its carpet-like continuity Chubarov's self-quotation and his mythologies existing in the collisions of dissimilar returns above the imagery and style of abstract expressionism, turning his heroic structures into archeological finds and ready-made objects. Both Jackson Pollack and Mark Toby as well as the German "New Wild" are impressed in Chubarov's intellectual energy much like film stars' names are on Hollywood plates. Post-historic handwriting reveals obvious legends in their contours of the remains of gilding, where respect borders on notions much broader than cultural memory, where irony alludes to the games in the labyrinths of time and space. In Einstein's shifted geometry with its "parallel" curvilinearity and relativity, these endless labyrinths bring to mind the abandoned caves and tunnels in Egyptian pyramids. Half-filled with crumbled stone, sand-drifts and excrescencies: they can be viewed both horizontally and vertically. Here you find forgotten and lost texts that were once declared revelations and prophecies. These multi-dimensional sign-bearing structures are being cleared and sorted out to be transformed into illuminations or oppositions like paradoxical tactile surfaces or jottings on the margins where the artist himself "archaeologizes" his mysterious verbalism weaving the fabric of a universal manuscript that

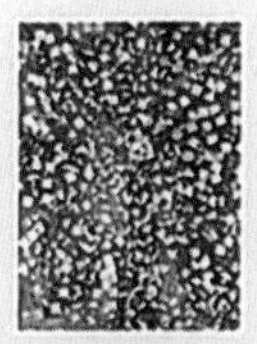

Псевдоживопись Зигмара Польке с использованием новых технологий возвращается в чистую «вещественность» и материальность Текста. Именно проблема «пространство как текст» сегодня становится общей для европейской и русской культуры, переходя из «ткани» смыслов в «ткань» пространства. Евгений Чубаров в этих текстовых слоях обретает свои собственные измерения.

Зигмар Польке, "Yggdrasil", 1984

Pseudo-painting of Sigmar Polke, based on new technologies, returns us to pure "objectivity" and materiality of the text as such. It is precisely the problem of "space as text" that has become common for European and Russian cultures, marked by the fact of transition from the texture of meaning to the texture of space. In these textual layers Evgeny Chubarov acquires his own dimension.

Sigmar Polke, Yggdrasil, 1984

Для Е.Чубарова, так же как и для А.Пенка, обращение к подсознанию давало выход к иным формам устойчивости, чем окружающая их тоталитарная реальность. «Динозавры» А.Пенка, символизирующие абсолютное сопротивление тоталитарной социальности, фактически те же змеевидные существа – знаки, наполняющие своей динамикой композиции Е.Чубарова.

А.Пенк, «G.B.I.», 1988

For Chubarov, the same as for A. Penk, address to the subconscious provided an outlet to other form of stability, distinct from those offered by the surrounding totalitarian reality. Penk's "Dinosaurs", symbolizing absolute resistance to totalitarian sociality, are in fact the same snake-like creatures-signs, which infuse with dynamism Chubarov's compositions as well.

A.R.Penk, "G.B.I." 1988

如果栏足够宽，文字足够小，那么两栏中的每一栏都将呈现出统一且可读的文本结构。整洁的文本设置可以配以其他信息，如图表或图片等

● 15.为功能而设计

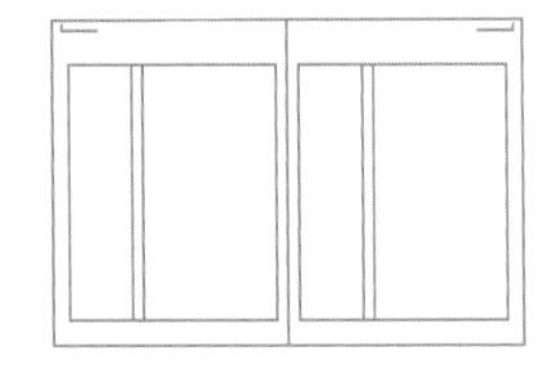

尽管双栏网格大部分使用的都是相同宽度的两栏，但两栏的宽度也可以不一样。当一个页面信息丰富，且要达到的效果是开放、可读和易于理解时，可以选择一个窄栏和一个宽栏的网格。宽栏适用于连续的文本，有助于作者（尤其是多个作者）传达连贯的信息，而窄栏可以用于容纳诸如图片说明、图片或表格等内容。

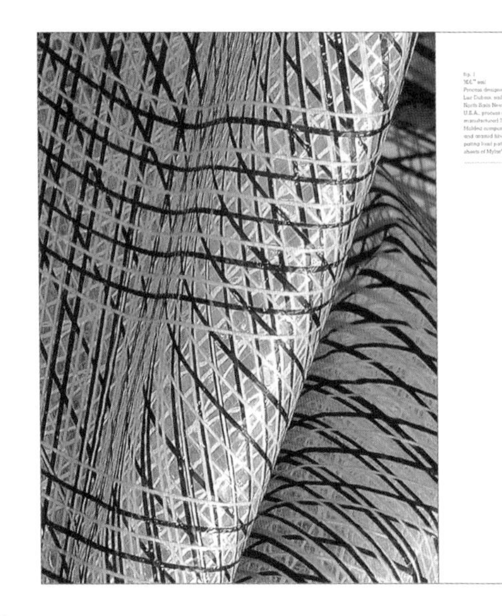

Susan Brown

TEXTILES:
FIBER, STRUCTURE,
AND FUNCTION

With the Mars Exploration Rovers (MER) six miles off the surface of Mars and traveling at twelve thousand miles per hour, the NASA team experiences "six minutes of terror": a rapid-fire series of high-stress events known as entry, descent, and landing. The cruise stage, which provides support for the voyage, is discarded. A parachute opens, slowing the descent of the craft to about 250 miles per hour. The heat shield is jettisoned, and, for a moment, the lander is hanging from a narrow, braided tether—the world on a string. Five seconds before touchdown, braking rockets fire and explosive gas generators inflate the four clusters of airbags attached to the lander. The bags hit the jagged surface at fifty miles per hour and bounce a hundred feet back up in the air, crashing down dozens more times before rolling to a stop on the rocky surface of Mars (fig. 2).[1]

The airbag system was first developed for *Pathfinder* in 1996 as part of a series of low-cost Discovery program missions, and was further refined for the Mars Exploration missions in 2003. While the animations of the projected landings are both amazing and amusing to watch, the bags are highly engineered by any standard, and performance fibers and textiles play an indispensable role in the successful design of the system (fig. 3).

无论图片说明是出现在章节开头还是文本页中，都可置于窄栏中，增加易读性。注意，比起普通页面，章节页的开头通常会在文本之前留出更多的空间（也被称作章头空白）

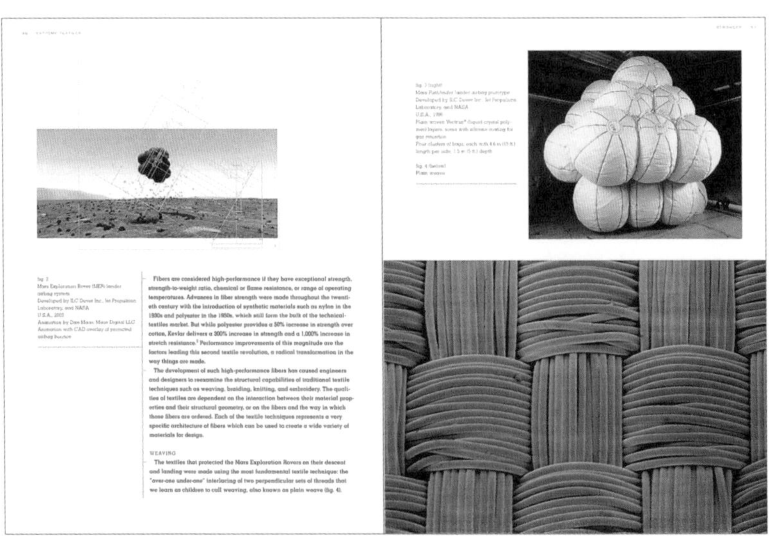

Fibers are considered high-performance if they have exceptional strength, strength-to-weight ratio, chemical or flame resistance, or range of operating temperatures. Advances in fiber strength were made throughout the twentieth century with the introduction of synthetic materials such as nylon in the 1930s and polyester in the 1950s, which still form the bulk of the technical-textiles market. But while polyester provides a 50% increase in strength over cotton, Kevlar delivers a 300% increase in strength and a 1,000% increase in stretch resistance.[2] Performance improvements of this magnitude are the factors leading this second textile revolution, a radical transformation in the way things are made.

The development of such high-performance fibers has caused engineers and designers to reexamine the structural capabilities of traditional textile techniques such as weaving, braiding, knitting, and embroidery. The qualities of textiles are dependent on the interaction between their material properties and their structural geometry, or on the fibers and the way in which those fibers are ordered. Each of the textile techniques represents a very specific architecture of fibers which can be used to create a wide variety of materials for design.

WEAVING

The textiles that protected the Mars Exploration Rovers on their descent and landing were made using the most fundamental textile technique: the "over-one under-one" interlacing of two perpendicular sets of threads that we learn as children to call weaving, also known as plain weave (fig. 4).

项目
《非常织品》（*Extreme Textiles*）展览名录

客户
美国国家设计博物馆库珀•休伊特设计博物馆（Cooper-Hewitt,National Design Museum）：非常织品展览（Extreme Textile Exhibition）

设计公司
曾•西摩设计工作室（Tsang Seymour Design）

设计总监
帕特里克•西摩（Patrick Seymour）

设计师
苏珊•布拉佐夫斯基（Susan Brzozowski）

展览名录根据材料的需要，将不同的形式组织在一起

在一个成功的、平衡的网格架构中，宽栏的宽度是窄栏的两倍。窄栏中的字体和正文字体相同，但较细。使用不同粗细的字体可以使页面构成更丰富

如果图片较少或者没有图片的话，也可以使用不对称的双栏网格，窄栏中不放任何内容

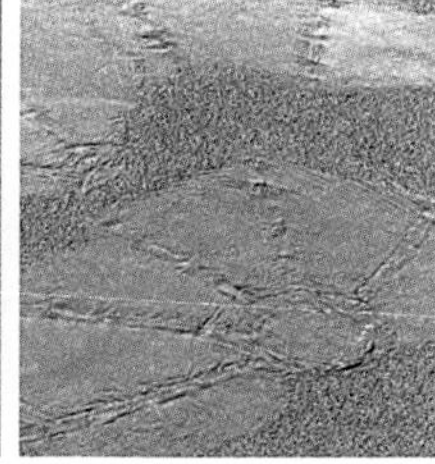

fig. 5
Impressions left by the airbags of the Mars Exploration Rover (MER) Opportunity in Martian soil, January 24, 2004

This classic plain weave has the greatest strength and stability of the traditional fabric structures. While no textiles survive from the earliest dates, impressions in clay of basic woven cloth demonstrate its use from at least 7000 BC.[3] Older than metal-working or pottery-making, perhaps even older than agriculture, cloth-weaving has a very primary relationship to the pursuits of humankind.[4]

It is fitting, then, that among the first marks made by man in the soil of Mars was that of a plain woven fabric: an impression made by the impact of the airbags (fig. 5).[5] Each bag has a double bladder and several abrasion-resistant layers made of tightly woven Vectran. Like most synthetic fibers, Vectran liquid crystal polymer is extruded from a liquid state through a spinneret, similar to a shower head, and drawn into filament fibers. The stretching of the fiber during the drawing process orients the polymer chains more fully along the fiber length, creating additional chemical bonds and greater strength. Vectran provides equal strength at one-fifth the weight of steel. Weight is of premium importance for all materials used for space travel, and Warwick Mills, the weaver of the fabric for the bags, achieved a densely woven fabric at a mere 2.4 ounces per square yard, but with a strength of 350 pounds per inch.[6]

The materials are also required to perform at severe temperatures. Because impact occurs two to three seconds after the inflation of the airbags, the fabrics endure their greatest stresses at both extremes of temperature: the explosive gasses that inflate the bags may elevate the temperature inside the bladder layers to over 212°F, but the temperature on the Martian surface is -117°F. Retraction of the airbags to allow the egress of the rovers required that the fabrics remain flexible at these very low temperatures for an extended period of time—about ninety minutes for the deflation and retraction process. Two other fiber types, aramid fibers (Kevlar 29 and Technora T-240) and ultra-high molecular weight polyethylene (UHMWPE) Spectra 1000, were also considered during the development of the Pathfinder airbags. Spectra, a super-drawn fiber, is among the strongest fibers known—fifteen times stronger than steel. However, it performs poorly at extreme temperatures, and so was eliminated early in the development process. Vectran was ultimately selected for the best performance at low temperatures, but Kevlar 129 was used for the tethers inside the bags because of its superior performance at higher temperatures.

The rovers themselves are also textile-based; they are made from super-strong, ultra-lightweight carbon-fiber composites, which are being widely used for aerospace components as well as high-performance sports equipment.[7] As composite reinforcements, textiles offer a high level of customization with regard to type and weight of fiber, use of combinations of fibers, and use of different weaves to maximize the density of fibers in a given direction. Fiber strength is greatest along the length. The strength of composite materials derives from the intentional use of this directional nature. While glass fibers are the most commonly used for composites, for high-performance products the fiber used is often carbon or aramid, or a combination of the two, because of their superior strength and light weight.

One advantage of composite construction is the ability to make a complex form in one piece, called monocoque construction. A woven textile is hand-laid in a mold; the piece is wetted out with resin and cured in an autoclave. The textile can also be impregnated with resin and cured without a wet stage. The same drape or hand that makes twill the preferred weave for most apparel is also desirable for creating the complex forms of boats, paddles, bicycle frames, and other sports equipment. The weft in a twill, rather than crossing under and over each consecutive warp, floats over more than one warp, and with each subsequent weft the grouping is shifted over one warp, creating the marked diagonal effect typical of twills (fig. 8).

Boat builders were among the first to experiment with carbon-reinforced composites. One early innovator, Edward S. ("Ted") Van Dusen, began making carbon-fiber composite racing shells in the 1970s (fig. 7). The critical factor in shell design is the stiffness-to-weight ratio, with greater stiffness meaning that more of the rower's power is translated into forward motion. Van Dusen found that all of the standard construction materials had about the same specific stiffness, or stiffness per unit weight, and began experimenting with glass, boron, and carbon fiber–reinforced composites.[8]

For his Advantage racing shells, Van Dusen uses glass fiber in a complex twill commonly known as satin weave. In a satin, each weft may float over

分隔线既可以用来分隔空间，又可以用来连接分栏

The numbers in these tables represent typical values of some important fiber properties; the actual behavior of fibers may differ as variants are produced for diverse end uses. These numbers were compiled from many different sources and are meant for illustration purposes only.

COMPARISON OF YARN STRENGTH

COMPARISON OF MODULI

CARBON

Thomas Edison first used carbon fiber when he employed charred cotton thread to conduct electricity in a lightbulb (he patented it in 1879). Only in the past fifty years, however, has carbon developed as a high-strength, high-modulus fiber.[8] Oxidized then carbonized from polyacrylonitrile (PAN) or pitch precursor fibers, carbon's tenacity and modulus vary depending on its starting materials and process of manufacture.[9]

Less dense than ceramic or glass, lightweight carbon-fiber composites save fuel when used in aerospace and automotive vehicles. They also make for strong, efficient sports equipment. Noncorroding, carbon reinforcements strengthen deep seawater concrete structures such as petroleum production risers.[10] Fine diameter carbon fibers are woven into sails to minimize stretch.[11] In outer apparel, carbon fibers protect workers against open flames (up to 1000°C/1,800°F) and even burning napalm: they will not ignite, and shrink very little in high temperatures.[12]

ARAMIDS

Aramids, such as Kevlar (DuPont) and Twaron® (Teijin), are famous for their use in bulletproof vests and other forms of ballistic protection, as well as for cut resistance and flame retardance. Initially developed in the 1960s, aramids are strong because their long molecular chains are fully extended and packed closely together, resulting in high-tenacity, high-modulus fibers.[13]

Corrosion- and chemical-resistant, aramids are used in aerial and mooring ropes and construction cables, and provide mechanical protection in optical fiber cables.[14] Like carbon, aramid-composite materials make light aircraft components and sporting goods, but aramids have the added advantages of impact resistance and energy absorption.

LIQUID CRYSTAL POLYMER (LCP)

Although spun from different polymers and processes, LCPs resemble aramids in their strength, impact resistance, and energy absorption, as well as their sensitivity to UV light. Compared to aramids, Vectran (Celanese), the only commercially available LCP, is more resistant to abrasion, has better flexibility, and retains its strength longer when exposed to high temperatures. Vectran also surpasses aramids and HMPE in dimensional stability and cut resistance: it is used in wind sails for America's Cup races, inflatable structures, ropes, cables and restraint-lines, and cut-resistant clothing.[15] Because it can be sterilized by gamma rays, Vectran is used for medical devices such as implants and surgical-device control cables.[16]

HIGH-MODULUS POLYETHYLENE (HMPE)

HMPE, known by the trade names Dyneema (Toyobo/DSM) or Spectra (Honeywell), is made from ultra-high molecular-weight polyethylene by a special gel-spinning process. It is the least dense of all the high-performance

DECOMPOSITION TEMPERATURE

DENSITY

fibers, and the most abrasion-resistant. It is also more resistant than aramids, PBO, and LCP to UV radiation and chemicals.[17] It makes for moorings and fish lines that float and withstand the sun, as well as lightweight, cut-resistant gloves and protective apparel such as fencing suits and soft ballistic armor. In composites, it lends impact resistance and energy absorption to glass- or carbon-reinforced products. HMPE conducts almost no electricity, making it transparent to radar.[18] HMPE does not withstand gamma-ray sterilization and has a relatively low melting temperature of 150°C (300°F)—two qualities that preclude its use where high temperature resistance is a must.

POLYPHENYLENE BENZOBISOXAZOLE (PBO)

PBO fibers surpass aramids in flame resistance, dimensional stability, and chemical and abrasion resistance, but are sensitive to photodegradation and hydrolysis in warm, moist conditions.[19] Their stiff molecules form highly rigid structures, which grant an extremely high tenacity and modulus. Apparel containing Zylon® (Toyobo), the only PBO fiber in commercial production, provides ballistic protection because of its high energy absorption and dissipation of impact. Zylon is also used in the knee pads of motorcycle apparel, for heat-resistant work wear, and in felt used for glass formation.[20]

PIPD

PIPD, M5 fiber (Magellan Systems International), expected to come into commercial production in 2005, matches or exceeds aramids and PBO in many of its properties. However, because the molecules have strong lateral bonding, as well as great strength along the oriented chains, M5 has much better shear and compression resistance. In composites it shows good adhesion to resins. Its dimensional stability under heat, resistance to UV radiation and fire, and transparency to radar expands its possible uses. Potential applications include soft and hard ballistic protection, fire protection, ropes and tethers, and structural composites.[21]

HYBRIDS

A blend of polymers in a fabric, yarn, or fiber structure can achieve a material better suited for its end use. Comfortable fire-retardant, anti-static clothing may be woven primarily from aramid fibers but feature the regular insertion of a carbon filament to dissipate static charge. Yarns for cut-resistant applications maintain good tactile properties with a wrapping of cotton around HMPE and fiberglass cores. On a finer level, a single fiber can be extruded from two or more different polymers in various configurations to exhibit the properties of both.

16.控制使用分隔线

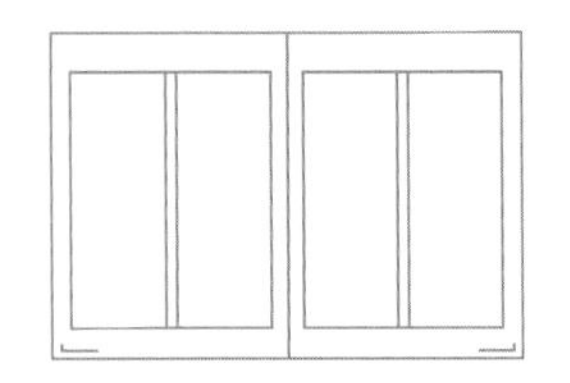

有时，说明性材料包含很多分散的信息块，这时，页面中的栏之间要有空间，以提高文章的可读性。在这样的情况下，垂直分隔线可以用来分隔栏，平行分隔线可以通过分隔正文与框形素材来分隔栏中的信息。另外，也可以用另一种平行分隔线将栏外注脚、页码与整个文本区域分隔开。

注意，过多的分隔线会让页面看上去呆板、无趣。

垂直分隔线将不同的信息块分布在各自的列中，这些信息块包含不同类型的字体和格式，如黑体、大写字母、斜体

NONFAT ROASTED GARLIC DRESSING

MAKES about 1 ½ cups
PREP TIME: 10 minutes
TOTAL TIME: 2 hours (includes 1 ½ hours roasting and cooling time)

To keep this recipe nonfat, we altered our usual technique for roasting garlic, replacing the oil we typically use with water.

2 large garlic heads
2 tablespoons water
Salt
2 tablespoons Dijon mustard
2 tablespoons honey
6 tablespoons cider vinegar
½ teaspoon pepper
2 teaspoons minced fresh thyme, or ½ teaspoon dried
½ cup low-sodium chicken broth

1. Adjust an oven rack to the upper-middle position and heat the oven to 400 degrees. Following the photos on page 000, cut ½ inch off the top of the garlic head to expose the tops of the cloves. Set the garlic head cut side down on a small sheet of aluminum foil, and sprinkle with the water and a pinch of salt. Gather the foil up around the garlic tightly to form a packet, place it directly on the oven rack, and roast for 45 minutes.

2. Carefully open just the top of the foil to expose the garlic and continue to roast until the garlic is soft and golden brown, about 20 minutes longer. Allow the roasted garlic to cool for 20 minutes, reserving any juices in the foil packet.

3. Following the photo on page 000, squeeze the garlic from the skins. Puree the garlic, reserved garlic juices, ¾ teaspoon salt, and the remaining ingredients together in a blender (or food processor) until thick and smooth, about 1 minute. The dressing, covered, can be refrigerated for up to 4 days; bring to room temperature and whisk vigorously to recombine before using.

LOWFAT ORANGE-LIME DRESSING

MAKES about 1 cup
PREP TIME: 10 minutes
TOTAL TIME: 1 hour (includes 45 minutes simmering and cooling time)

Although fresh-squeezed orange juice will taste best, any store-bought orange juice will work here. Unless you want a vinaigrette with off flavors make sure to reduce the orange juice in a nonreactive stainless steel pan.

2 cups orange juice (see note above)
3 tablespoons fresh lime juice
1 tablespoon honey
1 tablespoon minced shallot
½ teaspoon salt
½ teaspoon pepper
2 tablespoons extra-virgin olive oil

1. Simmer the orange juice in a small saucepan over medium heat until slightly thickened and reduced to ⅔ cup, about 30 minutes. Transfer to a small bowl and refrigerate until cool, about 15 minutes.

2. Shake the chilled, thickened juice with the remaining ingredients in a jar with a tight-fitting lid until combined. The dressing can be refrigerated for up to 4 days; bring to room temperature, then shake vigorously to recombine before using.

Test Kitchen Tip: **REDUCE YOUR JUICE**

Wanting to sacrifice calories, but not flavor or texture, we adopted a technique often used by spa chefs in which the viscous quality of oil is duplicated by using reduced fruit juice syrup or roasted garlic puree. The resulting dressings are full bodied and lively enough to mimic full-fat dressings but without the chemicals or emulsifiers often used in commercial lowfat versions. Don't be put off by the long preparation times of these recipes—most of it is unattended roasting, simmering, or cooling time.

Salads 65

项目
《美国实验厨房家庭烹饪书》（*America's Test Kitchen Family Cookbook*）

客户
美国实验厨房（America's Test Kitchen）

设计公司
BTDnyc

艺术总监
艾米•克里（Amy Klee）

水平分隔线放在页眉或者页脚，可以用来分隔信息或框住整个文本框

EASY JELLY-ROLL CAKE

MAKES an 11-inch log
SERVES 10
PREP TIME: 5 minutes **TOTAL TIME:** 1 hour

Any flavor of preserves can be used here. For an added treat, sprinkle 2 cups of fresh berries over the jam before rolling up the cake. This cake looks pretty and tastes good when served with dollops of freshly whipped cream (see page 000) and fresh berries.

- **¾ cup all-purpose flour**
- **1 teaspoon baking powder**
- **¼ teaspoon salt**
- **5 large eggs, at room temperature**
- **¾ cup sugar**
- **½ teaspoon vanilla extract**
- **1¼ cups fruit preserves**
- **Confectioners' sugar**

1. Adjust an oven rack to the lower-middle position and heat the oven to 350 degrees. Lightly coat a 12 by 18-inch rimmed baking sheet with vegetable oil spray, then line with parchment paper (see page 000). Whisk the flour, baking powder, and salt together and set aside.

2. Whip the eggs with an electric mixer on low speed, until foamy, 1 to 3 minutes. Increase the mixer speed to medium and slowly add the sugar in a steady stream. Increase the speed to high and continue to beat until the eggs are very thick and a pale yellow color, 5 to 10 minutes. Beat in the vanilla.

3. Sift the flour mixture over the beaten eggs and fold in using a large rubber spatula until no traces of flour remain.

4. Following the photos, pour the batter into the prepared cake pan and spread out to an even thickness. Bake until the cake feels firm and springs back when touched, 10 to 15 minutes, rotating the pan halfway through baking.

5. Before cooling, run a knife around the edge of the cake to loosen, and flip the cake out onto a large sheet of parchment paper (slightly longer than the cake). Gently peel off the parchment paper attached to the bottom of the cake and roll the cake and parchment up into a log and let cool for 15 minutes.

MAKING A JELLY-ROLL CAKE

1. Using an offset spatula, gently spread the cake batter out to an even thickness.

2. When the cake is removed from the oven, run a knife around the edge of the cake to loosen, and flip it out onto a sheet of parchment paper.

3. Starting from the short side, roll the cake and parchment into a log. Let the cake cool seam-side down (to prevent unrolling) for 15 minutes.

4. Unroll the cake. Spread 1¼ cups jam or preserves over the surface of the cake, leaving a 1-inch border at the edges.

5. Re-roll the cake gently but snugly around the jam, leaving the parchment behind as you go.

6. Trim thin slices of the ragged edges from both ends. Transfer the cake to a platter, dust with confectioners' sugar, and cut into slices.

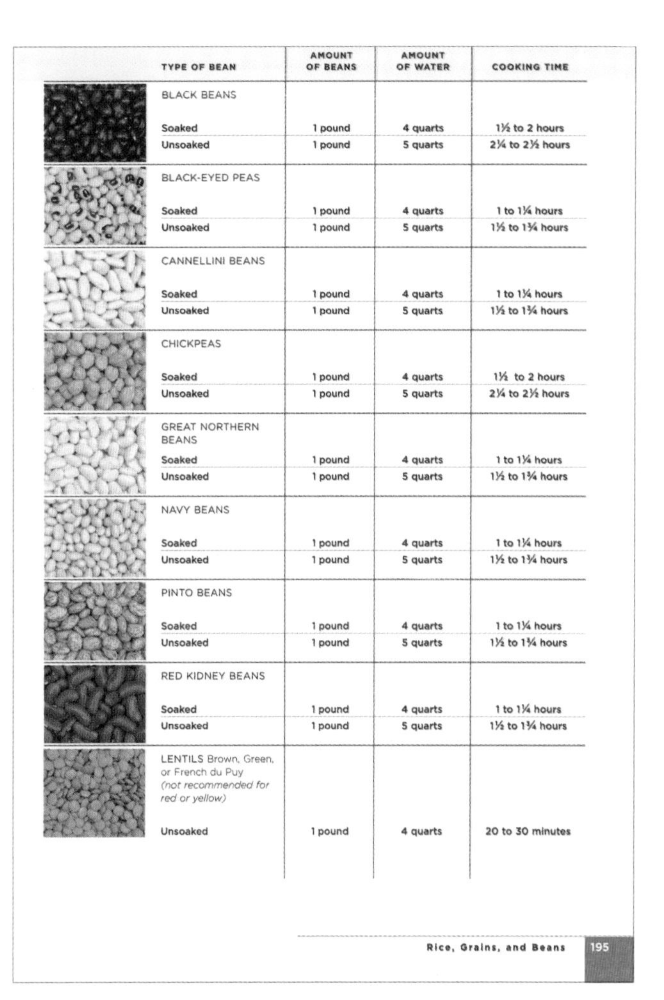

TYPE OF BEAN		AMOUNT OF BEANS	AMOUNT OF WATER	COOKING TIME
BLACK BEANS	Soaked	1 pound	4 quarts	1½ to 2 hours
	Unsoaked	1 pound	5 quarts	2¼ to 2½ hours
BLACK-EYED PEAS	Soaked	1 pound	4 quarts	1 to 1¼ hours
	Unsoaked	1 pound	5 quarts	1½ to 1¾ hours
CANNELLINI BEANS	Soaked	1 pound	4 quarts	1 to 1¼ hours
	Unsoaked	1 pound	5 quarts	1½ to 1¾ hours
CHICKPEAS	Soaked	1 pound	4 quarts	1½ to 2 hours
	Unsoaked	1 pound	5 quarts	2¼ to 2½ hours
GREAT NORTHERN BEANS	Soaked	1 pound	4 quarts	1 to 1¼ hours
	Unsoaked	1 pound	5 quarts	1½ to 1¾ hours
NAVY BEANS	Soaked	1 pound	4 quarts	1 to 1¼ hours
	Unsoaked	1 pound	5 quarts	1½ to 1¾ hours
PINTO BEANS	Soaked	1 pound	4 quarts	1 to 1¼ hours
	Unsoaked	1 pound	5 quarts	1½ to 1¾ hours
RED KIDNEY BEANS	Soaked	1 pound	4 quarts	1 to 1¼ hours
	Unsoaked	1 pound	5 quarts	1½ to 1¾ hours
LENTILS Brown, Green, or French du Puy *(not recommended for red or yellow)*	Unsoaked	1 pound	4 quarts	20 to 30 minutes

Rice, Grains, and Beans 195

信息单元间的空隙分隔了水平的元素，让页面看上去清晰干净

水平分隔线还可以用来组织元素。当有大批信息要处理时，水平分隔线可以将页码或栏外注脚与核心内容区分开

● 17.在秩序中增加流动感

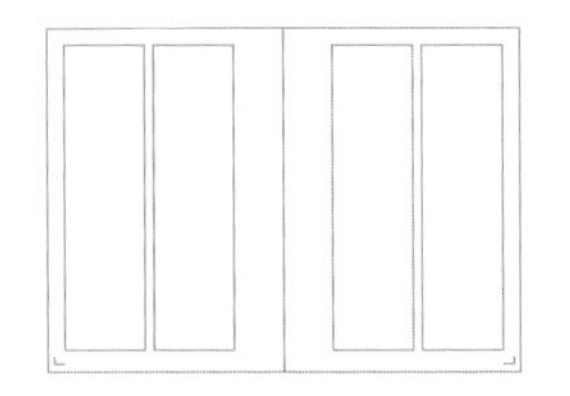

多样化的设计包括增加图片的宽度和使用不同宽度的字体

双栏网格是一种常见的框架，能使内容更有连贯性。图片可以轻松地插入网格内，图片上下可以增加图片说明。不仅如此，一旦基础框架建立起来，就有空间增加对开页面的多样性。比较宽的图片可以占两栏，也可以将图片说明放在页边空白处，这样可以使整个页面生动起来，既保持了秩序感，又增添了韵律感。

DESIGN

TO KICKSTART INCOMES

MARTIN FISHER

As an engineer, I am convinced that technology has a critical role to play in ending the scourge of global poverty. Finding and developing such technologies have been my life's work for more than twenty years, and in particular since 1991, when I teamed up with Nick Moon to establish KickStart. The tools and technologies our organization has designed have already helped more than 230,000 people in Africa escape from poverty, but we have only scratched the surface of what is needed and what is possible. It has been an exciting journey, with both failures and successes, and a lot of learning.

I started on this path in 1985, after finishing my Ph.D. in Mechanical Engineering at Stanford. I was uninspired by my prospects for employment—teaching at a university or conducting research for the military or big oil companies—so I went to Peru to hike in the Andes and ponder my future. It was there that I first encountered extreme poverty, and began to think about how every civilization in history had advanced through technology and invention. I returned from Peru determined to use my engineering skills to help the poor, and received a Fulbright fellowship to go to Kenya to study the Appropriate Technology Movement.

The Appropriate Technology Movement was inspired in the mid-1970s by E. F. Schumacher's book *Small Is Beautiful*. According to Schumacher, small-scale, locally made technologies in the hands of the poor would help bring an end to poverty, and this concept had quickly become all the rage in development circles. However, by the time I arrived in Africa in 1985, I discovered that the Appropriate Technology Movement had been declared dead. Once again, development agencies had spent millions of dollars with very little lasting impact.

I expected to be in Kenya for ten months; I stayed for seventeen years. I spent the first years studying what had gone wrong with the Appropriate Technology Movement and worked with a major nonprofit agency to implement "integrated rural-development programs." This had become the new fashion, and it involved providing a large amount of development resources to a small geographical area in the hopes that this would raise people out of poverty. We used state-of-the-art methods, but when we went back to the communities a few years later, we found that our work had few enduring effects. The community water systems we had built were crumbling. The income-generating groups we had trained had collapsed, and the things we had given away had created dependency rather than empowerment. Disappointed, I tried to find out what had gone wrong and how we could fix it, and also why forty years and trillions of dollars in development projects had actually left Africa deeper in poverty. During this period I met and worked closely with Nick. The lessons we learned during this time—and those we continue to learn—formed the foundation on which KickStart was built.

One of most important lessons was that the poorest people in the world are also among the most entrepreneurial—they have to be just to survive. They do not want handouts, they want opportunities. Nick and I created KickStart to address this fundamental fact.

1. Samuel Gichanu, a farmer from Kaharati of the Maragua District in Central Kenya, with his French beans for export.

2. A farmer on his Super MoneyMaker pump, outside of Nairobi, Kenya.

3. Compacting the blocks with the MoneyMaker Block Press in Soja, Uganda.

4. Teresa Mukeshi of Nairobi, Kenya, cooking yellow beans on the Kenya Ceramic Jiko Stove.

I hope that Design for the Other 90% will inspire others to address social issues through design and innovation. And to this end I would like to share a few of the most important lessons that we have learned about how to use technology to help abate poverty.

FIRST THINGS FIRST: INCOME

If I could offer one piece of advice, it would be this: What poor people need most is a way to make money. Today, everyone in the world lives in a cash economy, and we need money to feed our families, buy clothes, build our houses, pay for healthcare, and educate our children. Even in the poorest parts of Africa, a subsistence lifestyle is no longer viable, and without a way to make money, people simply cannot survive. Too often, designers for the developing world fail to understand this most basic fact, and focus instead on trying to make life a bit easier for the poor. But in order to have a significant impact in ending poverty, we need to focus on developing innovations that can be used directly by the poor to generate new income (fig. 2). In other words, the devices have to be used to produce goods or services that can be sold on the local market. In this way, whoever owns the device can use it to make money. This is KickStart's fundamental principle.

THE POOR DO NOT GENERALLY LACK TIME AND LABOR

It is tempting to design devices for the poor that will save them time and labor—after all, we pay dearly for such devices. However, time and labor are two things that many poor people have in abundance. Unless they have a way to make money with the saved time or labor, they are unlikely to invest much in such devices. For example, as long as they can get water from a free source, very few poor families will dig a well on their plot (which will cost them money) just to get drinking water—even if that source is many miles away. They make a decision that the time saved getting drinking water is not worth the initial costs of digging the well. For them, this money is better spent on basic food, clothing, shelter, healthcare, and education.

MONEY-SAVING DEVICES ONLY MAKE SENSE IF THEY ARE CHEAP

Some devices are promoted to help users save money. But the very poor live hand-to-mouth, and spend almost every penny as soon as they earn it. They have little money to save and even less to spend on money-saving devices. In Africa, the poorest people buy cooking oil and spices by the teaspoon and kales by the leaf. Cash flow is their biggest constraint, and, unless there is a way to offer them credit, it is difficult to get them to invest in a product that will save them money.

There are a few exceptions to this rule, and they are instructive to study. The Kenya Ceramic Jiko (KCJ) is a fuel-efficient charcoal cooking stove first designed in the early 1980s (fig. 4). It is forty percent more efficient than a common charcoal stove, and costs only about $2 more. Users of the KCJ save money because they need less charcoal, and at the same time the demand for charcoal—one of the biggest causes of deforestation—is reduced. The KCJ required millions of donor dollars to be spent on promotion over a fifteen-year period before it finally took off, but it now commands seventy percent of market share.

To an inventor who is working on an idea that will save poor people money, I offer this guideline: The product needs to sell for not much more than the price of a chicken in the local marketplace. A chicken is a luxury that even the very poor can afford from time to time. If the cost is this low, and the cost savings created are significant, you may have a successful product. If the cost is much higher (as it is for example with solar electric lanterns), then, unless financing can be provided, the only buyers will be the emerging middle class who already have enough money to cover their basic needs.

NOT INCOME OVER TIME, INCOME NOW

A new tool or technology for the poor must generate significant income in a very short period of time. So if an investment takes a long time to pay for itself, or goes bad altogether, they will go hungry for many months. Because the vast majority of the world's poor are rural farmers, we have to think in "farm time." Farmers are used to putting their money in the ground (in the form of seeds and fertilizer) and waiting three to six months to recoup their investment. This is another of the guiding design principles we use. Our tools have to completely pay for themselves in less than six months—preferably closer to three months.

UNDERSTAND THE POOR AND SOLVE THEIR PROBLEMS

In designing for the world's poor, there is too often a focus on developing things that "we" think "they" need. We design technologies that address a problem we have defined, without understanding the true needs of the people we are trying to help. Solar cookers are a good example. There are many clever designs for cookers which allow users to harness the power of the sun to prepare meals. The idea is that these cookers save forests from being cut down, slow global warming, and reduce the time required to collect firewood. These are laudable goals, but in most places, poor families prepare their largest meals in the evening, when the day's work is done. They prefer to cook inside so their neighbors cannot see what they are cooking, and often rely on a cooking fire for warmth, light, and to help repel mosquitoes. Moreover, poor families are unlikely to buy solar cookers unless they are very cheap and there is absolutely no other locally available cooking fuel.

THE POWER OF IRRIGATION

KickStart started by developing the means to make low-cost building materials, such as building blocks, out of soil and cement (fig. 3). Then we developed a technology to extract cooking oil from sunflower seeds. By 1997, we realized that to increase our impact we needed to focus on solutions for the eighty percent of Africa's poor who are small-scale farmers. They eke out an existence on small plots of land—often less than an acre per family. Their biggest asset is their land and their strongest skill is basic farming. We needed to help them leverage these assets, and irrigation was the missing piece. With irrigation, a farmer can grow crops all year round, instead of only once or twice a year when the rain comes. They can grow high-value crops such as fruits and vegetables, and, best of all, they can bring them to market in the dry season, when no one else has crops and the prices are high (fig. 1). But poor farmers in Africa cannot afford petrol pumps for irrigation, and almost none of them have electricity. So we developed a line of low-cost, manually operated irrigation pumps, appropriately named MoneyMaker Pumps. These lightweight, portable pumps can pull water from wells as deep as twenty feet or from rivers and ponds, and push it through a hose pipe under pressure to efficiently irrigate crops (fig. 6).

AVOID GIVEAWAYS—CREATE DIGNITY, NOT DEPENDENCE

From the very start, we at KickStart have sold our products. We do not give our pumps away, and this has caused some controversy, but we remain firm in this for a few reasons.

First, we want to create a sustainable solution to poverty, and no giveaway program can be sustainable. By selling our pumps, we create a sustainable supply chain. Every party has a vested interest in making our pumps, spares, and accessories available to anyone who wants one.

Second, we want to be as cost-effective as possible—moving as many people out of poverty for the least amount of money. When people buy something, they value it and are more likely to use it. These early adopters are more driven, more entrepreneurial, and more likely to succeed. Their success helps to inspire their more cautious neighbors and relatives. With giveaways, there are more likely to be a lot of early failures, and the technologies may never take off.

A great technology without a supply chain and promotion will have no impact. Perhaps the biggest reason that the Appropriate Technology Movement failed was because it tried to use a supply chain which ignored some of the basic rules of economics. It relied on the romantic notion that a new tool or technology could be made by individual end users or by local artisans spread across the countryside. The idea was that this would be more sustainable, create local jobs, and provide the valuable new technology. The idea was appealing, but fundamentally flawed. You and I are not asked to build our own automobiles, computers,

项目
《为另外的90%而设计》（*Design for the Other 90%*）展览名录

客户
史密森学会，美国国家设计博物馆库珀·休伊特设计博物馆

设计公司
曾·西摩设计工作室

设计总监
帕特里克·西摩

艺术总监/设计师
劳拉·霍威尔（Laura Howell）

流畅的布局能把一个故事讲得非常清晰

5. Felix Mururi with his MoneyMaker Hip Pump on his farm in Maragua District, Kenya

lawn mowers, and cell phones. They are made in large quantities in big factories. The economy of scale created by centralized manufacturing lowers the price, making the product affordable and ensuring higher quality and reliability. KickStart does the same thing. By centralizing our manufacturing in the most advanced factories available, we can produce high-quality, durable products at a lower cost (figs 7, 8). Wholesalers and middlemen move these goods from factory to marketplace, making a profit in the process. A network of more than 500 local retail shops in three countries stock and sell our pumps. This supply chain needs no artificial support, and will exist as long as there is consumer demand. KickStart also uses donor funds to market the new technologies and generate demand. As with any new product, this takes both time and money. When you are selling an expensive item to the poorest people in the world, it takes even longer and is more expensive (again, the KC) is a perfect example). But eventually, we will reach a point where we can end our marketing efforts and sell each pump at a profit, which we will then reinvest in developing new technology and expanding into new countries. This is a sustainable supply chain.

Third, there is a question of fairness. I have heard people say that it is not "fair" to ask poor people to invest in their own future, but is it fair to give one person or one village a gift when there are others just as needy? By making our products available through the marketplace, they are available to everyone, without patronage or favoritism. This is perhaps the hardest lesson for someone who wants to do good in the world. We see people in desperate need and want to alleviate their suffering. This spirit of generosity is human nature at its best. But as noble as this motivation is in the giving, it is demoralizing in the receiving. When people invest in themselves and their own futures, they have full ownership of their success, and that creates dignity.

INDIVIDUAL OWNERSHIP WORKS BEST

A good question to ask about any program is, Who will own the new technology? If the answer is unclear, or vague, then the program is unlikely to succeed in the long term. We have learned that individual ownership works better than group ownership. Africa is covered with failed community-owned technologies—tractors, water pumps, ambulances, water purification and irrigation systems, et cetera. The list goes on.

There is a common idea that poor people will come together for their collective benefit, or that "investing" in a community is more cost-effective or efficient than working with individuals. There are some situations where this works, like building roads or farmers' cooperatives. But it is much less likely to be effective with the joint ownership of a physical asset. The problem is that if everybody owns an asset, in reality, nobody owns it, and if nobody owns it, nobody will maintain it. Unless there is a way to extract a payment from everyone who uses the asset to cover the costs of maintenance, repair, and replacement, you have the classic free-rider problem.

It comes down to this. The poorest people in the world are just like you and me. No matter how community-minded we are, we will take care of the needs of our family first. And we value the most the items we had to work for.

DESIGN FOR AFFORDABILITY

Our best-selling Super MoneyMaker Pump can be used to irrigate more than two acres of land, and on average the users make $1,000 profit from selling fruits and vegetables in the first year of use. We continue to work to reduce the cost, but at $95, it is still too expensive for many families.

In response, we designed the Hip Pump, which can irrigate almost an acre and retails for less than $35. It looks like a bicycle-tire pump pivoted on a hinge at the end of a small platform. However, unlike a bicycle pump, it uses the operator's whole body. It is lightweight, portable, and extremely easy to use.

The Hip Pump has been a tremendous success: its initial production run of 750 units sold out almost immediately. One of them was bought by Felix Muruń, a young man from rural Kenya. He had a wife and three children to support, but they owned no land. Felix left his family to seek work in Nairobi, where he managed to earn $40 a month working in a restaurant in the city's slums, sending what he could home to his wife and children. When he saw the Hip Pump, he realized he could make more money farming back in his village. He saved his money, bought a pump, went home, and rented six small plots of land. He grew tomatoes, kale, baby corn, and French beans, which he sold to middlemen who took them to the city. Felix planted different crops on each of his small plots so he would have harvests at different times of the year. When we visited Felix three months after he started using his pump, he had already made $580 profit, and he and his wife were talking eagerly about buying land and building their own house. This small pump had enabled Felix to turn his own sweat and drive into cash, look after his family, and plan for his future (fig. 5).

MEASURE THE IMPACT OF WHAT YOU DO

Measuring real impact or outcome is where many would-be social entrepreneurs fail. The number of products you have sold or distributed tells the world nothing. You have to measure the change you are hoping to create with your invention. It is hard and expensive to do, but it is vital. We have learned a great deal from our impact-monitoring efforts. Not only does it enable us to measure ourselves against the goals we have set, it has also been hugely valuable in the design and improvement of our products and marketing efforts.

These are KickStart's core values, and they come together to create a very cost-effective and sustainable way to help people help themselves out of poverty. None of these principles are unique to KickStart or our technologies.

6. A farmer waters her French bean crop with water from a MoneyMaker pump, outside of Nairobi, Kenya.

7, 8. Kenya Vehicle Manufacturers (KVM), located in Thika, north of Nairobi, is one of the companies KickStart partners with to manufacture MoneyMaker Pumps.

They can be applied to many other technologies to make a real difference in the world. Each of these is important individually, but in our experience it is their combination that makes them truly effective.

Finally, for those people who are driven to innovate for the developing world (and also for those who are eager to fund such efforts), I offer this test. A truly successful program to develop and promote new technologies and/or business models needs to meet the following four criteria:

DOES THE PROGRAM CREATE MEASURABLE AND PROVEN IMPACT? This means that you need to carefully define the problem you are trying to solve, then carefully monitor and measure the actual impact you are having on that problem. In the case of KickStart, we are trying to bring people out of poverty by enabling them to earn more money. So we carefully measure how much more money the buyers of our technologies make as a result of owning them. If a program cannot create and prove real impact, then it is not worth implementing.

IS THE PROGRAM COST-EFFECTIVE? There are limited funds for developing and promoting new technologies, and we need to ensure that whatever is done uses these funds efficiently. "Cost-effective" is a subjective measure, so we offer this comparison. KickStart spends about $250 of donor funds to take an average family out of poverty, whereas a more traditional aid program claims on its Web site to do the same for $2,750.

IS THERE A SUSTAINABLE EXIT STRATEGY? One has to ensure that the benefits will continue to accrue for both the existing and new beneficiaries, even after the donor funds are depleted. Creating a program that continues to depend on donor funds forever is not a viable solution. There are four different ways that an effort can become sustainable: 1) build and leave in place a profitable supply chain to continue providing the goods/services, 2) hand over the program to a government which will fund it using tax money, 3) create a local situation that can continue to prosper without the injection of any new outside funds, for instance, establishing a local group savings and loan (merry-go-round) system, 4) completely eliminate the problem, such as eradicating a disease.

IS THE MODEL REPLICABLE AND SCALABLE? The problems we are trying to solve—poverty and climate change, among others—are very large in scale, and it is expensive to develop new technologies and new business models. So we want to ensure that the technologies themselves as well as the dissemination models are not too dependent on specific local conditions, and can be easily adapted to many different settings and locations.

Incorporating all of these guidelines into your work will be a challenge, but great inventors and designers enjoy a challenge. I can tell you that this experience has been an exciting, sometimes frustrating, often exhausting, and immensely satisfying journey. I wish you a fantastic journey of your own.

KICKSTART'S DESIGN PRINCIPLES:
Any tool or technology KickStart produces must meet all of the following design criteria:

INCOME-GENERATING — Any tool must have a profitable business model attached to it.

RETURN ON INVESTMENT — The business opportunity must be available to thousands of people, and the business must be profitable enough that the entrepreneur recoups his or her investment in six months or less.

AFFORDABILITY — We design our tools to retail at less than a few hundred dollars, ideally less than $100.

ENERGY-EFFICIENCY — All of our tools are human-powered, so they must be extremely efficient in converting human power into mechanical power.

ERGONOMICS AND SAFETY — Our products must be able to be used for long periods of time without injury.

PORTABILITY — Tools must be small and light enough to transport from store to home on foot, by bike, or by minibus.

EASE OF INSTALLATION AND USE — Tools must be easy to set up and use, without additional tools or training.

STRENGTH AND DURABILITY — Our tools are used in harsh conditions and will be pushed to their limits. They must be built to withstand abuse. We offer a one-year guarantee on all of our products.

DESIGN FOR AVAILABLE MANUFACTURING CAPACITY — Mass production keeps cost down, but locally available materials and processes can dictate the design.

CULTURAL ACCEPTABILITY — Local cultures will not change to adopt a new technology; the technology has to be adapted to local customs.

ENVIRONMENTAL SUSTAINABILITY — Our tools must not create a negative impact on the environment.

18.划分易于理解的区块

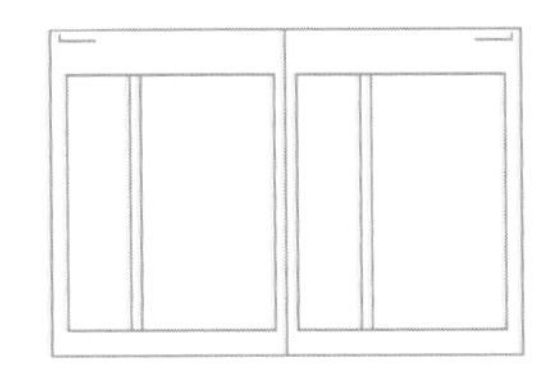

好的设计表现力强，能将素材与读者联系起来。无论出版物的用途是什么，成功的排版都需要以清晰易懂的方式划分区块。在一页或者一个故事里，区块可以是水平的，也可以是垂直的，同时保持有秩序感的整体性。排版的关键在于确保素材的相关度，尤其要确保读者能立刻理解基本内容；确保一个标题或者多个标题独立放置；确保图片说明位置准确，与图片联系紧密，便于读者阅读，尤其是一些说明性的材料。

あの人に会って、聞きたかった、大切なこと。

佐橋慶女さん、教えてください。

「消えつつある日本の生活文化、日本各地で出会った暮らしの知恵」

新聞記者だったときに日本各地を歩いて、おばあちゃんたちの知恵の豊かさに驚き、それを伝えるべく「伝承塾」として活動してきた佐橋さん。もちろん、自らの暮らしにも取り入れてきた。伝統食は？ 民間薬は？ 住まい方は？ いろいろなこと、全部、聞きたい。

木曽のお寺のお庫裏さんが、ゆべしを檀家の人たちと作り始めると、お寺が柚子の香りでいっぱいなんですって。

标题与正文分开，标题有时在页面的边缘，有时被置于页面中央。文本的各个部分被留白或分隔线分开，为图片说明留出醒目的位置

项目
《可颂》（*Croissant*）杂志

客户
《可颂》杂志社（*Croissant* Magazine）

艺术总监/设计师
马场诚子（Seiko Baba）

《可颂》是日本杂志，主要面向30岁以上的女性群体。这本杂志的版式既美观又清晰，它是一本杂志书（Mook），由《可颂》编辑部特别编辑出版，该书的标题是*Mukashi nagara no kurashi no chie*，意思是历久不衰的生活智慧

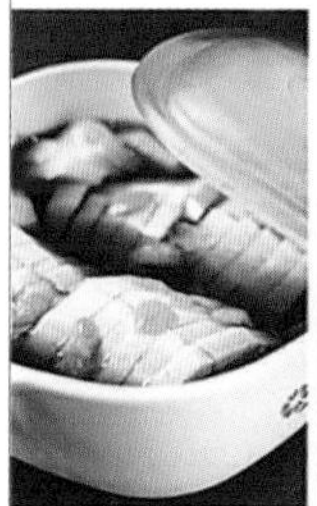

左・大根の二杯酢柚子巻き漬け。こうしておくと、いつまででももつ。いつ何どき人が来ても、慌てずに出せる。お茶とでもお酒とでもおいしい。
中・大根の皮はキンピラにする。「ちょっとだけ砂糖を入れるとおいしいのよ。ほんのちょっとだけですよ。で、唐辛子入れて——種取らないとダメよ。飛び上がっちゃうから。で、お醤油をタラーッとまわしかけて。味は自分の好みでね」
右・大根の葉は油揚げと炒める。味付けはショウガ醤油で。この葉っぱと皮の本体は、37ページで煮た大根。

大根は、葉っぱから尻尾まで全部食べられるのよ。皮はキンピラにして、ね。

薬用酒各種。「山椒の実、カリン、アロエ、ビワの葉、ニンニク、ナナカマド、クコ、クロモジ。毒草以外は、みんな薬用酒になります。効きます」。左から2つ目のアロエのお酒の作り方は、アロエの葉の棘を包丁でこそげ取り、1cmの厚さに切る。レモンも同じくらいの厚さの輪切りにする。広口瓶にアロエとレモンを入れ、果実酒用焼酎を入れて、冷暗所で保存。漬けて2～3か月したら飲める。

上左・押し寿司の押し器。上右・瓢箪のお弁当箱と、曲げわっぱのお弁当箱は「桐の、いい木を使わないと薄く切れないんですって」。手前・おつまみ入れ。「全部、お盆に伏せると平らになって、上にものがのっかるの」

右・お客さん用の靴袋。玄関に靴が並びきらず、思いついた。来た人はこの袋に自分の靴を入れ、廊下に。「間違わないように、袋も紐も、いろいろな色で作ったんです」。上・下駄箱の戸は空気が通るよう隙間がある。

「この畳は、いぐさの産地の九州は八代で作ってもらったの」。夏、マットの上に敷いて寝れば冷房要らず。「ヨガにも使うし。巻いとけば場所とらないしね」

薬用酒なんて、台所の納戸にい～っぱい！ 昔のものは、一つのものに効くんでなく、「効くんだとさ」、なんです。

会った木曽地方のおばあさんが作っていたもの。これを知人の恵那・明智の寺のお庫裡さまに紹介し、今ではその寺と檀家の人たちの産物になっています。作り方は、柚子をくり抜いてクルミやゴマなどと味噌を練り合わせて詰め、藁づとで1か月間、北側の風通しのいいところで、日干ししてね」

　佐橋さんは柚子のできない明智と四国の香り高い柚子を紹介し、知恵と産物を結びつけて、女性の産業にした。まさしく地域の産物と、知恵と文化の、交換と交流。

　撮影用に佐橋さんが切ってくれる。1切れ、2切れ、3切れ。あ、それで充分です。と言うと、

「3切れは身を切る、1切れは人を切る、と言ってね、5切れでめでたしとしましょうか」。

　野菜の二杯酢漬けと柚子味噌。これがあれば、心強いですよ。

「これは私が作った柚子味噌。今日は大根にのせます」

　さっきから、いい匂いがしていたのは大根を煮ている匂いだったのだ。

「お豆腐とかコンニャクにのせてもいいし、ねぎでもセロリでも細く切って、晒しておいて、柚子味噌で和えたりね。で、大根は、葉っぱから尻尾まで全部、食べられるんです。葉っぱは炒めて、皮はキンピラにするの。母たちは大根をたくさん煮たでしょ。皮はニンジンとキンピラにしてました」

　大根の皮はパリッとした歯ざわりもご馳走だ。

「大根、二杯酢に漬けておくと便利よ。カブでもキュウリでもね。二杯酢に漬けておくと、いつまでもおいしいんです。しなびないし。急なときに出せるでしょ。突然、電話がかかって、『これから行くけど、いい？』と言われても、いいわよ、何かあるでしょ、ってね」

　迎えられる。余裕である。

「だって、この間、アメリカのみやげにもらったカシューナッツを出したら、言うんですよ」

　やあだ、佐橋さんのところに来て、こんなの食べなくても～。

「ちっとも売れないの。エシャロットをひねショウガと漬けるのも、今年、たくさん作ったのに、みなさん、平らげちゃったの。にんにくをお醤油に漬けとくと、あらゆるお料理に使えますよ。お肉焼いたり、炒め物したりするときに、香り付けにいいです。にんにく特有の臭みはなくなって、でいて、旨味を出してくれるのね」

　使っては漬け、使っては漬けできるのも、ありがたい。

「でも、にんにくだって味噌に漬けるところもあって、東北はそうですね。鹿児島は焼酎漬けにするんです。35度の焼酎。でないともたないんでしょうね。暑いから。私は、ミョウガは二杯酢にも漬けるけど、焼酎と酢と氷砂糖にも漬けるんです。焼酎に漬けたほうが長くもつから」

　冷蔵庫のない時代、25度、35度の焼酎こそが支えだった。

「すべて冷暗所ですから。そうそう、昔は台所に床下の冷暗所があったんですよね」

　今、床下はなくて冷蔵庫はあるけれど、冷蔵庫には到底、入りきらない。一体どっちが便利やら。

　医者が「年ですね」ですますが、ビワの葉で治ったんです。

「薬用酒もたくさんありますからね。うちは。私ね、この間、脚が動かなくなっちゃって、腫れちゃって。寝てたんですけど、立つのも、寝返りもできないんです。枕元に電話があるんで」

　ちょっと来て！　いちばん近くに住む人にヘルプコールした。

「ビワの葉っぱで治ったんです」

　焼酎に漬けてあったビワの葉で湿布したのである。

「痛いところに葉っぱのせて、温めたコンニャクのせて。昔からの痛み止めなの。母が、よく、そう言ってました。そのときは、何言ってるのか、と思ったけど、ビワの葉の効用がこんなにあるとは思わなかった。お医者さんに行ったら、『年ですね』って言われるだけよ」

　昔はよく、どの家にもビワの木や柿の木が植わっていた。

「垣根の外に植えたものね。落ちるものは家運が落ちる、傾くということで。——薬用酒。私は今ね、山椒の実のがいっていいうんで。それ、飲んでもいいし。湿布にもいい」

　他にもたくさん。カリン、アロエ、ナナカマド、クロモジ。

「何でも。毒草以外はみんな効きます。マタタビもありますよ。猫を酔わせちゃうくらいで、人間にもいいらしいから」

　そんなアバウトな、と、思いました？　では、あらためて。

「山椒酒。冷えをとる。温めて飲めば風邪のときにいいです。カリンも体を温めるのと、あと、咳にいい。……私、これ、20年に漬けてるわね」

　アロエは神経的な痛みに、いちばん効く。また、精神的な不調にもいい。

「昔のものは、一つのものに効くんじゃないんです」

　自然は「一つのことにしか役に立ちません」なんて、そんなケチなことしない。昔の人は、それをよーく知っていた。

「昔の人って大学出てるわけじゃないし、小学校出ただけのおばあちゃんが、『効くんだとさ』ですもの。口伝ね。おばあちゃんが嫁いできたお嫁さんに伝えて、それを、その娘さんが聞いて。——そうなの！　最初にや

不同区块的字体标示着不同的信息。在这里，连续的正文与分步解说排列于不同的区域

19.连贯性与趣味性

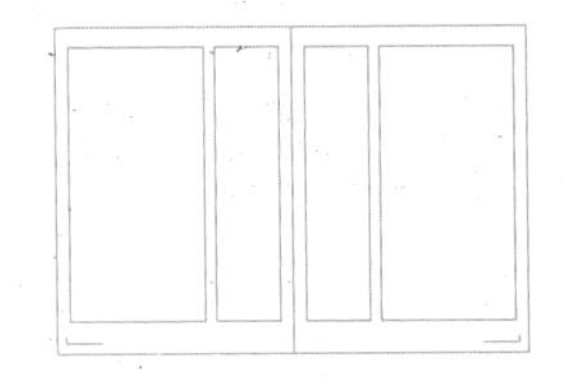

大多数成功的网格都具有连贯、有秩序感、清晰、结构性强的特点，在确保这些特点的基础上，网格还能够混搭各种元素。双栏网格可以设置不同宽度的列，在视觉上增加项目的张力和动感。即使采用一些不同寻常的变化让设计更鲜活，一个稳定的基础结构依然可以确保在设计具有戏剧感的同时还能架构明晰。

在大多数的项目中，连续的元素包括：

◎页面顶部的标题区域。

◎页面左端或右端同一个区域的连续文本框，作用是为读者提供一个有效的标志。

◎页脚的栏外脚注和页码，引导读者从头到尾阅读整篇文章。

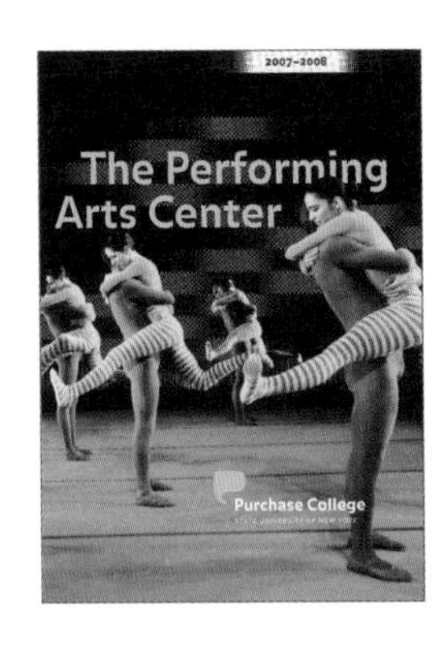

整个宣传册通过一个主模板来展示关键信息，关键的描述性文本和辅助信息很显眼。即使有着丰富的装饰细节，结构也依然清晰

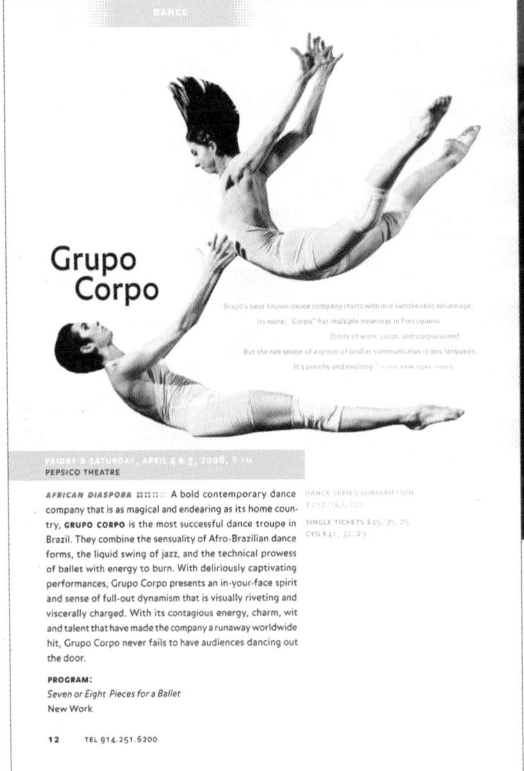

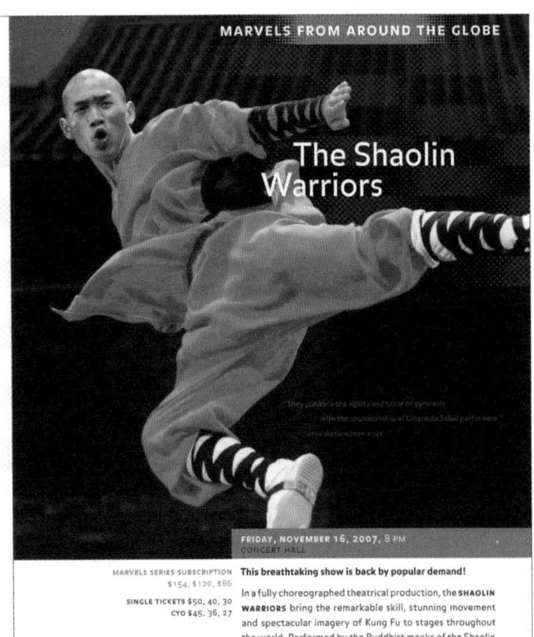

项目

《帕切斯学院表演艺术中心宣传册》

客户

美国纽约州立大学帕切斯学院（SUNY Purchase）

设计公司

Heavy Meta工作室

艺术总监

芭芭拉·格劳伯（Barbara Glauber）

设计师

希拉里·格林鲍姆（Hilary Greenbaum）

一个好的组织结构可以利用一些奇特的变化来让设计生动起来

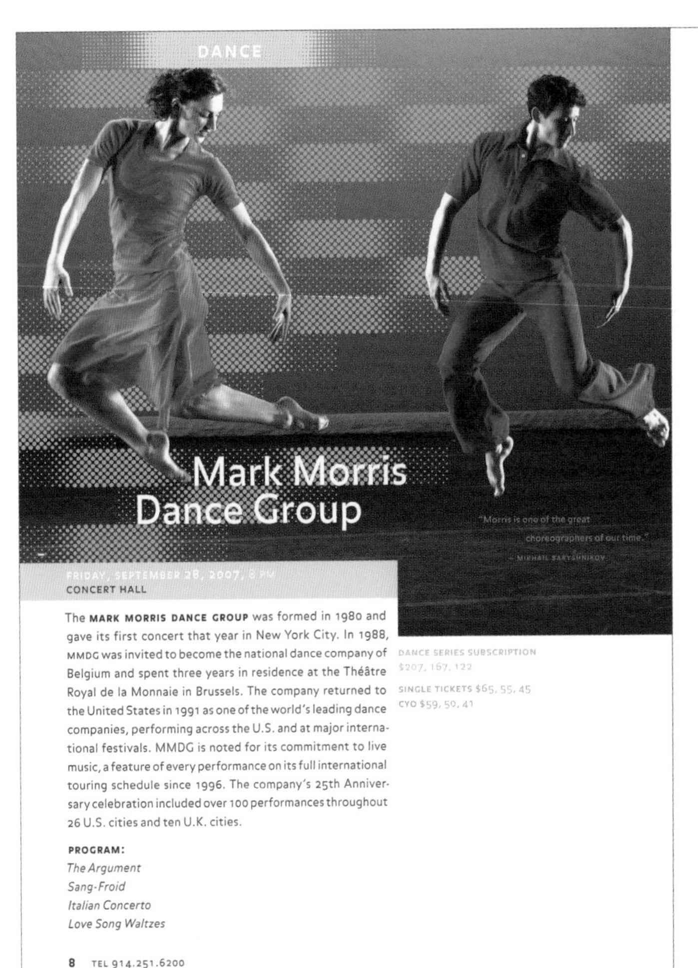

FRIDAY, SEPTEMBER 28, 2007, 8 PM
CONCERT HALL

The **MARK MORRIS DANCE GROUP** was formed in 1980 and gave its first concert that year in New York City. In 1988, MMDG was invited to become the national dance company of Belgium and spent three years in residence at the Théâtre Royal de la Monnaie in Brussels. The company returned to the United States in 1991 as one of the world's leading dance companies, performing across the U.S. and at major international festivals. MMDG is noted for its commitment to live music, a feature of every performance on its full international touring schedule since 1996. The company's 25th Anniversary celebration included over 100 performances throughout 26 U.S. cities and ten U.K. cities.

DANCE SERIES SUBSCRIPTION
$207, 167, 122
SINGLE TICKETS $65, 55, 45
CYO $59, 50, 41

PROGRAM:
The Argument
Sang-Froid
Italian Concerto
Love Song Waltzes

8 TEL 914.251.6200

FRIDAY & SATURDAY, OCT 19 & 20, 2007, 8 PM
PEPSICO THEATRE

Integrating China's traditional culture with influences from abroad and contemporary dance technique, **BEIJING LDTX** offers a unique and seamless blending of these three elements in a repertoire that shows off unsurpassed technical skill and choreographic excellence.

DANCE SERIES SUBSCRIPTION
$207, 167, 122
SINGLE TICKETS $45, 35, 25
CYO $41, 32, 23

FRIDAY'S PROGRAM: *The Cold Dagger* is the company's new full-evening work, choreographed by Li Han-zhong and Ma Bo. Based on the traditional Chinese game of Weigi, this intricately choreographed look at human confrontation juxtaposes incredible acrobatics with paired movement that would be otherwise impossible on a normal stage.

SATURDAY'S PROGRAM: A rep program that includes *All River Red*, a striking piece performed to Stravinsky's classic, *The Rite of Spring*; coupled with the company's newest commissioned work *Pilgrimage*, featuring music by the "father of Chinese rock," Cui Jian.

WWW.ARTSCENTER.ORG 9

上图：该项目的版式结构紧密，层次清晰。首次出现的标题比较大，之后出现的标题采用同样大小的文本框，但是字号更小。时间和地点用同样的颜色条装饰，但是采用了更直接的处理方法。设计师充分考虑了所有的关系，保证了层次的清晰

上页图：这些横向图覆盖了大半个页面，而插入图片中的文本框和颜色条为页面增加了动感和戏剧性。表演者的名字置于图片中各个显眼的位置，增加了页面的质感和喜剧感

Ronald K.Brown/ Evidence

Urban Bush Women/ Compagnie Jant-Bi

色彩与信息搭配和谐

右图：页面的轮廓和留白使阅读更有节奏

20.其他模板

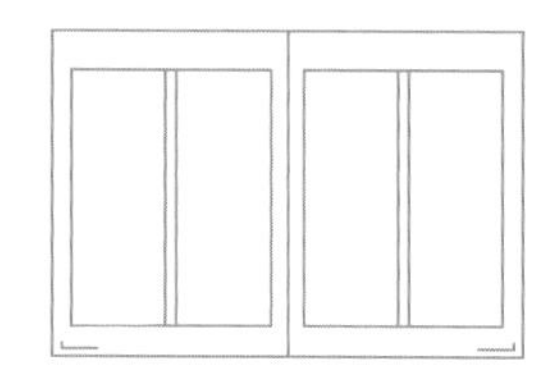

在一个作品中，混合使用不同的网格和排版系统是合理的。当页面存在不同的信息时，即使是清晰的双栏网格也需要稍作调整，以保证清晰和平衡的效果。

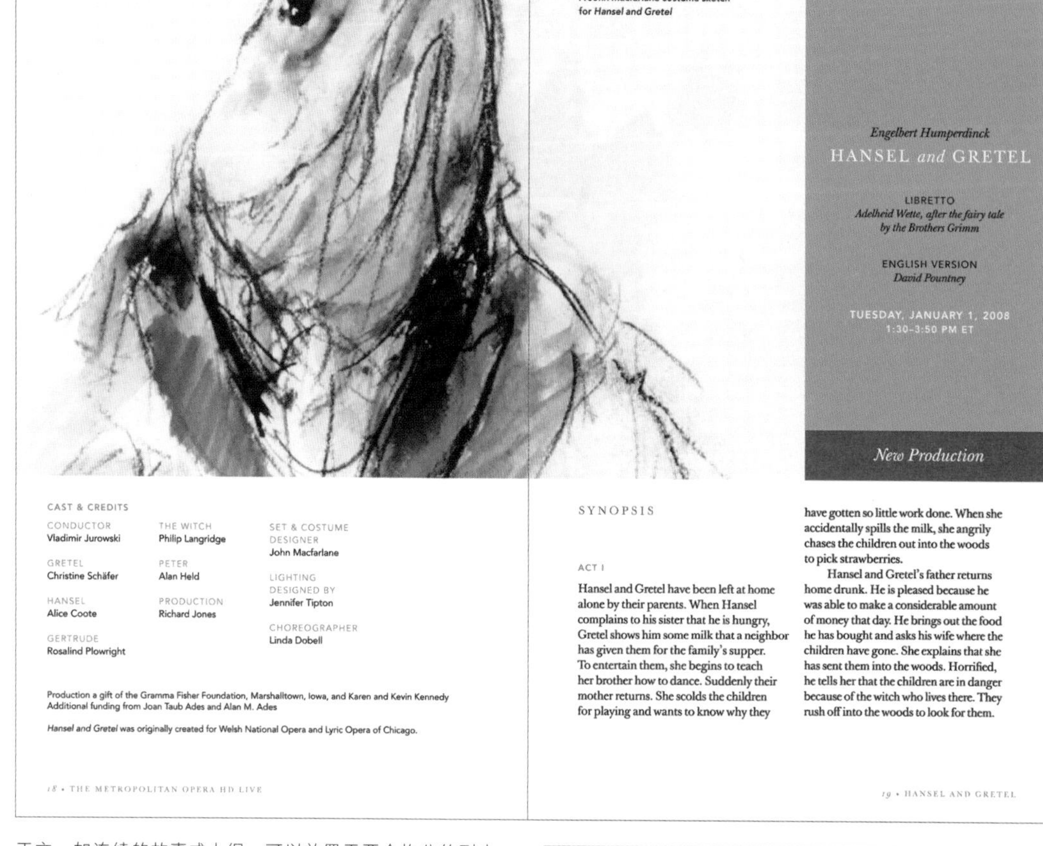

A John Macfarlane costume sketch for *Hansel and Gretel*

Engelbert Humperdinck

HANSEL *and* GRETEL

LIBRETTO
Adelheid Wette, after the fairy tale by the Brothers Grimm

ENGLISH VERSION
David Pountney

TUESDAY, JANUARY 1, 2008
1:30–3:50 PM ET

New Production

CAST & CREDITS

CONDUCTOR
Vladimir Jurowski

GRETEL
Christine Schäfer

HANSEL
Alice Coote

GERTRUDE
Rosalind Plowright

THE WITCH
Philip Langridge

PETER
Alan Held

PRODUCTION
Richard Jones

SET & COSTUME DESIGNER
John Macfarlane

LIGHTING DESIGNED BY
Jennifer Tipton

CHOREOGRAPHER
Linda Dobell

Production a gift of the Gramma Fisher Foundation, Marshalltown, Iowa, and Karen and Kevin Kennedy
Additional funding from Joan Taub Ades and Alan M. Ades

Hansel and Gretel was originally created for Welsh National Opera and Lyric Opera of Chicago.

18 • THE METROPOLITAN OPERA HD LIVE

SYNOPSIS

ACT I

Hansel and Gretel have been left at home alone by their parents. When Hansel complains to his sister that he is hungry, Gretel shows him some milk that a neighbor has given them for the family's supper. To entertain them, she begins to teach her brother how to dance. Suddenly their mother returns. She scolds the children for playing and wants to know why they have gotten so little work done. When she accidentally spills the milk, she angrily chases the children out into the woods to pick strawberries.

Hansel and Gretel's father returns home drunk. He is pleased because he was able to make a considerable amount of money that day. He brings out the food he has bought and asks his wife where the children have gone. She explains that she has sent them into the woods. Horrified, he tells her that the children are in danger because of the witch who lives there. They rush off into the woods to look for them.

19 • HANSEL AND GRETEL

正文，如连续的故事或大纲，可以放置于两个均分的列内。

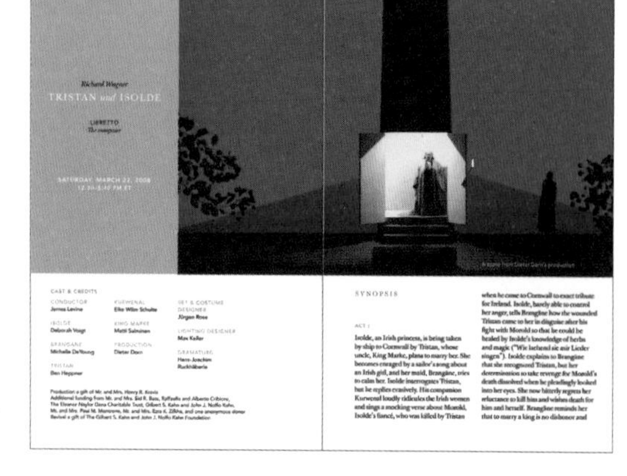

为每场表演设计的章节以大尺寸、戏剧性强的照片作为开头

项目
《2007—2008 HD项目指南》

客户
纽约大都会歌剧院

设计公司
亚当斯•莫瑞卡公司

创意总监
肖恩•亚当斯，诺林•莫瑞卡（Noreen Morioka）

艺术总监
莫妮卡•施劳格（Monica Schlaug）

设计师
莫妮卡•施劳格，克里斯•泰伦

一个内敛、经典又生动的设计为宣传册增添了年轻的活力，展现了一种历久弥新的艺术形式

Alice Coote and Christine Schäfer sing the title roles.

GRETEL WAKES HANSEL,
and the two find themselves in front of a gingerbread house.

ACT II

Gretel sings while Hansel picks strawberries. When they hear a cuckoo calling, they imitate the bird's call, eating strawberries all the while, and soon there are none left. In the sudden silence of the woods, the children realize that they have lost their way and grow frightened. The Sandman comes to bring them sleep by sprinkling sand on their eyes. Hansel and Gretel say their evening prayer. In a dream, they see 14 angels protecting them.

ACT III

The Dew Fairy appears to awaken the children. Gretel wakes Hansel, and the two find themselves in front of a gingerbread house. They do not notice the Witch, who decides to fatten Hansel up so she can eat him. She immobilizes him with a spell. The oven is hot, and the Witch is overjoyed at the thought of her banquet. Gretel has overheard the Witch's plan, and she breaks the spell on Hansel. When the Witch asks her to look in the oven, Gretel pretends she doesn't know how: the Witch must show her. When she does, peering into the oven, the children shove her inside and shut the door. The oven explodes, and the many gingerbread children the Witch had enchanted come back to life. Hansel and Gretel's parents appear and find their children. All express gratitude for their salvation.

20 • THE METROPOLITAN OPERA HD LIVE

IN FOCUS

Engelbert Humperdinck

HANSEL *and* GRETEL

PREMIERE: HOFTHEATER, WEIMAR, 1893

Originally conceived as a small-scale vocal entertainment for children, *Hansel and Gretel* outgrew its original design to become the most successful fairy-tale opera ever created. Like so many children's classics, *Hansel and Gretel* achieved greatness because it resonates with both adults and kids. The composer Engelbert Humperdinck was a protégé of the musical titan Richard Wagner, and the score of *Hansel and Gretel* is flavored with the sophisticated musical lessons he learned from his idol while maintaining a charm and a light touch that were entirely Humperdinck's own. The ancient tale of the young brother and sister who get lost in a dark forest and almost get eaten by an old witch became a classic of German literature in the famous collected stories of the Brothers Grimm. The opera acknowledges the darker features present in the story, yet presents them within a frame of grace and humor. Humperdinck's fellow composer Richard Strauss was delighted with this score from the start and conducted its world premiere. *Hansel and Gretel* has been internationally popular ever since and must be one of the very few operas that can boast equal acclaim from such diverse and demanding critics as children and musicologists.

THE CREATORS

Engelbert Humperdinck (1854–1921) was a German composer who began his career as an assistant to Richard Wagner in Bayreuth in a variety of capacities, including tutoring Wagner's son Siegfried in music and composition. Humperdinck even composed a few minutes of orchestral music for the world premiere of Wagner's *Parsifal* (1882) when extra time was needed to effect a scene change. (This music is not included in the printed score of *Parsifal* and is no longer performed). *Hansel and Gretel* was Humperdinck's first complete opera and remains the foundation of his reputation. The world premiere of his later opera *Königskinder* took place at the Met and was one of the sensations of the company's 1910–11 season, following less than three weeks after the world premiere of Puccini's *La Fanciulla del West*. *Hansel and Gretel*, however, is the only one of Humperdinck's works to remain in the repertory. The libretto was written by his sister, Adelheid Wette (1858–1916), and is based on the famous fairy tale from the Grimms' collection. The brothers Jacob (1785–1863) and Wilhelm (1786–1859) Grimm were German academics whose groundbreaking linguistic work revolutionized the understanding of language development. Today, they are best remembered for editing and publishing collections of folk tales.

THE SETTING

In the libretto, the opera's three acts move from Hansel and Gretel's home to the dark forest to the witch's gingerbread house deep in the forest. Put another way, the drama moves from the real, through the obscure, and into the unreal and fantastical. In this production, which takes the idea of food as its dramatic focus, each act is set in a different kind of kitchen, informed by a different theatrical style: a D.H. Lawrence-inspired setting in the first, a German Expressionist one in the second, and a Theater of the Absurd mood in the third.

THE MUSIC

The score of *Hansel and Gretel* successfully combines accessible charm with subtle sophistication. Like Wagner, Humperdinck assigns musical themes to certain ideas and then transforms the themes according to new developments in the drama. Much of this development occurs in the orchestra, like the chirpy cuckoo, depicted by the winds in Act II, which becomes

21 • HANSEL AND GRETEL

调整排版以区分信息，排版展示出衬线字体与无衬线字体之间的有趣对比

如果需要展示不同类型的信息，如问答形式，可以采用双栏网格，问题放在窄栏，回答放在宽栏

that Tristan is simply performing his duty. Isolde maintains that his behavior shows his lack of love for her, and asks Brangäne to prepare a death potion. Kurwenal tells the women to prepare to leave the ship, as shouts from the deck announce the sighting of land. Isolde insists that she will not accompany Tristan until he apologizes for his offenses. He appears and greets her with cool courtesy ("Herr Tristan trete nah"). When she tells him she wants satisfaction for Morold's death, Tristan offers her his sword, but she will not kill him. Instead, Isolde suggests that they make peace with a drink of friendship. He understands that she means to poison them both, but still drinks, and she does the same. Expecting death, they exchange a long look of love, then fall into each other's arms. Brangäne admits that she has in fact mixed a love potion, as sailors' voices announce the ship's arrival in Cornwall.

ACT II

In a garden outside Marke's castle, distant horns signal the king's departure on a hunting party. Isolde waits impatiently for a rendezvous with Tristan. Horrified, Brangäne warns her about spies, particularly Melot, a jealous knight whom she has noticed watching Tristan. Isolde replies that Melot is Tristan's friend and sends Brangäne off to stand watch. When Tristan appears, she welcomes him passionately. They praise the darkness that shuts out all false appearances and agree that they feel secure in the night's embrace ("O sink hernieder, Nacht der Liebe"). Brangäne's distant voice warns that it will be daylight soon ("Einsam wachend in der Nacht"), but the lovers are oblivious to any danger and compare the night to death, which will ultimately unite them. Kurwenal rushes in with a warning: the king and his followers have returned, led by Melot, who denounces the lovers. Moved and disturbed, Marke declares that it was Tristan himself who urged him to marry and chose the bride. He does not understand how someone so dear to him could dishonor him in such a way ("Tatest Du's wirklich?"). Tristan cannot answer. He asks Isolde if she will follow him into the realm of death. When she accepts, Melot attacks Tristan, who falls wounded into Kurwenal's arms.

ACT III

Tristan lies mortally ill outside Kareol, his castle in Brittany, where he is tended by Kurwenal. A shepherd inquires about his master, and Kurwenal explains that only Isolde, with her magic arts, could save him. The shepherd agrees to play a cheerful tune on his pipe as soon as he sees a ship approaching. Hallucinating, Tristan imagines the realm of night where he will return with Isolde. He thanks Kurwenal for his devotion, then envisions Isolde's ship approaching, but the Shepherd's mournful tune signals that the sea is still empty. Tristan recalls the melody, which he heard as a child. It reminds him of the duel with Morold, and he wishes Isolde's medicine had killed him then instead of making him suffer now. The shepherd's tune finally turns cheerful. Tristan gets up from his sickbed in growing agitation and tears off his bandages, letting his wounds bleed. Isolde rushes in, and he falls, dying, in her arms. When the shepherd announces the arrival of another ship, Kurwenal assumes it carries Marke and Melot, and barricades the gate. Brangäne's voice is heard from outside, trying to calm Kurwenal, but he will not listen and stabs Melot before he is killed himself by the king's soldiers. Marke is overwhelmed with grief at the sight of the dead Tristan, while Brangäne explains to Isolde that the king has come to pardon the lovers. Isolde, transfigured, does not hear her, and with a vision of Tristan beckoning her to the world beyond ("Mild und leise"), she sinks dying upon his body.

44 • THE METROPOLITAN OPERA HD LIVE

CLOSE-UP

SCALING THE HEIGHTS

Deborah Voigt and **Ben Heppner** on how they'll ascend opera's Mount Everest—the title roles of *Tristan und Isolde*—with a little help from Maestro James Levine.

Debbie, you've only sung Isolde on stage once before, several years ago. Why the long interval?

Deborah Voigt: I first sang the part in Vienna five years ago. It came along sooner than I anticipated, but the circumstances were right and I decided to go ahead and sing it. When you sing a role as difficult as Isolde, people are going to want you to sing it a lot, and I didn't want to have a lot of them booked if it didn't go well. So I didn't book anything until the performances were over. The first opportunity I had after Vienna are the Met performances.

Ben, what makes you keep coming back to Tristan?

Ben Heppner: Before it starts, it feels like I'm about to climb Mount Everest. But from the moment I step on the stage to the last note I sing it feels like only 15 minutes have gone by. There is something so engaging about this role that you don't notice anything else. It takes all of your mental, vocal, and emotional resources to sing. And I like the challenge of it.

The two of you appear together often, and you've also both worked a lot with James Levine.

DV: Maestro Levine is so in tune with singers—how we breathe and how we work emotionally. I remember I was having trouble with a particular low note, and in one performance, he just lifted up his hands at that moment, looked at me and took a breath, and gave me my entrance. The note just landed and hasn't been a problem since.
BH: He has this wonderful musicality that is so easy to work with. As for Debbie, we just love singing together and I think that is really its own reward.

This *Tristan* will be seen by hundreds of thousands of people around the globe. How does that impact your stage performance?

DV: None of us go out to sing a performance thinking that it is any less significant than another, so my performance will be the same. But when you are playing to a huge opera house, gestures tend to be bigger. For HD, some of the operatic histrionics might go by the wayside.
BH: When the opera house is filled with expectant listeners—that becomes my focus. The only thing I worry about is that it's a very strenuous role, and I'm basically soaking wet from the middle of the second act on! ■

45 • TRISTAN UND ISOLDE

● 21.设计要看起来简洁明了

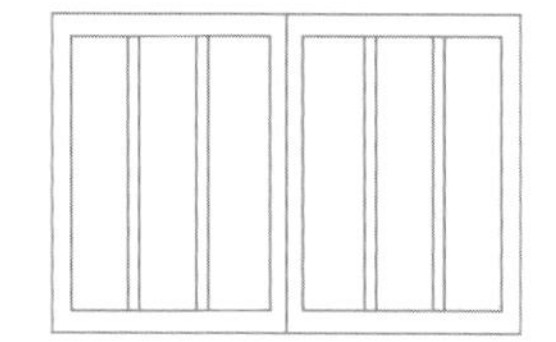

最成功的设计是看起来简单，却具有通用性的设计。一个看似开放和有很多空间的设计可以包含很多素材，尤其是在一本书或一个商品名录中。

如果项目同时包含文本和图片，要考虑两者之间的比例，并确定文本和图片各需要多少空间。当图片说明比较长，并包含大量额外的信息（如作者信息和附加信息）时，可以通过使用不同的字体，设置小一点儿的字号，或者调整元素之间的距离来区分图片说明和正文。

一种结构上的处理方式是采用三栏网格，但让它看上去是单栏或双栏设计。使用其中两栏放置单个文本，位于页面的右侧，这样能使正文看起来很干净，空间较大的左边空白处也可以用来放置较长的图片说明。

如果素材需要，双栏说明可以替代单独的文本栏，允许说明和图片位于同一页面上。使用三栏网格时，可以将图片的大小设置为一栏、两栏、三栏宽，甚至覆盖整个页面。

项目
《幸福的灵魂》（*Beatific Soul*）

客户
纽约公共图书馆（New York Public Library），斯卡拉出版社（Scala Publishers）

设计师
凯蒂•霍曼斯（Katy Homans）

作为一本指南，该书展示了杰克•凯鲁亚克（Jack Kerouac）的生活、职业生涯、日记及手稿，主要突出了他那本里程碑式的小说——《在路上》（*One the Road*）。三栏网格让版式有了更多灵活的变化，使页面看上去更有空间感，更平静，有一种简单的美

这个简单而通用的多栏网格可以容纳各种信息。正文较宽的行距和衬线字体使文字更易于阅读。说明位于左侧栏中，并设置为无衬线字体，以求清晰明确。页面结构可以适应文本的各种变化

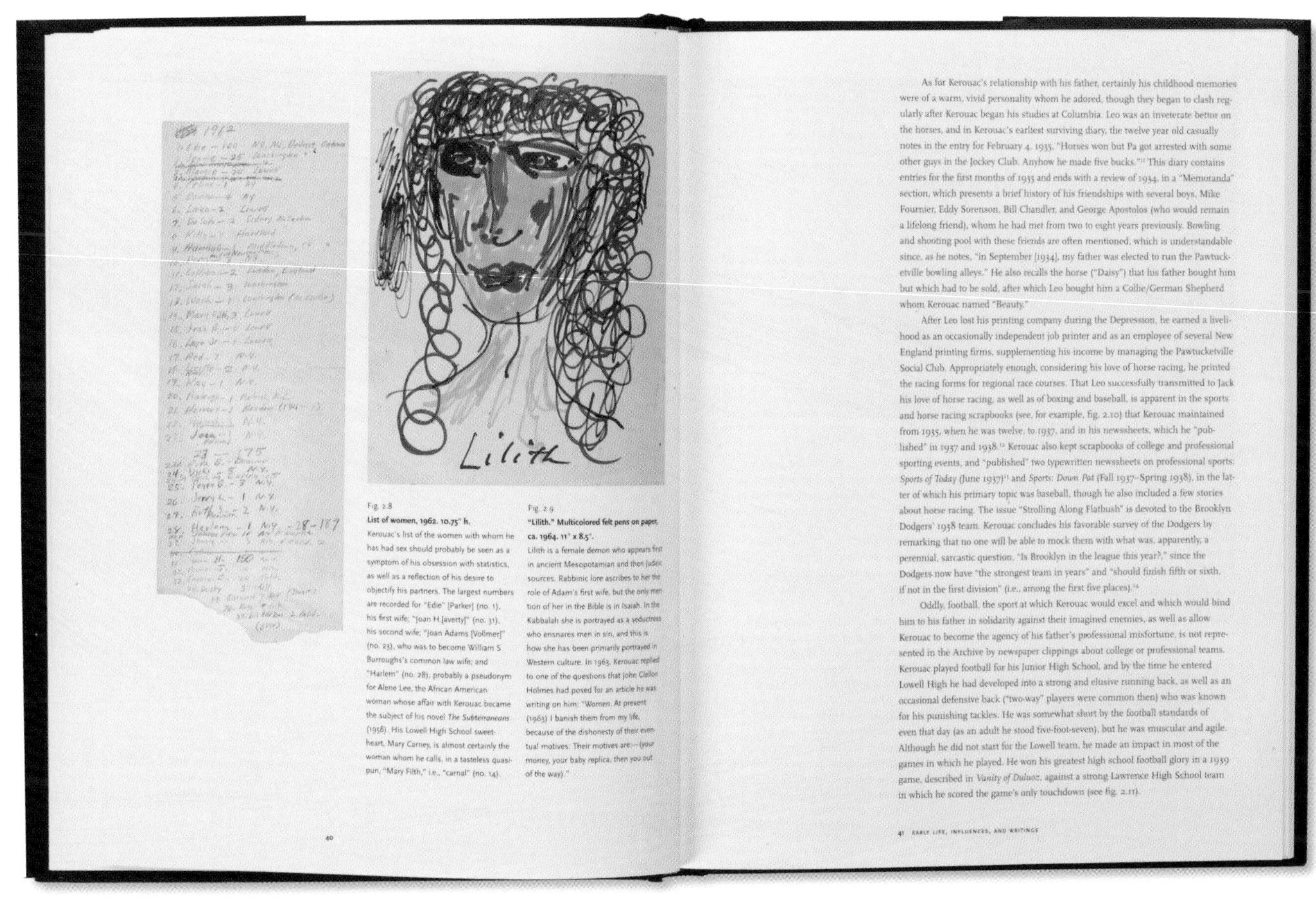

三栏网格为竖版的艺术作品和含有各项信息的图片说明提供了一个强大的框架。在左页中，说明取代了正文，一张细长的图片位于左侧栏，右页则全部留给文字

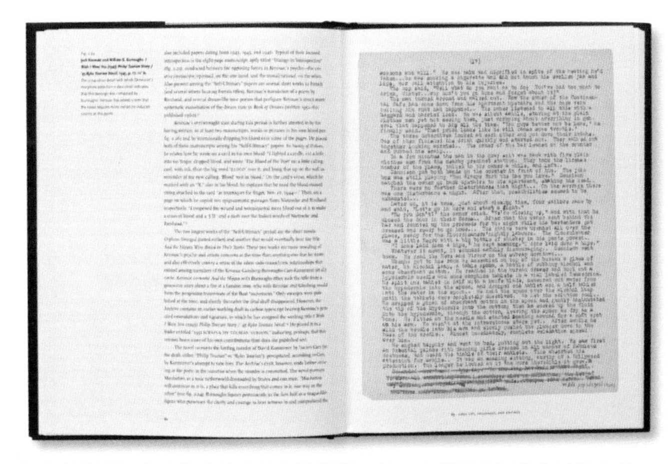

为了保持页面的节奏感和内容的清晰度，尺寸大的图片有时会单独占据一页。在这里，一张杰克·凯鲁亚克手写稿的照片占据了整个页面，与左页的文字形成对比

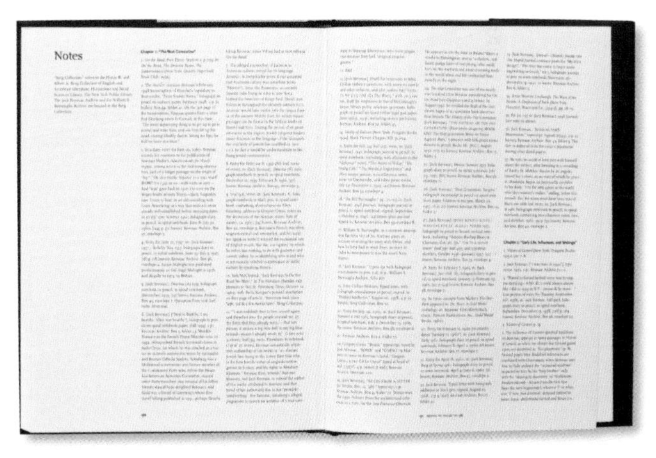

到了参考资料部分，如注释和索引，版式变成了三栏网格

● 22.通过排版确定分栏

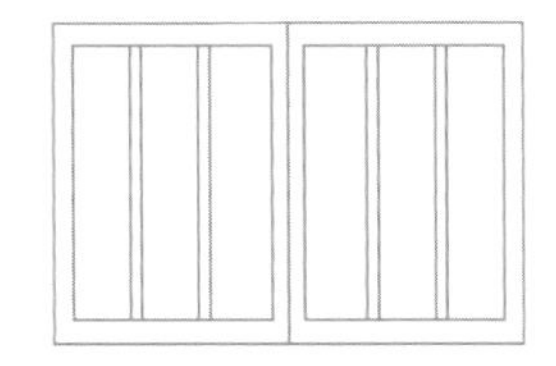

字体可以帮助分栏。使用粗细和大小不同的字体可以确定信息的顺序，给人一种秩序感，这种顺序既可以是横向的（标题、描述、相关内容），也可以是竖向的（栏目，从左至右）。不同的字体，如无衬线字体，可以用于列表、非正文文字或步骤介绍的信息。加粗的字体可以用于标题和步骤介绍中的数字，起到提示和增加页面趣味性的作用。较细的字体可以用于批注或其他辅助性的文本中。界限清晰的间距可以使字体排列整齐，看上去不会是乱糟糟的一团。

Soft and Chewy Chocolate Chip Cookies

A sugar- and butter-rich batter is the foundation for these cookie-jar classics. Just baked, they make a perfect snack on a chilly winter night -- or any time. MAKES ABOUT 3 DOZEN

- 2¼ cups all-purpose flour
- ½ teaspoon baking soda
- 1 cup (2 sticks) unsalted butter, room temperature
- ½ cup granulated sugar
- 1 cup packed light brown sugar
- 1 teaspoon coarse salt
- 2 teaspoons pure vanilla extract
- 2 large eggs
- 2 cups semisweet or milk chocolate chips, or a combination (about 12 ounces)

1. Preheat oven to 350°F. Whisk together flour and baking soda in a bowl. Put butter and sugars in the bowl of an electric mixer fitted with the paddle attachment. Mix on medium speed until pale and fluffy, about 2 minutes. Reduce speed to low. Add salt, vanilla, and eggs; mix until well blended, about 1 minute. Mix in flour mixture. Stir in chocolate chips.

2. Drop heaping tablespoons of dough onto baking sheets lined with parchment paper, spacing 2 inches apart. Bake cookies, rotating sheets halfway through, until edges turn golden but centers are still soft, 10 to 12 minutes. Let cool on sheets on wire racks 2 minutes. Transfer cookies to wire racks; let cool completely. Cookies can be stored between layers of parchment in airtight containers at room temperature up to 1 week.

58 • soft and chewy

Peanut Butter and Jelly Bars

This version of a well-loved combination from childhood concentrates the flavors into a sweet dessert that appeals to all ages. We like strawberry jam, but feel free to substitute any flavor you prefer. MAKES ABOUT 3 DOZEN

- 1 cup (2 sticks) unsalted butter, room temperature, plus more for pan
- 3 cups all-purpose flour, plus more for pan
- 1½ cups sugar
- 2 large eggs
- 2½ cups smooth peanut butter
- 1½ teaspoons salt
- 1 teaspoon baking powder
- 1 teaspoon pure vanilla extract
- 1½ cups strawberry jam, or other flavor
- 1 cup salted peanuts (5 ounces), roughly chopped

1. Preheat oven to 350°F. Butter a 9 by 13-inch baking pan, and line the bottom with parchment paper. Butter the parchment, dust with flour, and tap out excess.

2. Place butter and sugar in the bowl of an electric mixer fitted with the paddle attachment. Beat on medium speed until fluffy, about 2 minutes. With mixer running, add eggs and peanut butter; beat until combined, about 2 minutes. Whisk together flour, salt, and baking powder. Add to the butter mixture, and beat on low speed until combined. Add vanilla.

3. Transfer two-thirds of mixture to prepared pan; spread evenly with an offset spatula. Using offset spatula, spread jam on top of peanut-butter mixture. Crumble remaining third of peanut butter mixture on top of jam. Sprinkle evenly with peanuts.

4. Bake until golden, 45 to 60 minutes, rotating halfway through. Tent loosely with foil if bars are getting too dark. Transfer to a wire rack to cool. Run knife around edges and refrigerate, 1 to 2 hours. Cut into about thirty-six bars (about 1½ by 2 inches). Cookies can be stored in airtight containers at room temperature up to 3 days.

59

食品成分使用无衬线字体，步骤介绍使用衬线字体。加粗的无衬线字体表示强调

项目

《玛莎•斯图尔特的饼干宣传册》

客户

MSL 克拉克森•波特（MSL Clarkson Potter）

设计师

芭芭拉•德怀尔德（Barbara deWilde）

精致的摄影图片和排版准确地反映了一个国民品牌的优雅和品位

Coconut-Cream Cheese Pinwheels

Rich cream cheese dough, coconut-cream cheese filling, and a topper of jam make these pinwheels complex—chewy on the outside, creamy in the center. Create a variety of flavors by substituting different fruit jams for the strawberry. MAKES ABOUT 2½ DOZEN

for the dough:

- 2 cups all-purpose flour, plus more for work surface
- ⅔ cup sugar
- ½ teaspoon baking powder
- ½ cup (1 stick) unsalted butter, room temperature
- 3 ounces cream cheese, room temperature
- 1 large egg
- 1 teaspoon pure vanilla extract

for the filling:

- 3 ounces cream cheese, room temperature
- 3 tablespoons granulated sugar
- 1 cup unsweetened shredded coconut
- ¼ cup white chocolate chips

for the glaze:

- 1 large egg, lightly beaten
- Fine sanding sugar, for sprinkling
- ⅓ cup strawberry jam

1. Make dough: Whisk together flour, sugar, and baking powder in a bowl. Put butter and cream cheese into the bowl of an electric mixer fitted with the paddle attachment; mix on medium-high speed until fluffy, about 2 minutes. Mix in egg and vanilla. Reduce speed to low. Add flour mixture, and mix until just combined. Divide dough in half, and pat into disks. Wrap each piece in plastic, and refrigerate until dough is firm, 1 to 2 hours.

2. Preheat oven to 350°F. Line baking sheets with nonstick baking mats (such as Silpats).

3. Make filling: Put cream cheese and sugar into the bowl of an electric mixer fitted with the paddle attachment; mix on medium speed until fluffy. Fold in coconut and chocolate chips.

4. Remove one disk of dough from refrigerator. Roll about ⅛ inch thick on a lightly floured surface. With a fluted cookie cutter, cut into fifteen 2½-inch squares. Transfer to prepared baking sheets, spacing about 1½ inches apart. Refrigerate 15 minutes. Repeat with remaining dough.

5. Place 1 teaspoon filling in center of each square. Using a fluted pastry wheel, cut 1-inch slits diagonally from each corner toward the filling. Fold every other tip over to cover filling, forming a pinwheel. Press lightly to seal. Use the tip of your finger to make a well in the top.

6. Make glaze: Using a pastry brush, lightly brush tops of pinwheels with beaten egg. Sprinkle with sanding sugar. Bake 6 minutes. Remove and use the lightly floured handle of a wooden spoon to make the well a little deeper. Fill each well with about ½ teaspoon jam. Return to oven, and bake, rotating sheets halfway through, until edges are golden and cookies are slightly puffed, about 6 minutes more. Transfer sheets to wire racks; let cool 5 minutes. Transfer cookies to rack; let cool completely. Cookies can be stored in single layers in airtight containers at room temperature up to 3 days.

soft *and* chewy • 61

元素巧妙地堆叠在一起，创造出一种游戏的感觉。使用不同的字体强调重点会使版式更生动，同时也更具有趣味性和指导性

● 23.避免拥挤

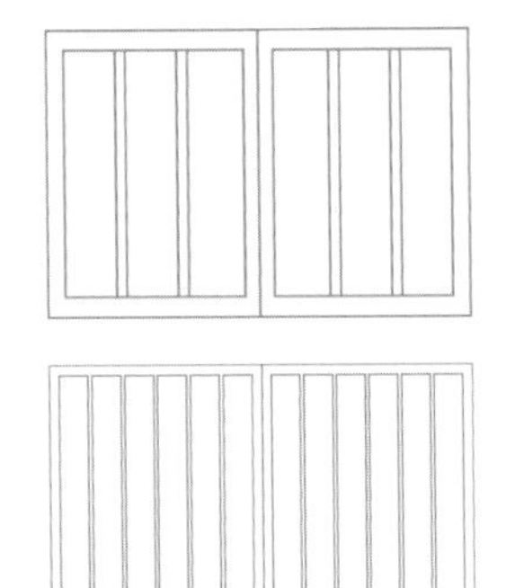

当设计多栏网格时，没有必要填满每一寸空间，最好的做法是在一些栏中留出空白。留白可以引导读者纵览整个页面，从中挑选特定的故事、图片或图标。不同粗细的分隔线可以控制页面并激活信息。

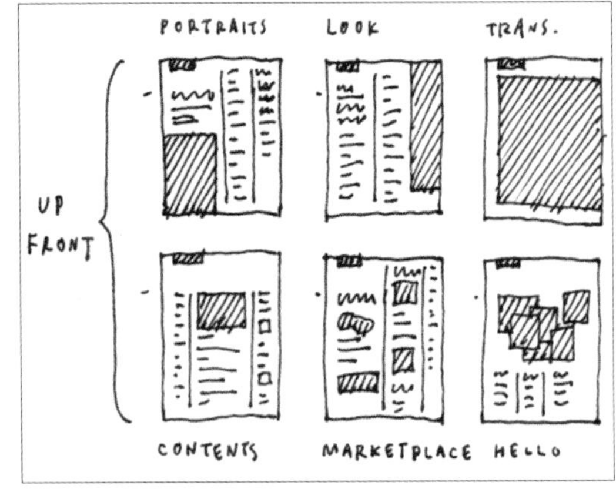

草图显示出一种空间感

项目
《好》杂志

客户
《好》杂志有限责任公司

设计公司
Open

设计总监
司各特·斯托厄尔

留白和大胆、前卫的设计帮助读者将注意力集中于页面中的核心观点——让地球变得更美好

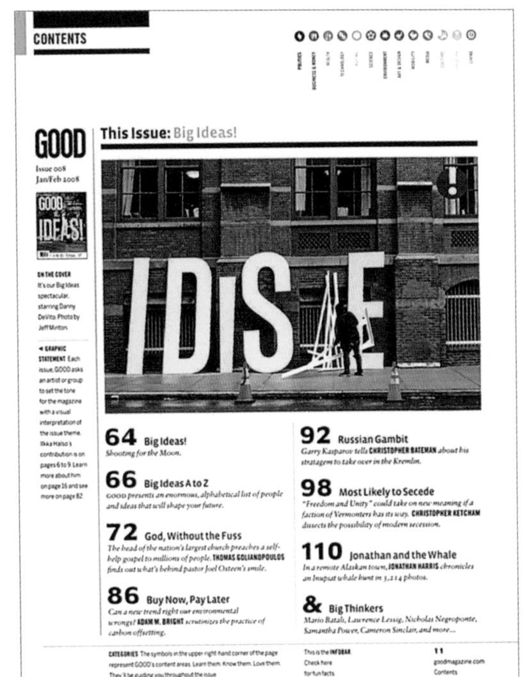

CONTENTS

GOOD

Issue 008
Jan/Feb 2008

ON THE COVER It's our Big Ideas spectacular, starring Danny DeVito. Photo by Jeff Minton.

◀ GRAPHIC STATEMENT Each issue, GOOD asks an artist or group to set the tone for the magazine with a visual interpretation of the issue theme. Ilkka Halso's contribution is on pages 6 to 9. Learn more about him on page 16 and see more on page 82

This Issue: Big Ideas!

64 Big Ideas!
Shooting for the Moon.

66 Big Ideas A to Z
GOOD presents an enormous, alphabetical list of people and ideas that will shape your future.

72 God, Without the Fuss
The head of the nation's largest church preaches a self-help gospel to millions of people. THOMAS GOLIANOPOULOS finds out what's behind pastor Joel Osteen's smile.

86 Buy Now, Pay Later
Can a new trend right our environmental wrongs? ADAM M. BRIGHT scrutinizes the practice of carbon offsetting.

92 Russian Gambit
Garry Kasparov tells CHRISTOPHER BATEMAN about his stratagem to take over in the Kremlin.

98 Most Likely to Secede
"Freedom and Unity" could take on new meaning if a faction of Vermonters has its way. CHRISTOPHER KETCHAM dissects the possibility of modern secession.

110 Jonathan and the Whale
In a remote Alaskan town, JONATHAN HARRIS chronicles an Inupiat whale hunt in 3,214 photos.

& Big Thinkers
Mario Batali, Lawrence Lessig, Nicholas Negroponte, Samantha Power, Cameron Sinclair, and more...

CATEGORIES The symbols in the upper right hand corner of the page represent GOOD's content areas. Learn them. Know them. Love them. They'll be guiding you throughout the issue

This is the INFOBAR. Check here for fun facts.

11
goodmagazine.com
Contents

目录页通常很难设计。这个页面没有杂乱的感觉，让读者很容易就能找到杂志想表达的内容。不同大小和粗细的字体增加了页面的趣味性，也使页面布局更平衡。右上角的图标样式贯穿了整本杂志

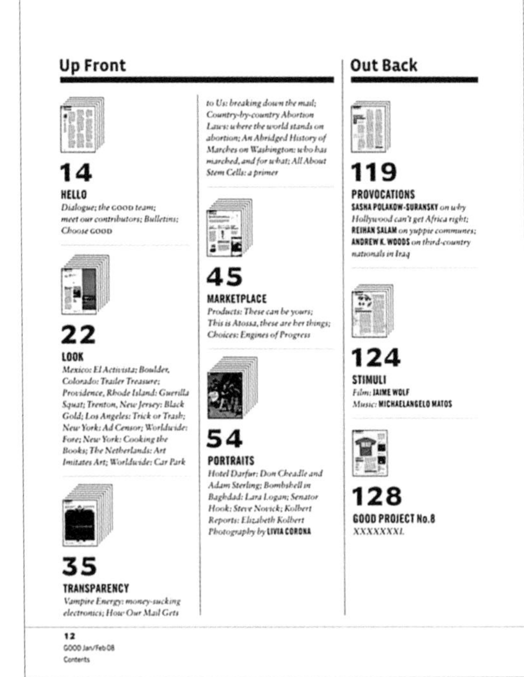

Up Front

14
HELLO
Dialogue; the GOOD team; meet our contributors; Bulletins; Choose GOOD

22
LOOK
Mexico: El Activista; Boulder, Colorado: Trailer Treasure; Providence, Rhode Island: Guerilla Squat; Trenton, New Jersey: Black Gold; Los Angeles: Trick or Trash; New York: Ad Censor; Worldwide: Fore; New York: Cooking the Books; The Netherlands: Art Imitates Art; Worldwide: Car Park

35
TRANSPARENCY
Vampire Energy: money-sucking electronics; How Our Mail Gets to Us: breaking down the mail; Country-by-country Abortion Laws: where the world stands on abortion; An Abridged History of Marches on Washington: who has marched, and for what; All About Stem Cells: a primer

45
MARKETPLACE
Products: These can be yours; This is Atossa, these are her things; Choices: Engines of Progress

54
PORTRAITS
Hotel Darfur: Don Cheadle and Adam Sterling; Bombshell in Baghdad: Lara Logan; Senator Hook: Steve Novick; Kolbert Reports: Elizabeth Kolbert
Photography by LIVIA CORONA

Out Back

119
PROVOCATIONS
SASHA POLAKOW-SURANSKY on why Hollywood can't get Africa right; REIHAN SALAM on yuppie communes; ANDREW K. WOODS on third-country nationals in Iraq

124
STIMULI
Film: JAIME WOLF
Music: MICHAELANGELO MATOS

128
GOOD PROJECT No.8
XXXXXXXL

12
GOOD Jan/Feb 08
Contents

这个页面包含了五个层次的信息，紧凑的排版和充足的留白使这些信息清晰易读

分隔线和巧妙的排版设定了一系列不同类别的信息

左侧简洁有力的排版贯穿页面，右页上是同样充满力量的插画

BIG IDEAS!

QUANTUM HIPPIES

Quantum mechanics is all about the relationship between matter and energy. So it's not hard to imagine why the science has been co-opted by a subculture of bong-toting academe—expand your mind, anyone? Scholars (cough) like Daniel Pinchbeck have struck mono-atomic gold with a prophetic philosophy that marries quantum theory and Ayahuasca-induced hallucination, all in an effort to come to grips with what rational materialism neglects: the inexplicable nature of being.

We see evidence every day—at every grade level, and in urban and rural communities all across the country—that when children facing the challenges of poverty are given the opportunities they deserve, they excel. This is the truth, and yet those who believe that it is impossible for schools to overcome the challenges of poverty consider it a radical idea. This is the idea I'd like to see our nation's leaders embrace—the idea that with a new approach to education, we can ensure that all of our nation's children, regardless of where they are born, have the opportunity to attain an excellent education.

BIG THINKER:
WENDY KOPP

Wendy Kopp is the director of Teach For America (one of GOOD's nonprofit partners).

RUSSIAN DEMOCRACY

R

Russian Gambit

Garry Kasparov, the Russian presidential candidate and former chess grandmaster, is trying to keep Vladimir Putin in check.

interview by
CHRISTOPHER BATEMAN
illustrations by
DARREN BOOTH

92
GOOD Jan/Feb 08
Big Ideas!

BIO Christopher Bateman is an editorial associate at Vanity Fair. He lives in New York.

93
goodmagazine.com
Big Ideas!

观点很重要？那就用大号的字体！标题中大字号的首字母下沉，暗示着故事的开始。目录页中的标志连续出现在多页的同一位置，也就是页面的右上角，确保在合适的地方使用了适当的标志

● 24.下移栏目

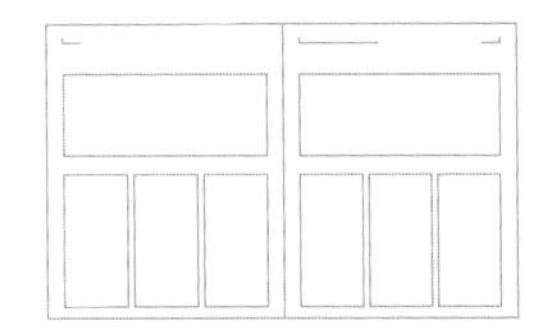

全页都是三栏网格的页面可能会显得拥挤。要想保持读者的注意力，让他们有耐心阅读，一个有效的办法是下移页面上的分栏，让页面保持干净，并赋予页面动感。

下移的分栏也利于创建一片明确的区域，用于放置导览信息，如页首标题、页码、分页标题、批注和照片。

8

Pew Environment Group

Halloween 1948 was all trick and no treat in Donora, Pennsylvania. In the last week of October, this town of 14,000 in the western part of the state underwent a weather event called a "temperature inversion," trapping at ground level the smog from local metal factories.

Nearly half of Donora's residents experienced breathing problems, hundreds suffered permanent heart and lung damage, and some 50 deaths were attributed to the disaster. Sixty years ago, public policy gave Americans relatively little protection from industrial accidents. However, Donora and similar disasters helped focus national attention on the government's responsibility to protect the population from environmental hazards. Eventually, the Donora catastrophe led to the Air Pollution Control Act of 1955, the United States' first piece of federal legislation on this issue and an early step in what has become an ongoing effort to save the environment, for the sake of the natural world as well as public health.

A related development in 1948 produced no fatalities but was a harbinger of a situation that was ultimately even more serious. As energy demand and prices soared in the postwar boom and Western companies discovered vast oil fields in the Middle East, the United States for the first time became a net importer of oil.

Sixty years later, this country—indeed, the world—faces unprecedented environmental challenges. Changes to terrestrial and marine environments resulting from climate change, overfishing, agriculture, grazing and logging are already transforming the planet in ways that impair its ability to be hospitable to life—both ours and that of the countless other species that occupy it with us. The Pew Environment Group is focused on reducing the scope and severity of three major global environmental problems:

- Dramatic changes to the Earth's climate caused by the increasing concentration of greenhouse gases in the atmosphere;
- The erosion of large wilderness ecosystems that contain a great part of the world's remaining biodiversity;
- The destruction of the world's oceans, with a particular emphasis on global marine fisheries.

Climate change. To reduce the threat of climate change, we are urging the adoption of a mandatory national policy to reduce greenhouse gas emissions. While its centerpiece is a market-based cap and trade system, complementary measures are needed to create additional incentives to invest in less polluting technologies in key sectors, particularly transportation.

Early in 2007, we launched the Pew Campaign for Fuel Efficiency to promote legislation to increase fuel-efficiency standards for passenger vehicles to 35 miles per gallon by 2020. Nationwide, vehicles account for two-thirds of oil consumption and one-third of greenhouse gas emissions, with light-duty passenger vehicles—cars, pick-ups, minivans and SUVs—producing about 60 percent of transportation-related emissions. Globally, U.S. transportation accounts for about 8 percent of all greenhouse gas pollution and 17 percent of an increasingly tight and volatile world oil market. Higher standards would reduce our country's dependence on foreign oil, enhance national security, save consumers money and reduce global warming pollution.

Pew Prospectus 2008 IMPROVING PUBLIC POLICY
Pew Environment Group

9

Seeking protection from the smog in Donora, Pa., in 1948

Car emissions, a leading contributor to climate change

Sharks, sought for their fins, among the sea's endangered creatures

Wilderness protection. Due to the spread of human civilization, habitat destruction and, increasingly, climate change, scientists estimate that we may be losing as many as 30,000 species each year. To slow or stop this loss, many conservation biologists say, we need to create new parks, wildlife refuges and protected areas where extractive activity and development are prohibited. Pew has played a critical role in the permanent protection of more than 200 million acres of wilderness in the United States and Canada since 1990. More recently, we have launched a joint initiative with The Nature Conservancy to establish new national parks and indigenous protected areas in Australia. Together, these three countries contain more than 30 percent of the world's remaining old growth forests and an even larger share of pristine wilderness areas.

Ocean conservation. Overfishing, chemical and nutrient pollution, habitat alteration, introduction of exotic species and climate change are taking what may be an irreversible toll on the world's marine environment. The Pew Environment Group has helped lead the way in bringing about many of the major improvements in fisheries management and marine conservation in the United States since the mid-1990s. In recent years, we have expanded our oceans work internationally and are working in various other regions of the world to curtail overfishing, protect critical marine habitat and reduce the amount of unintended bycatch—the fish, seabirds, sharks, whales and other species that are routinely thrown back into the sea, either dead or dying.

Pew today is in a stronger position to address all of these problems as the result of the merger of our Environment Program with the National Environmental Trust. The consolidated team has a domestic and international staff of more than 100, making us one of the nation's largest environmental scientific and advocacy organizations with a presence across not only the United States but also Australia, Canada, Europe, the Indian Ocean, Latin America and the Western Pacific.

Society has historically invested little time, thought or effort in protecting the environment for posterity. Sixty years ago, once the smog in Donora had cleared, most people simply assumed that things would return to the way they had been. We can no longer afford to make that mistake.

Joshua S. Reichert

Managing Director
Pew Environment Group

项目
《2008年皮尤慈善信托基金招股说明书》

客户
皮尤慈善信托基金会（The Pew Charitable Trusts）

设计公司
Iridium集团（Iridium-Group）

编辑
马歇尔•A. 莱杰（Marshall A. Ledger）

助理编辑/项目经理
桑德拉•萨尔曼（Sandra Salmans）

一家非盈利性机构的项目得以严肃又优雅地展示了出来

变化是设计的“调味料”，所以，最好的办法是加宽放有引导性内容的文本框，增强页面的对比度。对附加的结构来说，批注的文字可以使用与其他内容完全不同的字体

Culture

Change was sweeping the arts scene in 1948, with an impact that would not be fully realized for years. American painters led the way into abstract expressionism, reshaping both the visual arts and this country's influence on the art world.

Long-playing records, enthralling the public in 1948

The Village of Arts and Humanities, revitalizing North Philadelphia

Development workshop for Bill Irwin's *The Happiness Lecture*

Meanwhile, technology was setting the stage for revolutions in music and photography. The LP record made its debut, and the Fender electric guitar, which would define the rock 'n roll sound in the next decade and thereafter, went into mass production. Both the Polaroid Land camera, the world's first successful instant camera, and the first Nikon went on sale.

In New York, the not-for-profit Experimental Theatre, Inc., received a special Tony honoring its path-breaking work with artists such as Lee Strasberg and Bertolt Brecht. But in April it was disclosed that the theatre had run up a deficit of $20,000—a shocking amount, given that $5,000 had been the maximum allocated for each play—and in October *The New York Times* headlined, "ET Shelves Plans for Coming Year."

Apart from its miniscule budget, there is nothing dated about the travails of the Experimental Theatre. The arts still struggle with cost containment and tight funds. But if the Experimental Theatre were to open its doors today, it might benefit from the power of knowledge now available to many nonprofit arts organizations in Pennsylvania, Maryland and California—and, eventually, to those in other states as well. Technology, which would transform music and photography through inventions in 1948, is providing an important tool to groups that are seeking to streamline a grant application process that, in the past, has been all too onerous.

That tool is the Cultural Data Project, a Web-based data collection system that aggregates information about revenues, employment, volunteers, attendance, fund-raising and other areas input by cultural organizations. On a larger scale, the system also provides a picture of the assets, impact and needs of the cultural sector in a region.

The project was originally launched in Pennsylvania in 2004, the brainchild of a unique collaboration among public and private funders, including the Greater Philadelphia Cultural Alliance, the Greater Pittsburgh Arts Council, The Heinz Endowments, the Pennsylvania Council on the Arts, Pew, The Pittsburgh Foundation and the William Penn Foundation. Until then, applicants to these funding organizations had been required to provide similar information in different formats and on multiple occasions. Thanks to the Pennsylvania Cultural Data Project, hundreds of nonprofit arts and cultural organizations throughout the state can today update their information just once a year and, with the click of a computer mouse, submit it as part of their grant applications. Other foundations, such as the Philadelphia Cultural Fund, the Pennsylvania Historical and Museum Commission and the Independence Foundation, have also adopted the system.

So successful has the project been that numerous states are clamoring to adopt it. In June, with funding from multiple sources, Maryland rolled out its own in-state Cultural Data Project. The California Cultural Data Project, more than five times the size of Pennsylvania's with potentially 5,000 nonprofit cultural organizations, went online at the start of 2008, thanks to the support of more than 20 donors. Both projects are administered by Pew.

As cultural organizations in other states enter their own data, the research will become exponentially more valuable. Communities will be able to compare the effects of different approaches to supporting the arts from state to state and city to city. And the data will give cultural leaders the ability to make a fact-based case that a lively arts scene enriches a community economically as well as socially.

The Cultural Data Project is not the first initiative funded by Pew's Culture portfolio to go national or to benefit from state-of-the-art technology. For example, the system used by Philly-FunGuide, the first comprehensive, up-to-date Web calendar of the region's arts and culture events, has been successfully licensed to other cities.

In addition to the Cultural Data Project, another core effort within Pew's Culture portfolio is the Philadelphia Center for Arts and Heritage and its programs, which include Dance Advance, the Heritage Philadelphia Program, the Pew Fellowships in the Arts, the Philadelphia Exhibitions Initiative, the Philadelphia Music Project and the Philadelphia Theatre Initiative. Since the inception of the first program in 1989, these six initiatives have supported a combined total of more than 1,100 projects and provided more than $48 million in funding for the Philadelphia region's arts and heritage institutions and artists.

Through its fellowships, Pew nurtures individual artists working in a variety of performing, visual and literary disciplines, enabling them to explore new creative frontiers that the marketplace is not likely to support. The center also houses the Philadelphia Cultural Management Initiative, which helps cultural groups strengthen their organizational and financial management practices.

Almost from the time it was established, Pew was among the region's largest supporters of arts and culture. While it continues in this role, committed to fostering nonprofit groups' artistic excellence and economic stability, and to expanding public participation, Pew—like the arts themselves—has changed its approach with the times.

Marian A. Godfrey

Managing Director
Culture and Civic Initiatives

2007 Milestones

Each year, we join with excellent organizations to produce work that exemplifies exactly what we mean in stating that Pew serves the public interest. On these pages, we highlight the results of some of the Pew-supported work that made a difference in 2007.

Environment

Health and Human Services

Pew Center on the States

● 25.改变形状

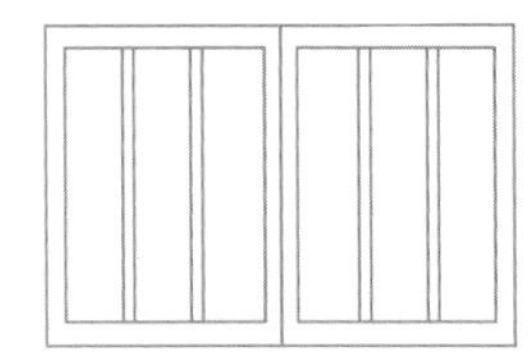

改变插图的形状可以让指南类的文本更生动，更具启发性。如果所有内容都是同样的大小和宽度，虽然整体版式会很清晰，但也会很无趣。更好的办法是使用不同样式的图片组合。

Handbook How-Tos

HOW TO WASH, DRY, AND STORE LETTUCE

1. Fill a clean basin or a large bowl with cold water, and submerge the lettuce leaves completely. (For head lettuce, first discard the outer leaves; they're most likely to harbor bacteria. Chop off the end, and separate the remaining leaves.) Swish the leaves around to loosen dirt.

2. Once sediment has settled, lift out the lettuce, pour out the dirty water, and refill the bowl with clean water. Submerge the lettuce again, and continue swishing and refilling until there are no more traces of dirt or sand in the bowl. You may need to change the water 2 or 3 times.

3. Dry the lettuce in a salad spinner until no more water collects at the bottom of the bowl. Alternatively, blot the leaves between layered paper towels or clean dish towels until no water remains.

4. If you plan to store the lettuce, arrange the dry leaves in a single layer on paper towels or clean dish towels, roll up, and seal inside a plastic bag. Lettuce can be stored this way in the refrigerator for 3 to 5 days. To prevent it from browning rapidly, don't tear the leaves into smaller pieces until you're ready to use them.

SOAK AND SPIN THE LEAVES

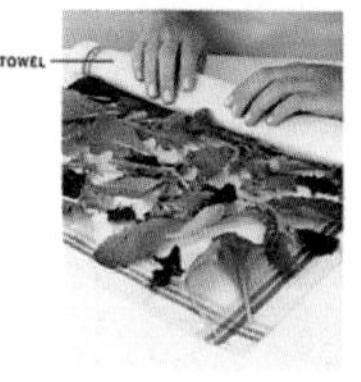

STORE IN A TOWEL

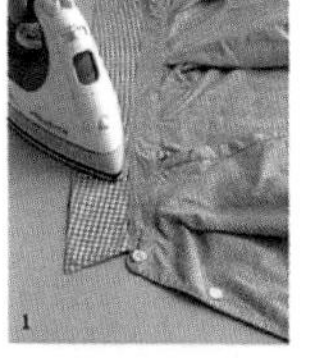

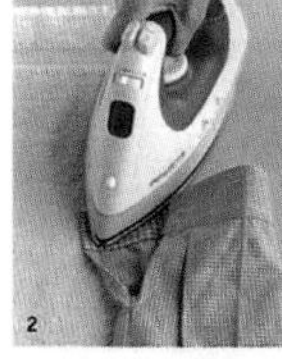

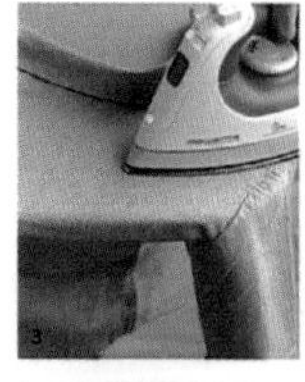

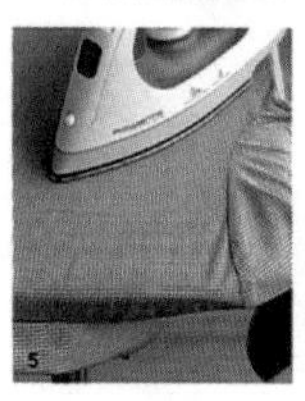

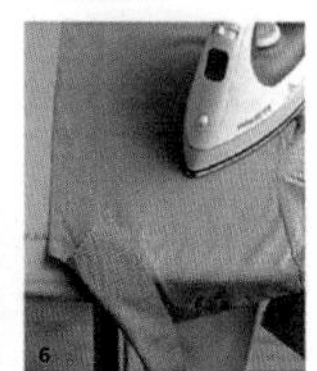

HOW TO IRON A BUTTON-FRONT SHIRT

For easier ironing and the best results, start with a thoroughly damp shirt. Mist the shirt with water using a spray bottle, roll it up, and keep it in a plastic bag for 15 minutes or up to a few hours. (If you can't iron the shirt sooner, refrigerate it in the bag so the shirt won't acquire a sour smell.) Most of the ironing will be on the wide end of the board. If you're right-handed, position the wide end to your left; if you're left-handed, it should be on your right.

1. Begin with the underside of the collar. Iron, gently pulling and stretching the fabric to prevent puckering. Turn the shirt over, and repeat on the other side of collar. Fold the collar along seam. Lightly press.

2. Iron the inside of the cuffs; slip a towel under the buttons to cushion them as you work. Iron the inside of the plackets and the lower inside portion of the sleeves, right above the cuffs. Iron the outside of the cuffs.

3. Drape the upper quarter of the shirt over the wide end of the board, with the collar pointing toward the narrow end of the board, and iron one half of the yoke. Reposition, and iron the other half.

4. Lay 1 sleeve flat on the board. Iron from shoulder to cuff. (If you don't want to crease the sleeve, use a sleeve board.) Turn the sleeve over, and iron the other side. Repeat with the other sleeve.

5. Drape the yoke over the wide end of the board, with the collar facing the wide end, and iron the back of the shirt.

6. Drape the left side of the front of the shirt over the board, with the collar pointing toward the wide end; iron. Repeat with the right front side, ironing around, rather than over, buttons. Let the shirt hang in a well-ventilated area until it's completely cool and dry, about 30 minutes, before hanging it in the closet.

62

想要使文本或说明清晰明了，一种方法是加上步骤解析的插图以及最终完成品的照片。这些不同大小的图片使页面不至于有呆板的感觉

项目
《玛莎•斯图尔特生活》（*Martha Stewart Living*）杂志

客户
玛莎•斯图尔特生活全媒体公司

设计公司
《玛莎•斯图尔特生活》杂志

首席创意官
盖尔•塔伊（Gael Towey）

清晰的步骤解析图片和完成品的照片放置在一个明确又灵活的版式里

下页图： 这个作品的排版功能性强，注重细节，精美又不烦琐。加框的侧边栏提醒读者这里有不同于菜谱的重要信息

SAUTÉED SOLE WITH LEMON

SERVES 2

Gray sole is a delicately flavored white fish. You can substitute flounder, turbot, or another type of sole.

- ½ cup flour, preferably Wondra
- 1 teaspoon coarse salt
- ½ teaspoon freshly ground pepper
- 2 gray sole fillets (6 ounces each)
- 2 tablespoons unsalted butter
- 2 tablespoons olive oil
- 2 tablespoons sliced almonds
- 1½ tablespoons chopped fresh parsley
- Finely chopped zest and juice from 1 lemon, plus wedges for garnish

1. Combine flour, salt, and pepper in a shallow bowl. Dredge fish fillets in flour mixture, coating both sides, and shake off excess.

2. Melt butter with oil in a sauté pan over medium-high heat. When butter begins to foam, add fillets. Cook until golden brown, 2 to 3 minutes per side. Transfer each fillet to a serving plate.

3. Add almonds, parsley, zest, and 2 tablespoons juice to pan. Spoon over fillets, and serve with lemon wedges.

HARICOTS VERTS

SERVES 2

- Coarse salt, to taste
- 8 ounces haricots verts
- 2 tablespoons extra-virgin olive oil
- Freshly ground pepper, to taste
- 1 bunch chives, for bundling (optional)

1. Bring a pot of salted water to a boil. Add haricots verts, and cook until bright green and just tender, 3 to 5 minutes. Drain, and pat dry. Transfer to a serving bowl.

2. Toss with oil, salt, and pepper. Tie into bundles using chives.

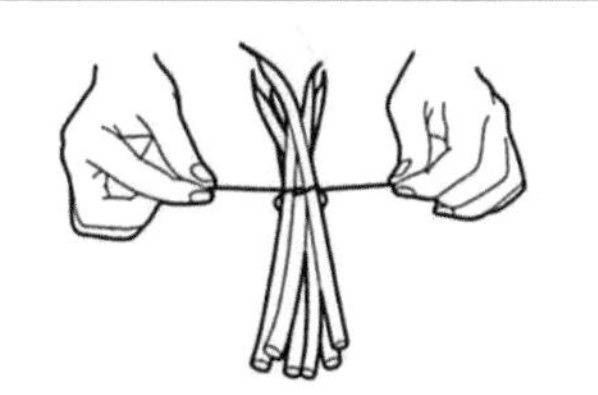

HOW TO BUNDLE GREEN BEANS

1. Cook haricots verts. Drain, and pat dry. Let stand until cool enough to handle.

2. Lay a chive on a work surface. Arrange 4 to 10 haricots verts in a small pile on top of chive. Carefully tie chive around bundle. Trim ends of chive if desired.

QUICK-COOKING CLASSIC Seared sole fillets glisten beneath a last-minute pan sauce made with lemon, parsley, and almonds. The resulting entrée, served with blanched haricots verts, is satisfyingly quick yet sophisticated.

● 26.切分节奏

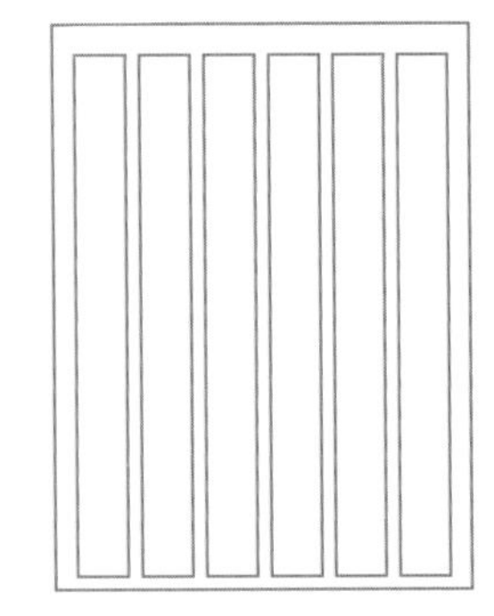

尽管页面或屏幕的整洁和秩序感非常关键，但重复且无变化的元素也会让读者感到无聊。让文本栏沿着小号的形状排列（如下图）可以避免排版僵化。多样化可以凸显（而不是破坏）核心信息。

这个网格包含了大量的信息。阶梯形的分栏沿着小号的形状排列，进一步强化了漂亮生动的列表。从排版设计的角度来看，这个节目单是一个具有平衡性、节奏感和工艺感的杰作

项目
林肯中心爵士乐团（Jazz at Lincoln Center）节目单

客户
林肯中心爵士乐团

设计师
博比•马丁

清晰的布局使大量信息变得生动活泼

柱状网格为文本框提供了一个清晰的框架，包括了诸多功能。文本框中容纳了多项素材，在照片上创建一层平面的形式增加了节目单的维度，使这些文本框有节奏地在页面上“流动”

27.混搭

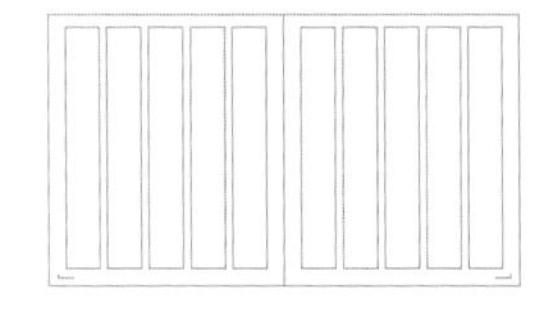

粗细、大小、质感、形状、数量、间距、色彩，把多种元素的变化结合在一起可以使版式看上去富有生机，风格多变又统一。紧凑的网格好比一个基础框架，使一个页面可以容纳许多图片和标题，以便为另外一个甚至多个页面腾出空间。在不牺牲基本信息可读性的前提下，字体的粗细和大小，以及图片大小和形状的动态变化都需要注意。

大胆的五栏网格连续出现在这本杂志上，成为页面的基础框架，也容纳了各种形状和大小的素材。页面结构强大，图片周围还有留白

REGIONAL CRAFTS

A sampling of people and programs supporting the work of traditional artisans

PILCHUCK GLASS SCHOOL

Over three decades, the center has fostered a glass-blowing industry in the Pacific Northwest.

A former New York designer returns home to create a unique partnership with gifted local artisans

ALABAMA CHANIN

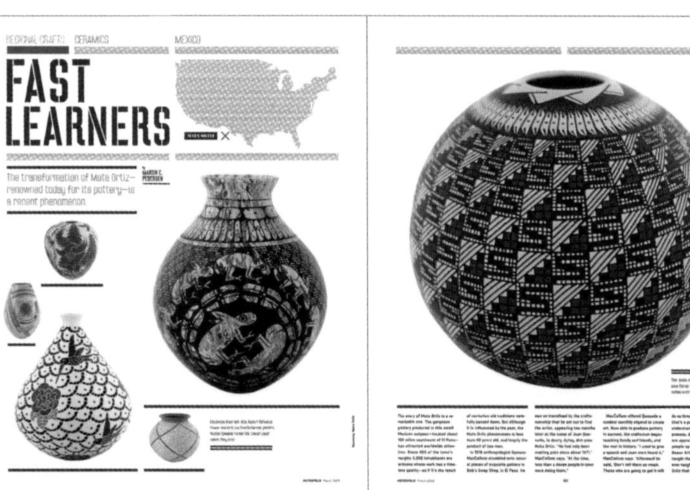
FAST LEARNERS

The transformation of Mata Ortiz—renowned today for its pottery—is a recent phenomenon.

尽管版式中的文本只采用了黑色和一种强调色，但这种版式通过变换黑体和模板式的字体，削弱了较小且细的字体的存在感，凸显了页面的颜色和质感。粗细不等的分隔线也使页面的结构更加清晰

项目
《大都会》（*Metropolis*）杂志

客户
《大都会》杂志

创意总监
克里斯威尔·拉宾（Criswell Lappin）

设计精良的网格使这个具有地方特色的作品焕发光彩。页面底部的多列网格通过一系列文本的不同大小、粗细和色彩，为想要展示的内容提供了坚实的基础

下页图：分隔线成了摇椅侧面的打底元素

REGIONAL CRAFTS HOUSEWARES + ARCHITECTURE NORTH CAROLINA

HANDMADE HOME

by BELINDA LANKS

A crafts group enlists local artisans to create a one-of-a-kind dwelling.

AKIRA SATAKE
CERAMICS
Satake produces functional ceramic pieces—from vases, platters, and bowls to decorative tiles—with a refined Japanese aesthetic.

FATIE ATKINSON
FURNITURE
Employing a steam-bending technique, Atkinson can make this chair out of any open-pored wood, including hickory, ash, and white or red oak (shown).

BARBARA ZARETSKY
TEXTILES
Zaretsky creates earth-toned patterns using natural fibers, plant dyes, and textile paints.

HandMade in America has been fervently promoting craft in Western North Carolina since 1993, but this year marks the nonprofit's first foray into real estate. In a novel collaboration, the group has partnered with private developer Biltmore Farms to construct the HandMade Home, a 3,700-square-foot model in Asheville showcasing the work of 100 local craftspeople. The house, which broke ground last September, is expected to meet the green-building standards of North Carolina's Healthy Built Homes program and fetch $2.25 million when it makes its debut in October as part of the city's annual "Parade of Homes."

Founding executive director Becky Anderson hopes the project will spur other developers, architects, and homeowners to tap the region's greatest resource: the 4,500 resident artisans making everything from furniture and lighting fixtures to tableware and rugs (examples shown above). "We want to become the center of handcrafted homes," she says. To make it easy, HandMade in America has produced directories featuring the work of and contact information for the craftspeople in its network. But Ben Brown, the project's publicist, recommends that people considering such an undertaking think smaller. "This is the first project of its kind, and it will probably be the last," Brown says. "With one hundred independent-minded artists involved, people are ready to shoot each other." ○

Courtesy HandMade in America

PEWABIC
The designs for Brookside School at Cranbrook (top), in Bloomfield Hills, and Detroit's Comerica Park stadium (above) were custom-made by the pottery's in-house team.

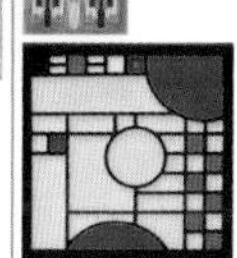

MOTAWI
The Frank Lloyd Wright Collection includes Avery (left) and Confetti (below). Also shown: Montrose (above) and Amaryllis (right), an adaptation of a Louis Sullivan design.

DAVID ELLISON
Sold by Country Floors, the Apollo Plaque (above) is a reinterpretation of historic details found on buildings in New York's Flatiron District.

Eastern Michigan is home to one of the most active crafts movements in the country.

REGIONAL CRAFTS TILE MICHIGAN

MOTOR CITY GLAZE

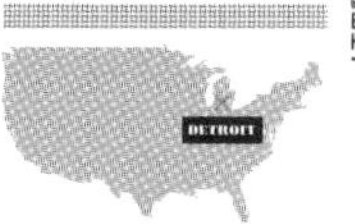

by EVA HAGBERG

"We'd start doing these tile shows that were just tile, and we'd think, How could anyone make a living at this?" says Marcia Hovland, part of a loose-knit group of Michigan-based tile-makers, reminiscing about the good old days before the tile industry took off. "And now everyone is doing really well."

Hovland is one of the artisans who came up through Detroit's famed Pewabic Pottery—a tile factory, exhibition space, and educational facility. She studied there with David Ellison—a name that comes up again and again in conversation with these eastern-Michigan tile fiends—and realized that she could turn her painting and design background into a whole new bag of (ceramic) chips.

Karim Motawi runs Motawi Tileworks out of Ann Arbor with his sister, Nawal. The company makes historically influenced pottery in line with the types of things that were produced in the earliest days of Pewabic in the 1900s. "We're literally plowing through the history books and the source books, the old catalogs," he says. "We're trying to re-create the lost craft." As the official Frank Lloyd Wright licensee, it's reproducing just fine.

Motawi Tileworks operates on a relatively tiny scale—it produces 18,000 square feet of tile a year, a drop in the bucket—and so do many of its local cohorts, which is why they're so happy to know Joseph Taylor, president of the Tile Heritage Foundation, which works to raise the historic craft's profile. "They are like tile cheerleaders," Motawi says. ○

Courtesy the manufacturers

REGIONAL CRAFTS SEATING NEW YORK

BROOKLYN'S OWN

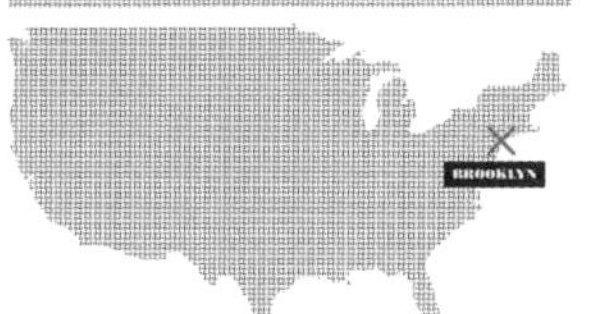

A crafty, DIY-inspired furniture movement emerges in New York's most creatively vibrant borough.

ELUCIDESIGN
DANISH CHAIR
Inspired by the Scandinavian classics, this Chris Jondle–designed piece is made of maple and uses a hand-silkscreened linen for the back and seat.

WÜD
WÜD CHAIR
The dining-room chair, designed by Corey Springer and Eric Ervin in 2006, comes in a variety of woods, including cherry (shown), walnut, and maple.

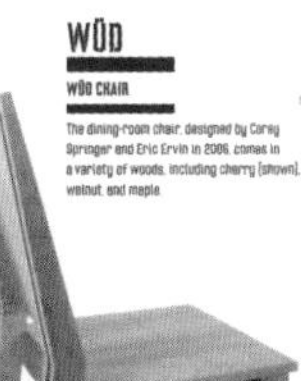

UHURU DESIGN
OK METAL ARMCHAIR
Designed by Jason Horvath, this lounge chair consists of a one-inch-by-two-inch steel frame and upholstered cushions available in custom colors and patterns.

Courtesy the designers

CITY JOINERY
WEDGE CHAIR
This dining-room chair was designed by Jonah Zuckerman in 2007. Pictured in black walnut, it's available in a variety of woods.

SCRAPILE
PROTOTYPE I
Designed by Bart Bettencourt, the chair is made of repurposed wood scraps that were bound for a landfill. The process makes the materials unique to each piece.

PAUL SAMKO
ROCKING CHAIR
The walnut rocker is composed of 15 different pieces. Created by Samko in 2007, the chair can be customized using different types of wood or upholstery.

by EVA HAGBERG

Far from the maddening crowds of the contemporary-furniture scene, a small group of intrepid designers is sprouting like trees in Brooklyn. Aesthetically, they're all over the map. Scrapile (from Greenpoint) is known for the pun it's named after: a scrap pile of locally sourced wood that designers Bart Bettencourt and Carlos Salgado turn into a building material; each block incorporates everything from walnut to plywood and is then processed through a labor-intensive layering method. Uhuru, founded by Bill Hilgendorf and Jason Horvath, offers a line of sleek, multimaterial pieces, all of which, if viewed through a larger lens, are just as sustainable.

Those firms got started about four years ago, and they join the older guard Elucidesign, founded in 2001, and City Joinery, which set up shop in 1996. Elucidesign's Redpoint collection is a beautifully spare series of pared-down pieces; City Joinery's range and look is broader and heavier.

These firms may not share a look, but they do share a sensibility shaped by their size, scale, and voluntary outsider status in the design world. "We're in this straddling position," City Joinery's Jonah Zuckerman says. "We care a lot about design, but we also care a lot about craft." Horvath brings up a similar tension: "We don't want to be this big furniture company that does production overseas, but we don't want to be just building furniture in Red Hook." He shouldn't worry too much. His company and his compatriots are part of a new phenomenon—the rise of the artisan designer, Brooklyn division. ○

● 28.简化复杂的元素

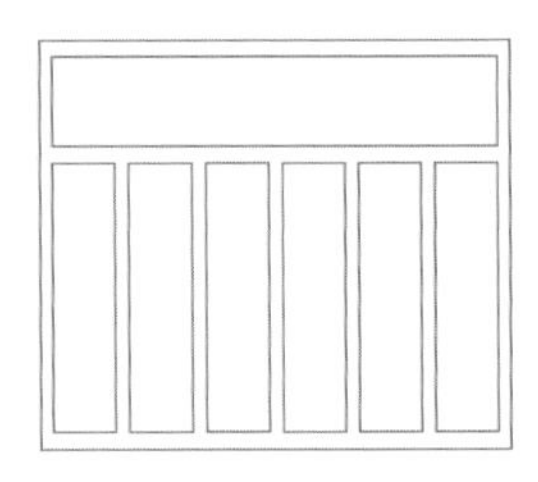

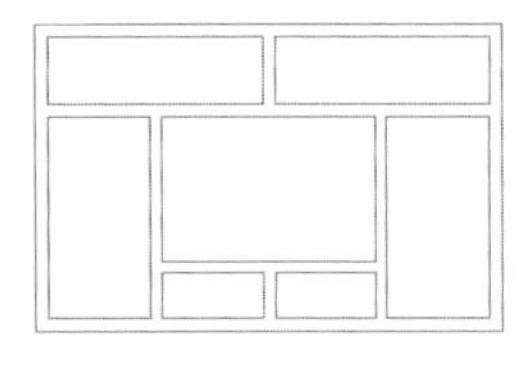

多栏网格非常适合主题严肃的文本的排版，清晰的规划可以通过多种方式将信息分块。不同大小、字体和颜色的文本、列与分隔线共同传达着技术类的信息。

下页图（上）：粗分隔线上方的模块用于放置标题、作者、地点和标识。有时，标题下的粗分隔线会断开，以标记不同分栏之间的间隙

不同的字号和不同的间距使研究信息区别于被放大的研究结论。标题用了对比鲜明的无衬线字体，简要地重复了一遍事实。垂直分隔线将分栏中各个文本段落分隔开，进一步明确了信息

Figure 2: The ICE probe is placed in the right heart for imaging during PFO closure and pulmonary vein isolation.

The ICE probe can be advanced into the inferior vena cava (IVC), enabling high quality imaging of the abdominal aorta (Figure 3).

项目（上图）
海报

客户
纽约大学医学中心（NYU Medical Center）

设计公司
卡拉佩鲁齐设计公司（Carapellucci Design）

设计师
贾尼斯·卡拉佩鲁齐（Janice Carapellucci）

这张纽约大学医学中心的海报是一个教科书式的案例，展示了一个清晰的信息结构。海报中的数据和研究结果的部分易于阅读。每类信息都被区分开，元素之间的间距和空白比例完美，使页面可读性更强。尽管海报的信息量很大，但即使是对非医学从业者而言，读起来也很容易

项目（下图）
SXSW工作室垫纸

客户
smith & beta

设计师
苏珊娜·戴尔·奥尔托

这个为制造商设计的分步指南包括了不同的元素，如图标、贴士、层级、检查表

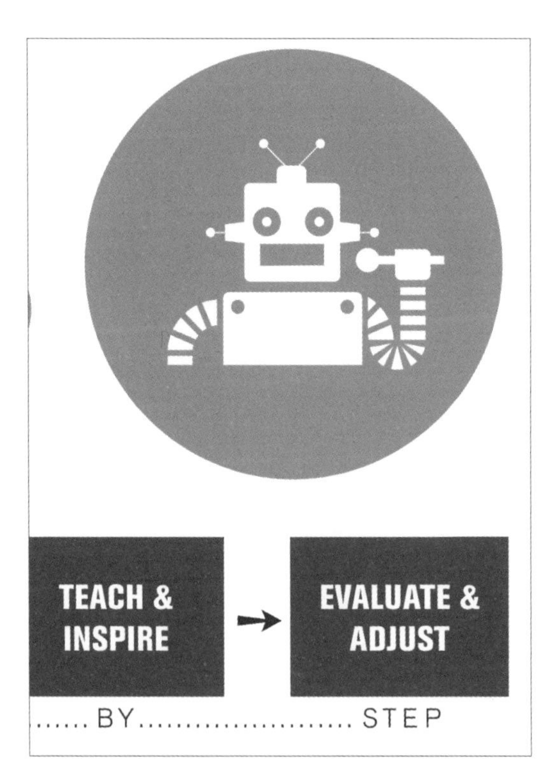

下页图（下）：顶部从左至右排列的文字解释了提纲挈领的观点，中间区域放置了碎片化的信息，粗分割线之下的字体区别于中间区域的列表、流程图及右侧文本框内的字体

Evaluation of the Abdominal Aorta and its Branches Using an Intravascular Echo Probe in the Inferior Vena Cava

Carol L. Chen, MD
Paul A. Tunick, MD
Lawrence Chinitz, MD
Neil Bernstein, MD
Douglas Holmes, MD
Itzhak Kronzon, MD

New York University School of Medicine New York, NY USA

Background

Ultrasound evaluation of the abdominal aorta and its branches is usually performed transabdominally. Not infrequently, the image quality is suboptimal. Recently, an intracardiac echocardiography (ICE) probe has become commercially available (Acuson, Mountain View CA, Figure 1). These probes are usually inserted intravenously (IV) and advanced to the right heart for diagnostic and monitoring purposes during procedures such as ASD closure and pulmonary vein isolation (Figure 2). Because of the close anatomic relation between the abdominal aorta (AA) and the inferior vena cava (IVC), we hypothesized that these probes would be useful in the evaluation of the AA and its branches.

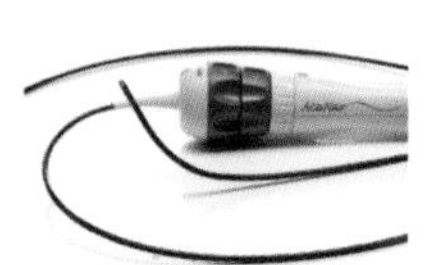

Figure 1: ICE probe (AcuNav, Acuson)

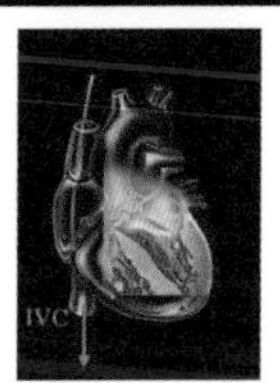

Figure 2: The ICE probe is placed in the right heart for imaging during PFO closure and pulmonary vein isolation.

The ICE probe can be advanced into the inferior vena cava (IVC), enabling high quality imaging of the abdominal aorta (Figure 3).

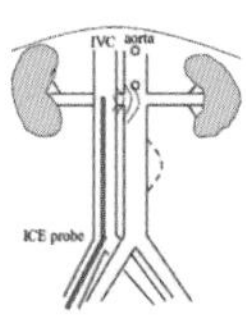

Figure 3: The position of the ICE probe in the IVC allows for excellent imaging and Doppler flow interrogation of the abdominal aorta and its branches (renal arteries, SMA, celiac axis) and the diagnosis of diseases such as renal artery stenosis and abdominal aortic aneurysm.

Methods

Fourteen pts who were undergoing a pulmonary vein isolation procedure participated in the study. In each pt, the ICE probe was inserted in the femoral vein and advanced to the right atrium for the evaluation of the left atrium and the pulmonary veins during the procedure. At the end of the procedure, the probe was withdrawn into the IVC.

Results

High resolution images of the AA from the diaphragm to the AA bifurcation were easily obtained in all pts. These images allowed for the evaluation of AA size, shape, and abnormal findings, such as atherosclerotic plaques (2 pts) and a 3.2 cm AA aneurysm (1 pt). Both renal arteries were easily visualized in each pt. With the probe in the IVC, both renal arteries are parallel to the imaging plane (Figure 4), and therefore accurate measurement of renal blood flow velocity and individual renal blood flow were possible.

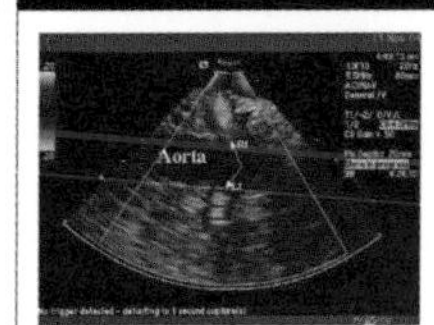

Figure 4: Two-dimensional image with color Doppler, of the abdominal aorta at the level of the right (Rt) and left (Lt) renal ostia. Note visualization of the laminar renal blood flow in the right renal artery, toward the transducer (red) and the left renal artery, away from the transducer (blue).

Calculation of renal blood flow: The renal blood flow in each artery can be calculated using the cross-sectional area of the artery ($\pi r2$) multiplied by the velocity time integral (VTI, in cm) from the Doppler velocity tracing, multiplied by the heart rate (82 BPM in the example shown).

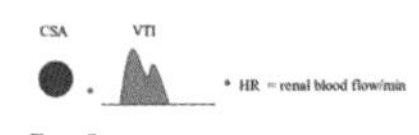

Figure 5

Figure 6: Pulsed Doppler of the right renal artery blood flow. The diameter of the right renal artery was 0.65 cm, and the VTI of the right renal blood flow was 0.19 meters (19 cm). Therefore the right renal blood flow was calculated as 516 cc/minute.

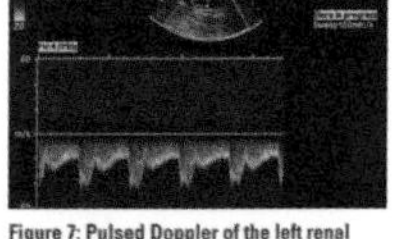

Figure 7: Pulsed Doppler of the left renal artery blood flow. The diameter of the left renal artery was 0.51 cm, and the VTI of the left renal blood flow was 0.2 meters (20 cm). Therefore the left renal blood flow was calculated as 334 cc/minute.

The total renal blood flow (right plus left) in this patient was therefore 850 cc/min. (average normal = 1200 cc/min.)

Conclusions

High resolution ultrasound images of the AA and the renal arteries are obtainable using ICE in the IVC. The branches of the abdominal aorta can be visualized and their blood flow calculated. Renal blood flow may be calculated for each kidney using this method. This may prove to be the imaging technique of choice for intra-aortic interventions such as angioplasty of the renal arteries for renal artery stenosis, fenestration of dissecting aneurysm intimal flaps, and endovascular stenting for AA aneurysm.

Why making? *Are you a maker? We hope so.*

It is a particularly critical time to put intelligent, ethical thought into "things." Perhaps you are shaping products that move markets…or, knitting fluffy hats. Do you recognize, in an antique chair, its narrative… The hand of the artisan who reshaped a tree to offer comfort?

Have you ever optimistically pulled apart broken electronics with hope of resuscitation? Confused by new car engines? *(You are not alone.)* Since the onset of the Anthropocene Age, humans have been obsessed with things. We have allowed them to help us, crowd us, amuse us, comfort us, etc. ***Joy, sustainability, curiosity and purpose*** are some of the keywords for 21c making manifestos. One must have trust in invisible electronic worlds yet remember the many paths we have traveled.

Why make makers?

Since making can be manifested in so many ways—software, toys, or an epicurean meal, it is essential to recognize the elements of a making processes that transcend media.

Materials, meaning-making, and mastery come together as a guide for companies who value creative processes and courageous individuals.

I HEAR AND I FORGET. I SEE AND I REMEMBER. I DO AND I UNDERSTAND."
—Confucius

Making Makers Who Fearlessly Make

SXSW 2015

Lori Kent, Ed. D.
Allison Kent-Smith
Catherine McGowan

10 tips

1. Know what your team makes. Know their skills.
2. Design learning experiences that engage the senses…have emotional meaning and connect to everyday work.
3. Define common terminology around making. Acknowledge team's existing knowledge.
4. Manage people so that their inner imaginations soar. *Tell them that what they know recombines as "creativity."*
5. Encourage everyone around you to have pride in their craft and continue to grow over a lifetime.
6. Design learning experiences that support multiple learning styles and configure complementary teams.
7. Making EXPLICIT a vision and your provisional goals.
8. AND…create a work (making) process that is shared and iterative.
9. Get people to connect with their inner child to lift creative blocks. Take makers to unexpected places.
10. This workshop is a beginning. You have a specific culture, individual needs…***take a first step.***

GATHER TEAM & VISION

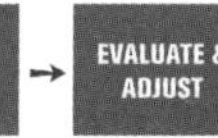

PROGRAM……BUILDING……STEP……BY……STEP

Resources

Texts

Shopcraft as Soulcraft: An Inquiry into the Value of Work by Matthew B. Crawford
The Courage to Create by Rollo May
Spark: How Creativity Works by Julie Bernstein (Studio 360)
Makers: The New Industrial Revolution by Chris Anderson
The Craftsmen by Richard Sennett
Ten Faces of Innovation by Kelley & Littman (IDEO)

WWW

http://aeon.co/magazine/being-human/
https://dschool.stanford.edu/groups/dhandbook/
http://edge.org/
http://makerfaire.com/
http://dx.cooperhewitt.org/lesson-plans/
http://www.fixerscollective.org/
http://www.techshop.ws/

Thanks to

Strawbees, SparkFun, Sally Oetttinger, Meredith Olsen, Grace Borchers, and the s & b teacher collective.
Designed by Suzanne Dell'Orto.

3 ELEMENTS OF MAKING : MATERIALS, MEANING-MAKING, & MASTERY

Materials

- Materials tell you what to do.
- "Functional fixedness" is seeing a "thing" or material as having a specific use…rethink.
- Ordinary materials can inspire, transform….

Meaning-Making

- Be a generative thinker…able to sort, filter, bifurcate, combine and expand.
- Your experience gives you an incredibly rich "well" for making.
- Develop wonder. Think too much.

Mastery

- Mastery? What do you do best?
- How can you deconstruct process to teach mastery?
- How do you support individual and team mastery?

smith&beta.
www.smithandbeta.com

SXSW Evaluation Link: sxsw.feedogo.com/fdbk.do?sid=IAP36301

29.让DIY指南便于理解

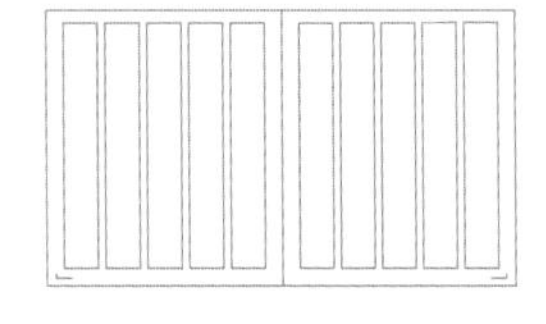

指南类的文本应该方便读者阅读。即使内容是由读者不理解的外文写成的，清晰的版式也可以在一定程度上让文本更易读。给每个步骤和图片编号的方式有助于清楚地表达内容。在版式设计中，要结合照片内容和照片的清晰度，使页面达到既赏心悦目又易于阅读的效果。

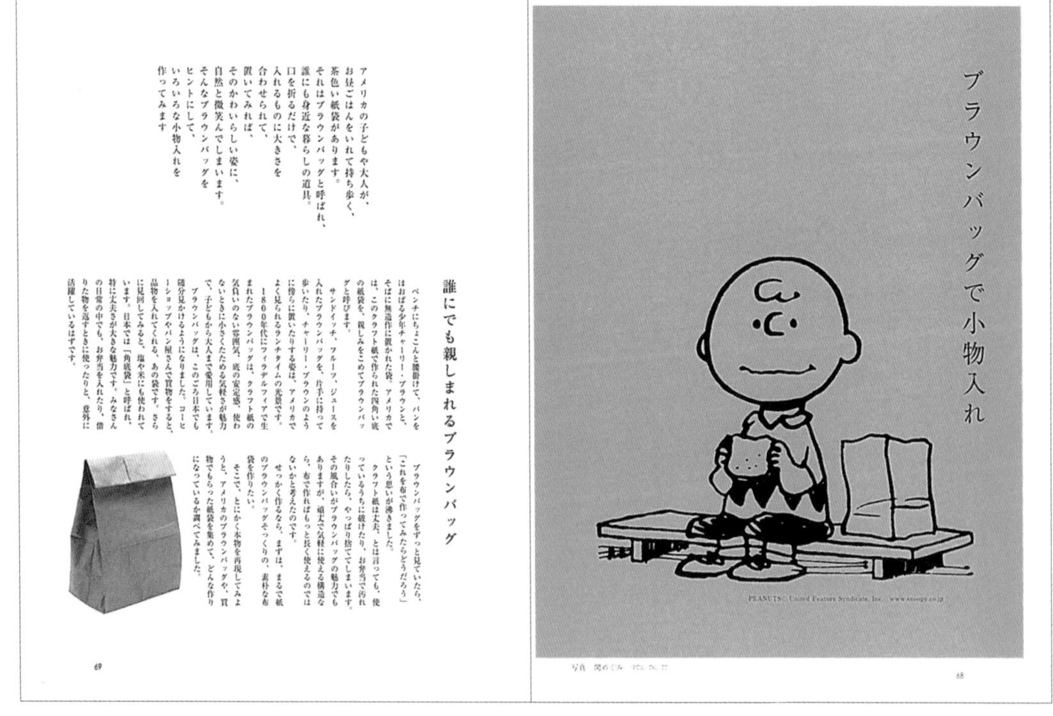

ブラウンバッグで小物入れ

アメリカの子どもや大人が、
お昼ごはんをいれて持ち歩く、
茶色い紙袋があります。
それはブラウンバッグと呼ばれ、
誰にも身近な暮らしの道具。
口を折るだけで、
入れるものに大きさを
合わせられて、
置いてみれば、
そのかわいらしい姿に、
自然と微笑んでしまいます。
そんなブラウンバッグを
ヒントにして、
いろいろな小物入れを
作ってみます

誰にでも親しまれるブラウンバッグ

ベンチにちょこんと腰掛けて、パンをほおばる少年チャーリー・ブラウンと、そばに無造作に置かれた袋。アメリカでは、このクラフト紙で作られた四角い底の紙袋を、親しみをこめてブラウンバッグと呼びます。

サンドイッチ、フルーツ、ジュースを入れたブラウンバッグを、片手に持って歩いたり、チャーリー・ブラウンのように傍らに置いたりする姿は、アメリカでよく見られるランチタイムの光景です。

1800年代にフィラデルフィアで生まれたブラウンバッグは、クラフト紙の気負いのない雰囲気、底の安定感、使わないときに小さくたためる気軽さが魅力で、子どもから大人まで愛用しています。

ブラウンバッグは、このごろ日本でも随分見かけるようになりました。コーヒーショップやパン屋さんで買物をすると、品物を入れてくれる、あの袋です。さらに見回してみると、塩や米にも使われています。日本では「角底袋」と呼ばれ、特に丈夫さが大きな魅力です。みなさんの日常の中でも、お弁当を入れたり、借りた物を返すときに使ったりと、意外に活躍しているはずです。

ブラウンバッグをずっと見ていたら、「これを布で作ってみたらどうだろう」という思いが沸きました。

クラフト紙は丈夫、とは言っても、使っているうちに破けたり、お弁当で汚れたりしたら、やっぱり捨ててしまいます。その風合いがブラウンバッグの魅力でもありますが、頑丈で気軽に使える構造なら、布で作ればもっと長く使えるのではないかと考えたのです。

せっかく作るなら、まずは、まるで紙のブラウンバッグそっくりの、素朴な布袋を作りたい。

そこで、とにかく本物を再現してみようと、アメリカのブラウンバッグや、買物でもらった紙袋を集めて、どんな作りになっているか調べてみました。

69

68

空白的地方可以用来放置介绍性文字。一幅卡通画可以跨越多种文化

项目
《每日笔记》（*Kurashi no techo*）杂志

客户
《每日笔记》杂志社

设计师
林修三（Shuzo Hayashi），柳政明（Masaaki Kuroy-anagi）

这篇指南型的文章将西方虚拟的人物形象——查理•布朗（Charlie Brown）和他的午餐包与东方的空间感完美结合

方框中的箭头标出了从一个要点到另一个要点的步骤方法，页面上的每个部分都清晰地显示在了规整的网格上

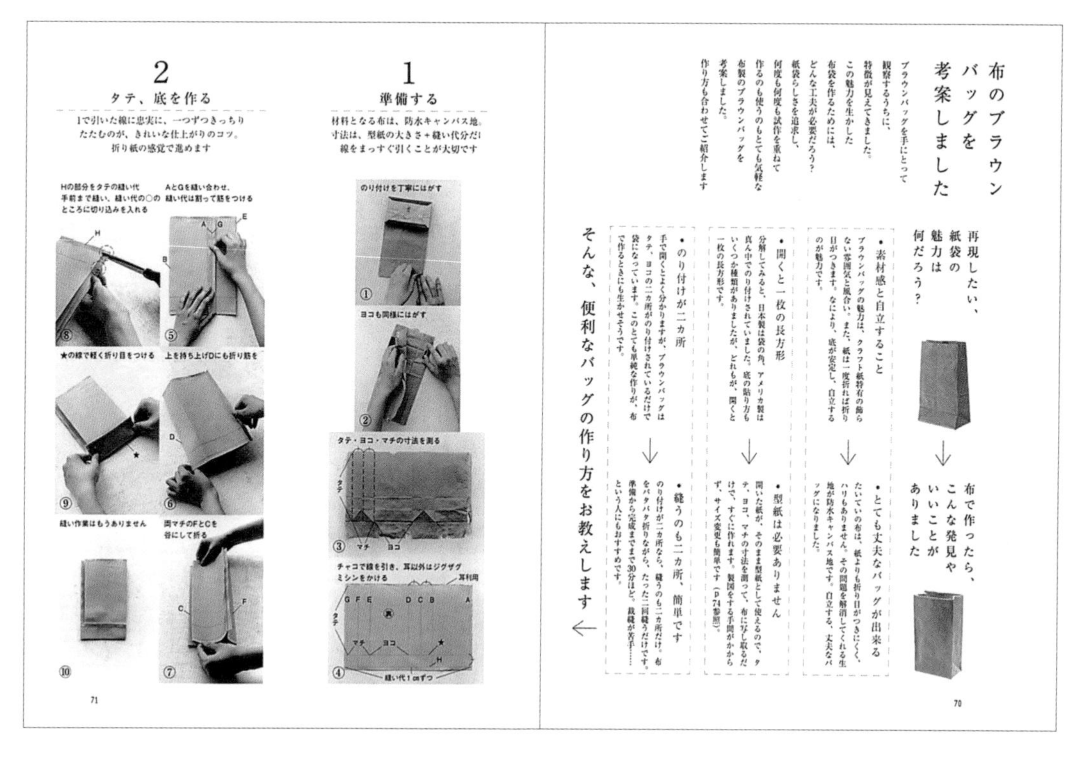

数字标明了流程的每个步骤，圆圈里的小数字标明了每一个细分步骤。各种元素排列有序。图表非常清晰，即使是不懂该语言的手工制作者，也可以完成这个作品。每个元素的相对尺寸适合，照片精美，这使详细的说明看起来不至于令人望而却步

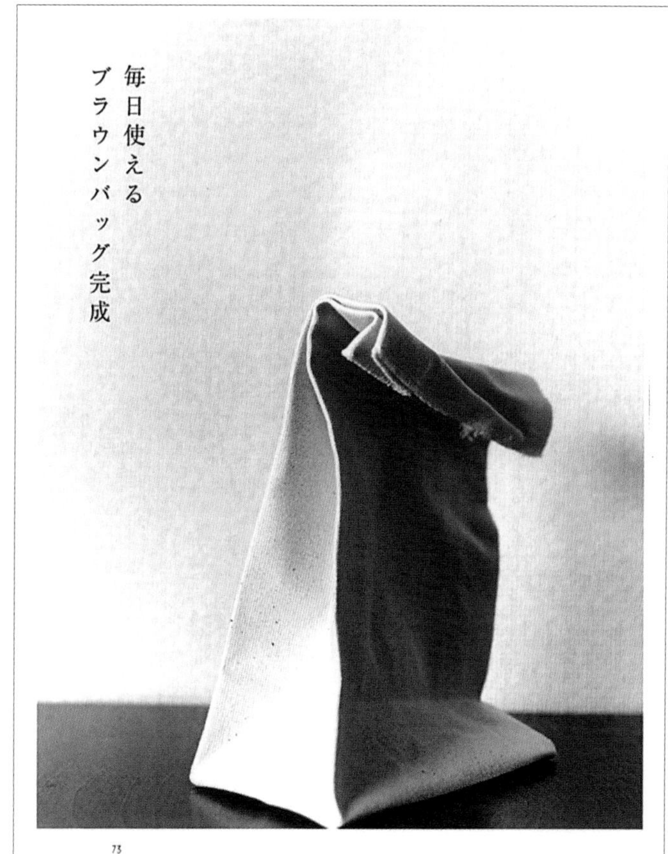

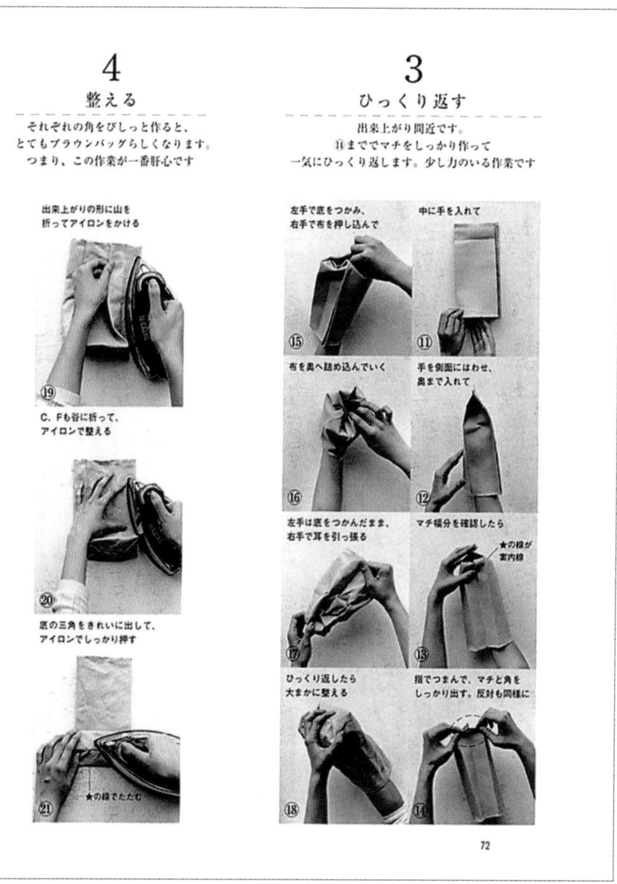

● 30.熟悉网页基础知识

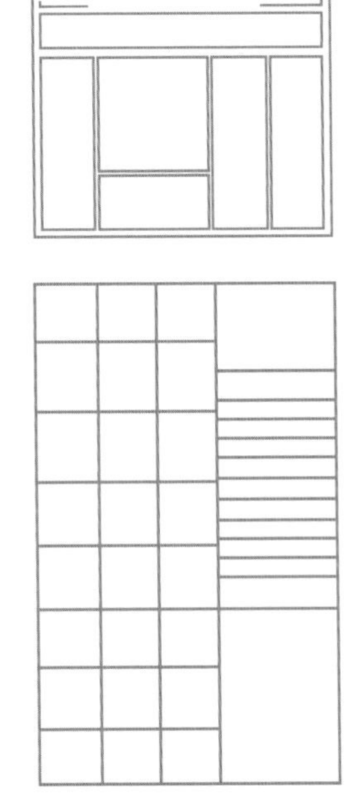

为了容纳大量的信息，大型网页也会使用网格来搭建。网页空间被划分为不同版块，以容纳信息。设计之前，首先要检查网页是否有明确的限制条件，然后考虑屏幕的宽度和工具栏，如屏幕和浏览器的导航栏。与印刷品一样，网页设计需要考虑内容所占的空间，大多数网页中需要考虑的元素包括广告、一系列复杂的标题和副标题、作者名字、列表和链接。所以，选择合适的版式至关重要。

屏幕尺寸

用户的屏幕有不同的尺寸，因此，大多数设计师会设定一块活动区域，使其宽度和长度适合小屏幕。近些年，电脑屏幕的尺寸越来越大，而掌上设备的屏幕尺寸也千变万化，因此，版式更简单，层次更清晰。

SECTIONS SEARCH SUBSCRIBE NOW LOG IN

ENGLISH 中文 (CHINESE) ESPAÑOL

Friday, July 6, 2018 | Today's Paper | Video | 77°F | Nasdaq +1.18% ↑

World U.S. Politics N.Y. Business Opinion Tech Science Health Sports Arts Books Style Food Travel Magazine T Magazine Real Estate ALL

Trade War With China Becomes Reality as Tariffs Take Effect

By ANA SWANSON 9:21 AM ET

- The Trump administration followed through with its threat to impose tariffs on $34 billion worth of Chinese products, setting up a clash between the world's two largest economies.
- The penalties prompted quick retaliation by Beijing. China said it immediately put its own similarly sized tariffs on an unspecified clutch of American goods.

322 Comments

China Strikes Back at Tariffs, but Its Consumers Worry

By RAYMOND ZHONG 20 minutes ago

While many maintain a defiant tone, Chinese shoppers could get hit as the trade war hurts exports and makes even basic types of food more

David Moskowitz

Who's Afraid of the Big Bad Wolf Scientist?

Rob Wielgus was one of America's pre-eminent experts on large carnivores. Then he ran afoul of the enemies of the wolf.

By CHRISTOPHER SOLOMON

Your Friday Briefing

By CHRIS STANFORD 8:34 AM ET

Here's what you need to know to start your day.

Listen to 'The Daily': A Reunification Story

Since President Trump ended the practice of

Opinion

We'll All Be Paying for Scott Pruitt for Ages

By THE EDITORIAL BOARD

A hotter planet, shortened life spans and diminished trust in public service are the cost of the E.P.A. chief's service.

Pruitt's Resignation Is Just the Beginning

By NORMAN L. EISEN and NOAH BOOKBINDER

The damage he caused to the E.P.A., the federal government and the Trump administration will linger for years.

- Bruni: The Moist Mystery of Scott Pruitt's Resignation

OP-ED COLUMNISTS

- Brooks: Fred Rogers and the Loveliness of the Little Good
- Cohen: America Never Was, Yet Will Be
- Krugman: Big Business Reaps Trump's Whirlwind

READER CENTER »
The Times at Gettysburg, July 1863: A Reporter's Civil War Heartbreak

What Nelson Mandela Lost

By TAYARI JONES

The South African freedom fighter's letters from prison remind us that the separation of families is the ultimate expression of state power.

'Hope Is a Powerful Weapon': Unpublished Mandela Prison Letters

By THE EDITORS

A new volume offers insight into the personal and political life of one of the 20th century's most influential freedom fighters.

- Egan: Down and Out in San Francisco on $117,000
- Making America Unemployed Again
- The Déjà Vu of Mass Shootings
- Adrian Piper Speaks! (for Herself)
- Sign Up for Our World Cup Newsletter »

THE CROSSWORD »
Play Today's Puzzle

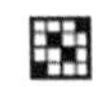

项目
《纽约时报》（*The New York Times*）网站

客户
《纽约时报》

设计公司
《纽约时报》

创意总监
汤姆•博德金（Tom Bodkin）

这个网页的设计将严肃的信息和干净、漂亮又传统的版式结合在一起，通过使用变化的无衬线字体辅助衬线字体，用不同的颜色标注不同的故事和时间线，使网页品质得以提升

紧凑的结构使页面能够容纳导航栏、故事内容、不同尺寸和边距的图片、广告及视频

前《纽约时报》设计总监科伊•荣（Khoi Vinh）认为，“一个个单元格仿佛是搭建网格系统的‘砖块’，而分栏则由一组单元格组成，为页面打造视觉结构”。科伊•荣提到，在理想的情况下，单元格的个数应该是3或4的倍数，以12为最佳，因为12是3和4的乘积。尽管人们在视觉上看不出单元格的个数，但这样的形式为网站打造了强大的结构，使网站的各单元格和分栏的形式井然有序

一旦设计师将单元格聚合成分栏，最重要的就是要在文字左右加入额外的空间，这样前后排版才能整齐，无论分栏中包含的是图片、文字还是文本框

World »

Thai Cave Rescue Will Be Murky, Desperate Ordeal, Divers Say

English City, Stunned, Tries to Make Sense of New Poisonings

Japan Executes Cult Leader Behind 1995 Sarin Gas Subway Attack

Business Day »

U.S. Hiring Stayed Strong in June Despite Trade Strains

The Unemployment Rate Rose for the Best Possible Reason

China Strikes Back at Trump's Tariffs, but Its Consumers Worry

Opinion »

What Nelson Mandela Lost

'Hope Is a Powerful Weapon': Unpublished Mandela Prison Letters

We'll All Be Paying for Scott Pruitt for Ages

REAL ESTATE »

THE HUNT

Trading Chelsea Clatter for Greenpoint Calm

By JOYCE COHEN

Living on Eighth Avenue was fun, but after six years Emery Myers wanted some peace and quiet — not to mention a garden. Walls were optional.

- Search for Homes for Sale or Rent
- Mortgage Calculator

U.S. »

A Black Oregon Lawmaker Was Knocking on Doors. Someone Called the Police.

Migrant Shelters Are Becoming Makeshift Schools for Thousands of Children

Trump Administration in Chaotic Scramble to Reunify Migrant Families

Technology »

Tech Giants Win a Battle Over Copyright Rules in Europe

State of the Art: Employee Uprisings Sweep Many Tech Companies. Not Twitter.

The New New World: Why Made in China 2025 Will Succeed, Despite Trump

Arts »

If It's on 'Love Island,' Britain's Talking About It

Four Musicals on Three Continents: An Australian Company's Big Bet

The Art of Staying Cool: 10 Can't-Miss Summer Shows in New York

MOST EMAILED | MOST VIEWED | RECOMMENDED FOR YOU

1. What Can You Do About a Hammertoe?

2. Mom, I Need a Break

3. Countdown to Retirement: A Five-Year Plan

4. A Cult Show's Recipe for Success: Whiskey, Twitter and Complex Women

5. Facebook Removes a Gospel Group's Music Video

6. A Cult Show's Recipe for Success: Whiskey, Twitter and Complex Women

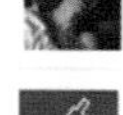

7. Airline Crew Have Higher Cancer Rates

8. London Mayor Allows 'Trump Baby' Blimp for President's Trip to U.K.

9. When a Vegan Gets Gout

10. Statue of Liberty Stamp Mistake to Cost Postal Service $3.5 Million

Log in to discover more articles based on what you've read.

LOG IN | REGISTER NOW

What's This? | Don't Show

Politics »

Amy McGrath Set Her Sights on the Marines and Now Congress. Her Mother Is the Reason.

Trump Assails Critics and Mocks #MeToo. What About Putin? 'He's Fine'

Brett Kavanaugh, Supreme Court Front-Runner, Once Argued Broad Grounds for Impeachment

Fashion & Style »

Modern Love: This Is What Happens When Friends Fall in Love

The Secret Price of Pets

Fashion Review: A Declaration of Independence at Valentino and Fendi

Movies »

Lakeith Stanfield Is Playing Us All

Review: 'Ant-Man and the Wasp' Save the World! With Jokes!

Review: 'Sorry to Bother You,' but Can I Interest You in a Wild Dystopian Satire?

New York »

The Rise of the Stressed-Out Urban Camper

A City Founded by Alexander Hamilton Sets the Stage for Its Next Act

Culture of Fear and Ambition Distorted Cuomo's Economic Projects

Sports »

Neymar and the Art of the Dive

Garbiñe Muguruza and Marin Cilic Join the Wimbledon Exodus

On Pro Basketball: Finally Free From LeBron's Reign, the N.B.A. East Has No Reason to Change

Theater »

Four Musicals on Three Continents: An Australian Company's Big Bet

Review: 'The Royal Family of Broadway,' This Time in Song

Critic's Notebook: Orlando Bloom and Aidan Turner Are Drenched in Blood in London

Science »

Trilobites: Never Mind the Summer Heat: Earth Is at Its Greatest Distance From the Sun

The Lost Dogs of the Americas

Rhino Embryos Made in Lab to Save Nearly Extinct Subspecies

Obituaries »

Ed Schultz, Blunt-Spoken Political Talk-Show Host, Dies at 64

Michelle Musler, Courtside Perennial in the Garden, Is Dead at 81

Claude Lanzmann, Epic Chronicler of the Holocaust, Dies at 92

Television »

If It's on 'Love Island,' Britain's Talking About It

'Sharp Objects,' a Mesmerizing Southern Thriller, Cuts Slow but Deep

On Comedy: A Netflix Experiment Gives Deserving Comics Their 15 Minutes

Health »

Global Health: In a Rare Success, Paraguay Conquers Malaria

Trilobites: Lots of Successful Women Are Freezing Their Eggs. But It May Not Be About Their Careers.

Voices: When a Vegan Gets Gout

Travel »

The Getaway: Looking for a Weekend Escape? Here Are 5 Family-Friendly Options

Carry On: What W. Kamau Bell Can't Travel Without

The Rise of the Stressed-Out Urban Camper

Books »

Profile: Attention, Please: Anne Tyler Has Something to Say

Books of The Times: When It Comes to Politics, Be Afraid. But Not Too Afraid.

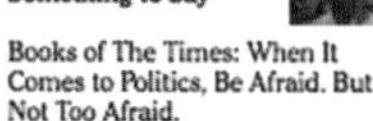

Captain America No. 1, by Ta-Nehisi Coates, Annotated

Education »

'Access to Literacy' Is Not a Constitutional Right, Judge in Detroit Rules

Colleges and State Laws Are Clamping Down on Fraternities

In the Age of Trump, Civics Courses Make a Comeback

Food »

Wines of The Times: American Rosés Without Clichés

Hungry City: A Filipino Specialty Best Paired With a Brew in the East Village

Australia Fare: Yatala Pies Has Served Nostalgia for More Than 130 Years. Arguably.

Magazine »

Feature: Can the A.C.L.U. Become the N.R.A. for the Left?

Letter of Recommendation: Letter of Recommendation: 'The Totally Football Show With James Richardson'

Feature: Who's Afraid of the Big Bad Wolf Scientist?

Real Estate »

New Buildings Rise in Flood Zones

Right at Home: Buried in Paperwork

The Hunt: Trading Chelsea Clatter for Greenpoint Calm

The Upshot »

The Unemployment Rate Rose for the Best Possible Reason

There Isn't Much the Fed Can Do to Ease the Pain of a Trade War

Americans Are Having Fewer Babies. They Told Us Why.

Times Insider »

Outsmarted by a Smart TV? Not This Reporter.

With Our World Cup App, Fans Are Part of the Action

The Times at Gettysburg, July 1863: A Reporter's Civil War Heartbreak

● 31.素材分解

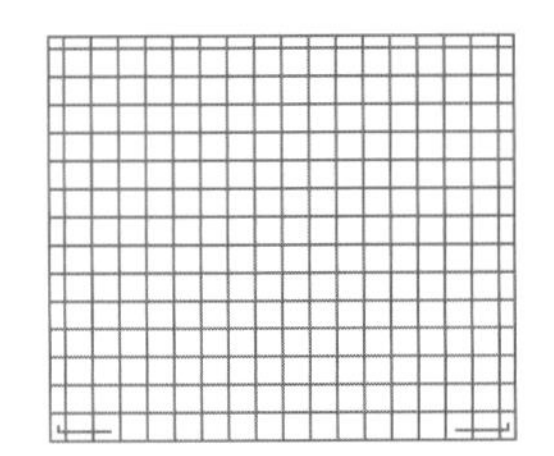

有时，信息会在图表和模块之间交叉。当展示复杂的信息时，需要考虑到清晰度、可读性、留白和多样化。将复杂的信息分解为可管理的模块，可以使布局更清晰。

在以下情况中可以使用模块化网格：
◎有很多独立的信息块，且没有必要连续阅读或者不可能连续阅读时；
◎当需要所有的材料都填入形状相似的块状区域时；
◎当需要一个前后一致的（或者几乎一致的）版式时；
◎各个信息单元的开头由数字或日期组成，各单元素材长度相似时。

为了让版式服务于内容，素材有时候还需要分解。用不同大小和粗细的字体衬托解释性文本，这样可以帮助读者理解文本内容。正如其他原则里提到的，有规律地使用不同的字体可以区分不同的信息，比如，清楚但单调的信息和另类的信息。

项目
《每日笔记》杂志

客户
《每日笔记》杂志社

设计师
林修三，柳政明

这是一本指南类杂志中的一个专题，教授读者生活中的小窍门，版式设计展现出了一种克制的时尚风格

下页图： 在列出的贴士中，文本周围留有同样大小的空间，文字的多少决定了文本框的大小。分隔线在不会遮挡文本框中材料的情况下，可以分隔各个贴士，形成了包含子信息的侧边栏

在任何语言中，项目编号都会起到提示的作用，字体的大小和粗细表示信息的等级顺序

对于编号的模块来说，正如字体的大小和粗细能改变版式的外观一样，阿拉伯数字和日文中的汉字也能给有用的信息带来变化和舒适的感觉

今日はなにを

●暮らしのヒント集

ここにならんでいる
いくつかのヒントのなかで、
ふと目についた項目を
読んでみてください。
たぶん、ああそうだったと
いうことになるでしょう

1 テーブルにコップを置くときは、静かに置くことを心がけましょう。やさしいしぐさが気持ちをやわらげます。

2 組み立て式の椅子やテーブルのネジは、意外とゆるんでいるものです。締めなおしておきましょう。

3 暮らしには笑顔が大事です。いろいろあっても、にっこり笑顔を忘れずに。

4 一年使った枕を新しいものに替えてみましょう。新しい気持ちで眠りにつけるでしょう。

5 今日こそゆるんだ水道のパッキンを取替えましょう。家中の蛇口をチェックします。

6 毎日の暮らしのなかで見て見ぬふりはやめましょう。そういう癖を身につけてはいけません。

7 空気が乾燥してきます。外から帰ったらすぐにうがいができるように、洗面所のコップをきれいにしておきましょう。

8 朝、目が覚めたら、ベッドの中で今日一日、何をするかを考えます。することがたくさんあれば、うかうかしていられず、すぐ起きるでしょう。

9 どんなことでもまずはお金を使わずにできるかを考えてみましょう。それが工夫の一歩になります。

10 言いたいことを言った後は、笑顔で接することが大事です。険悪にならないように、まわりに気を使いましょう。

11 日曜日の朝、天気が良かったら、外でご飯にしませんか。ごく簡単なお弁当を近所の公園などで食べるのです。散歩もかねて気分も変わります。

12 風邪をひいて、お風呂に入れないときは、足だけでも洗って、温めましょう。さっぱりして気分がよくなります。

13 今日は一歩ゆずってみましょう。その一歩がそのまま新しい一歩を進めるちからになるものです。

14 裁縫箱を整理しましょう。さびた針やよれた糸は処分して、新しいものに取替えます。

15 今夜は粗食デーにしましょう。味噌汁にお漬物とか、ありあわせのおかずで間に合わせます。明日は今夜の分もごちそうにしましょう。

16 冷蔵庫が夏の設定になっていませんか。気温も下がったし、あけての回数も減ってきたので、調節しておきます。

17 虫歯があったら、いますぐ治しておきましょう。年末年始のお医者さんが休みのときに痛くなったら大変です。

18 手紙ばさみを買ってみましょう。とても便利なので、毎日届く郵便をさっさと片づけられます。

19 今日は一日、お年寄りのお相手をつとめましょう。お茶を飲みながら、ゆっくりと昔話を聞いてあげたり、一緒に出かけたりします。

20 毎日を心地よく過ごすには、あまりに潔癖すぎてもいけません。よごれやけがれも受け入れてこそ暮らしがあるのです。人との関係も同様です。

21 きびしい肌寒さをおぼえる夜になりました。ことにお年寄りにはひざ掛けか、肩掛けを一枚、早めに用意してあげましょう。

22 しめきりの窓をあけて、敷居のゴミを払いましょう。アルミサッシの溝など、ほこりがつまっているものです。

23 洋服ダンスの防虫剤は大丈夫でしょうか。においはしていても、中身はもうなくなっていることが案外多いものです。

24 新しいチャレンジは自分で決めるものです。ひとに惑わされて後悔しないように。

25 ガス台の下やすきまを掃除しましょう。意外に汚れているものです。きれいになると気持ちよく料理ができるでしょう。

● 32.留出一些呼吸的空间

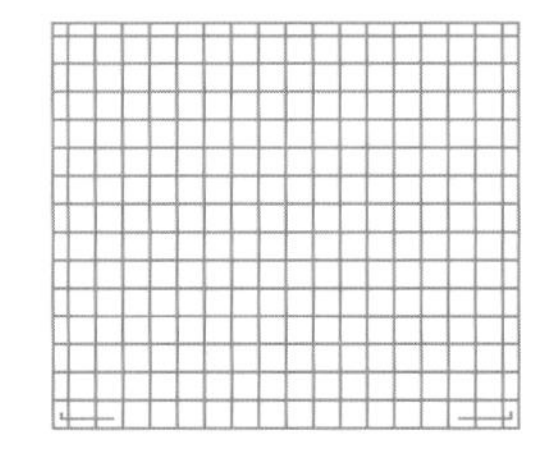

不是所有模块都需要填满。一个模块化的网格能精确地确定各部分的尺寸，从而让设计师可以分配和管理更多细节。模块可以是有形的，也可以是无形的。模块可大可小，它们形成了一个稳定的结构，容纳文字、字母、色彩或装饰。而且，它们还可以是完全空白的。

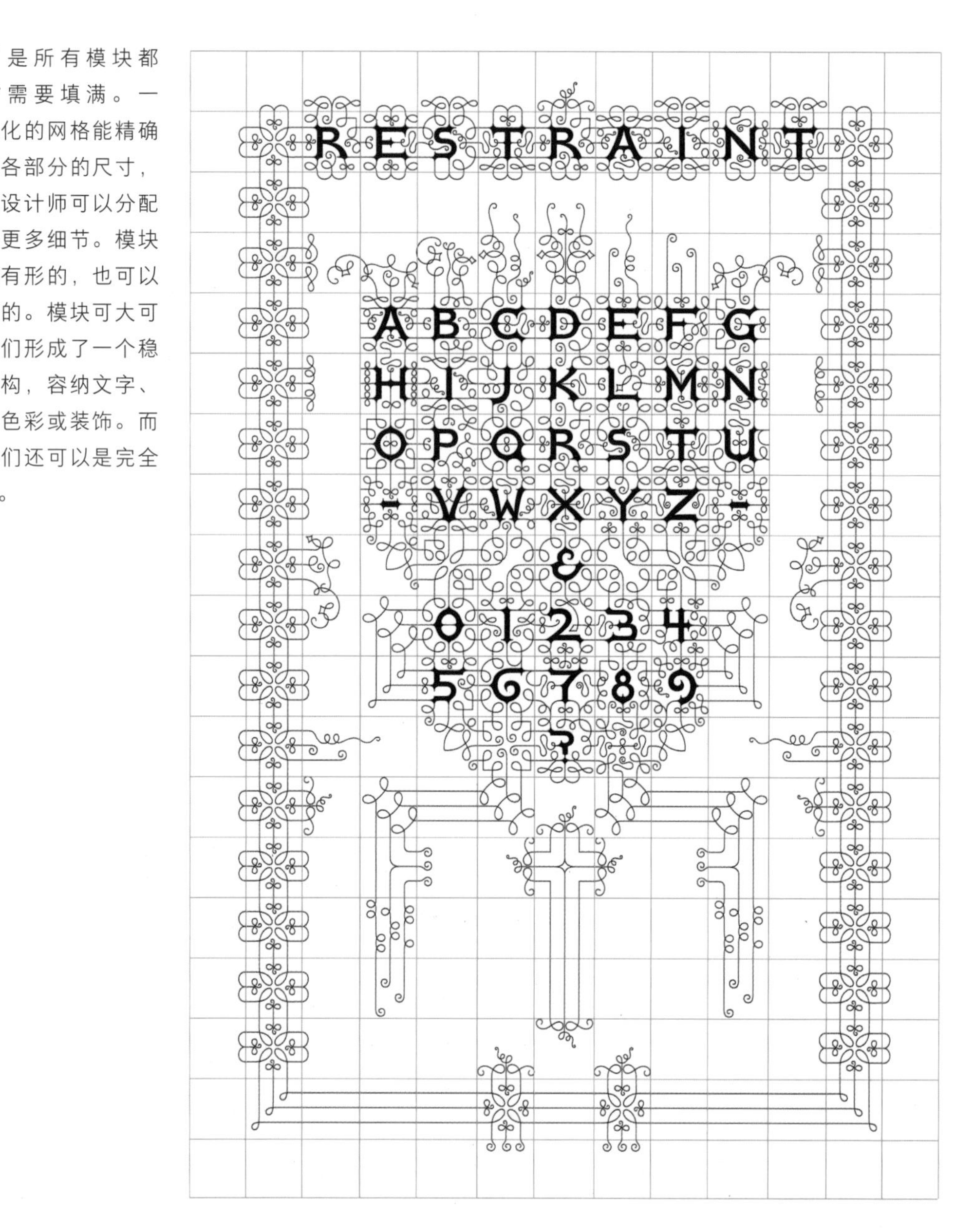

项目
Restraint字体

客户
玛丽安•班杰斯（Marian Bantjes）

设计师
玛丽安•班杰斯，罗斯•米尔斯（Ross Mills）

手工制作的印刷格式将数字数码化

展示性字体是专门为标题设计的，不适用于正文。因为当字号较小时，字体显著的特征就会消失，变得很难阅读

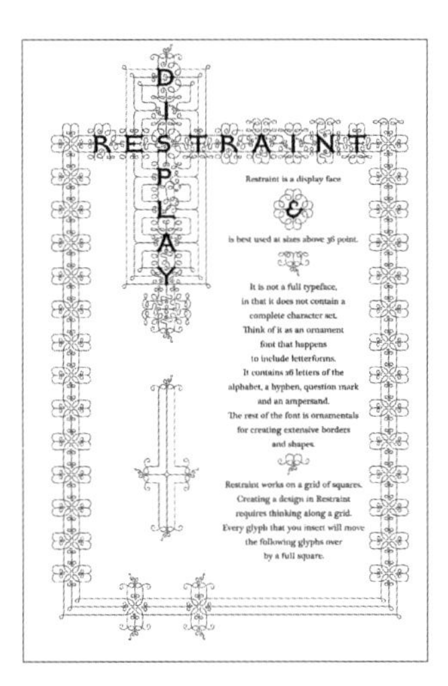

填充页面中心的模块，并在边缘留出空间，这样可以将外部的空间变成一个框架

另一个方法是用模块作为框架，在中心留出空间

RESTRAINTS

Font Software Product License
End-User License Agreement (EULA)

(page 1 of 2)

❖ PLEASE READ ❖

Some restrictions apply to the use of this software

The 'Restraint' typeface (Font Software) and designs contained therein is protected by copyright laws and international copyright treaties, as well as other intellectual property laws and treaties. The Font Software is licensed, not sold. This license is only valid when the licensee has been listed below and this agreement is signed by a representative of Tiro Typeworks. Please retain copies of this agreement.

Whereas 'Tiro Typeworks' is represented by one or both of the following individuals:
William Ross Mills of Galiano Island, British Columbia, Canada. DBA Tiro Typeworks and
John Hudson of Gabriola Island, British Columbia, Canada. DBA Tiro Typeworks

Subject to the foregoing, Tiro Typeworks grants (hereafter the 'licensee') :
M E Tondreau
611 Broadway
Room 511
New York, NY 10012
United States
a perpetual non-exclusive license to use the Restraint Font Software with the following terms and conditions:

1. ACCEPTANCE OF TERMS
Installation and use of this Font Software constitutes acceptanceof the terms of this licence agreement.

1.1 You acknowledge that the Font Software is the intellectual property of Tiro Typeworks and/or designers represented by Tiro Typeworks and contains copyrighted material authored by Tiro Typeworks and/or designers represented by Tiro Typeworks. The term Font Software shall also include any updates, upgrades, additions, modified versions, and development copies of the Font Software licensed to you by Tiro Typeworks. The media itself is and shall remain the property of Tiro Typeworks. Expanded versions, subsets or other derivatives of this design may also exist under other names and be distributed by Tiro Typeworks or other licensed Distributors.

2. GRANT OF LICENSE.
This document grants you the following rights:

2.1 INSTALLATION AND USE.
You may install and use the Font Software on up to five computer hard drives or other storage devices and up to two physical output devices (e.g. printers, imagesetters) based at one single geographical location stipulated by the licensee (laptops may be considered 'based' at a single location). The Font Software may not be used by more than five users on a network. Extended licenses may also be purchased, in which case a new license agreement will be drafted to reflect the new conditions.

For the sole purpose of data backup, additional backup copies of the Font Software may be made.

2.2 FAIR USE.
You may use the Font Software in most personal and commercial applications. However, under this license, you may not use the font software:

a) for the creation of logos or identities (including movie titles)

b) for the creation of signage or architectural details.

c) for the creation of advertising campaigns which include outdoor advertising (billboards, bus shelters, etc.) or television advertising, wherein the designs contained in the Font Software comprises the sole or major design element.

d) to manufacture products wherein the designs contained in the Font Software comprises the sole or major design element, including but not limited to T-shirts, jewellery, fridge magnets, greeting cards, ceramics, posters for sale, etc.

If you wish to use the Font Software for any of the above, please contact us at restraint@tiro.nu for additional licensing or royalty fees. If in doubt, ask.

2.3 MODIFICATION.
You are not allowed to without written approval granted by Tiro Typeworks:

a) modify and/or recompile the Font Software: this includes generating or re-compiling the Font Software from any font design program. (where a 'font design' program is any piece of software capable of reading and re-compiling any standard font format),

b) adapt modules, produce sub-sets or supersets or alter any internal font data thereof for your own developments,

c) put the software solutions embodied in the Font Softwareto any commercial use other than operating your own computer(s) or output device(s), or

d) merge, ship or embed the Font Software with other software programs.

PLEASE CONTACT TIRO TYPEWORKS OR A LICENSED DISTRIBUTOR IF THERE ARE SPECIFIC MODIFICATIONS THAT YOU REQUIRE.
We acknowledge that no typeface can solve all problems and accept that some clients may wish to have modifications made to suit their particular needs. We would be happy to help with this and no one knows better the typefaces you are licensing, so please ask first.

这个最终用户许可协议展示了漂亮的版式以及关于使用Restraint字体的规定

33.融合模块

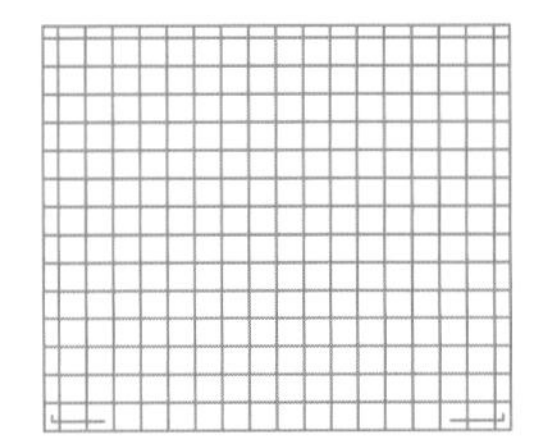

作为图表时，模块化网格可能看起来很复杂，但实际上并非如此，而且没有必要把每一个模块都填满。根据需要决定放入空间的信息量的多少，可以使用几个大的文本框建立一个模块，模块里放置图片，更重要的是要放置关键信息，如目录和索引信息。

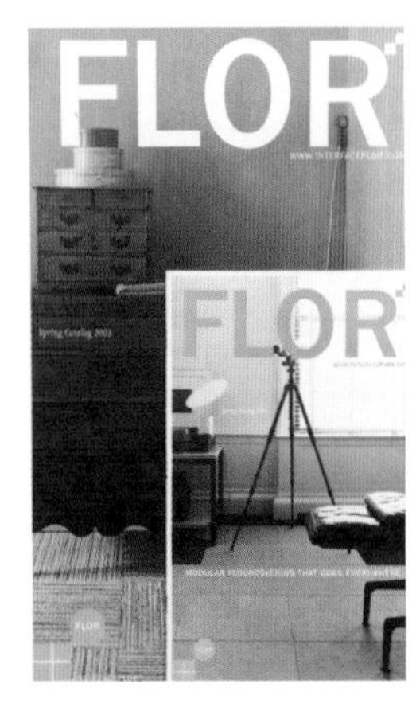

模块被嵌入图片中，页面左下角还嵌入了一个带有Flor标志的模块

项目
Flor产品目录

客户
Flor地板集团

设计公司
华伦天奴集团（The Valentine Group）

模块化网格最适合用于理性的内容。它可以将一个页面分解成步骤式的视觉指南，正如这个目录，它将地板的介绍模块化了

这个目录页被分解为数个小格子，并使用了不同的色彩，易于阅读和浏览

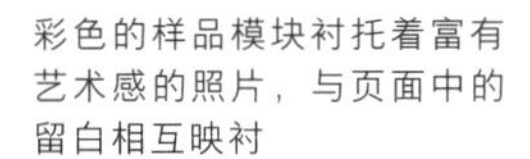

彩色的样品模块衬托着富有艺术感的照片，与页面中的留白相互映衬

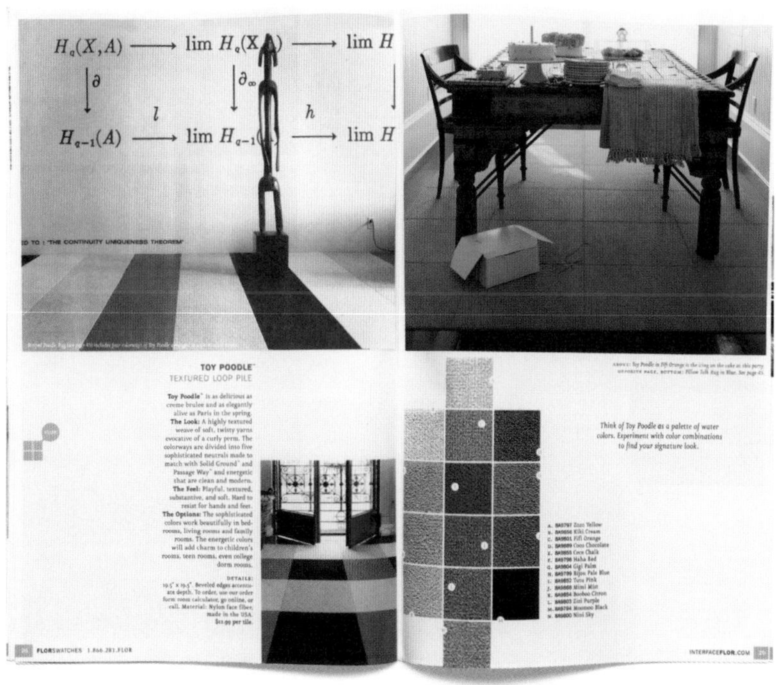

ROOM FEET APPROX	7'	9'	11'	12'	13'	15'	17'	18'	20'	22'	23'	25'	27'
4'	12 TILES	16	19	21	22	26	29	30	34	37	39	42	45
5'	15	19	23	26	28	32	36	38	42	46	48	52	56
7'	21	27	32	35	38	44	50	53	58	64	67	73	78
9'	27	34	41	45	49	56	64	67	75	82	86	93	100
11'	32	41	50	55	59	68	77	82	91	100	104	113	122
12'	35	45	55	60	65	75	84	89	99	109	114	124	133
13'	38	49	59	65	70	81	91	97	107	118	123	134	144
15'	44	56	68	75	81	93	105	111	124	136	142	154	167
17'	50	64	77	84	91	105	119	126	140	154	161	175	189
18'	53	67	82	89	97	111	126	133	148	163	170	185	200
20'	58	75	91	99	107	124	140	148	165	181	189	205	222
22'	64	82	100	109	118	136	154	163	181	199	208	226	244
23'	67	86	104	114	123	142	161	170	189	208	217	236	255
25'	73	93	113	124	134	154	175	185	205	226	236	256	277 TILES

Flor地板集团的计算表本质上就是一个模块表格

34.合理规划空间

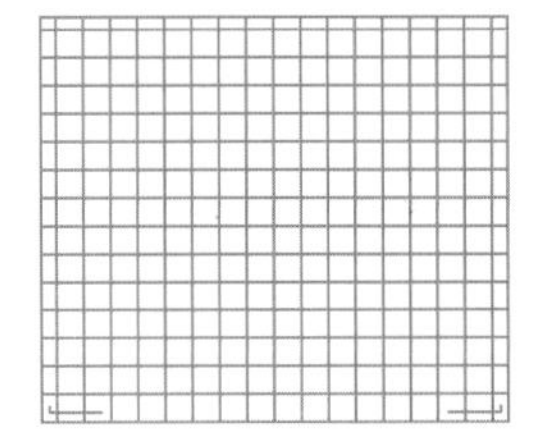

编辑复杂信息的第一步是设计出简洁有力的网格。根据呈现的信息规划每个模块的比例，这样即使是可能让人混淆的材料也会很清晰。

海报之所以能作为一个引人注意的载体，得益于它足够大的面。标题最好大一些，以便人们在远处也能看清，并吸引读者继续阅读其中的细节。

A COMMUNICATIONS MA
AMERICAN VOTER'S EX

EDUCATION

LEARNING ABOUT VOTING RIGHTS AND DEMOCRACY

WORD-OF-MOUTH

Families are a primary source of civics education, but this method of voter education is inadequate.

HIGH SCHOOL CIVICS CLASSES

We learn about voting rights in high school civics classes, which are disappearing from U.S. education.

CITIZENSHIP CLASSES

Laborious self-study books are replacing the traditional citizenship classes required for naturalization.

REGISTRATION

SIGNING UP TO BECOME A REGISTERED VOTER

PAPER REGISTRATION FORMS

Complicated, badly-printed voter registration forms are common in most states.

ONLINE REGISTRATION FORMS

Oregon Voter Registration

Download a Voter Registration Form now!

Many states are testing on-line registration systems. To minimize fraud, most states still print out a paper form.

MOTOR VOTER APPLICATIONS

Many states allow for voter registration on the driver's license application, but the check boxes can be hard to find.

VOTER ROLLS

Many voters are turned away from the polls because their registration is incomplete or inaccurate.

DESIGN PROBLEM
DISAPPEARING CIVICS CLASSES

DESIGN PROBLEM
FORMS THAT ARE BARRIERS TO PARTICIPATION

DESIGN TO THE RESCUE

ALL KINDS OF DESIGNERS CAN PARTICIPATE IN VOTER REFORM. HERE'S WHO

GRAPHIC DESIGNERS

项目
海报——通过设计投票

客户
明尼苏达大学设计学院（Design Institute, University of Minnesota）

编辑/项目总监
珍妮特•艾布拉姆斯（Janet Abrams）

艺术总监/设计师
西尔维亚•哈里斯（Sylvia Harris）

这张海报的内容是对投票中的关键步骤的分解，海报充分利用了每一寸空间，使用网格引导人们的阅读

下页图：虽然海报包含了很多信息，但将投票过程分解成步骤的方法使其更易于阅读

VOTING BY DESIGN

The century began with an electoral bang that opened everyone's eyes to the fragility of the American voting system. But, after two years of legislation, studies and equipment upgrades, major problems still exist. Why?

Voting is not just an event. It's a complex communications process that goes well beyond the casting of a vote. For example, in the 2000 presidential election, 1.5 million votes were missed because of faulty equipment, but a whopping 22 million voters didn't vote at all because of time limitations or registration errors. These and many other voting problems can be traced not just to poor equipment, but also to poor communications.

Communicating with the public is what many designers do for a living. So, seen from a communications perspective, many voting problems are really design problems. That's where you come in.

Take a look at the voting experience map below, and find all the ways you can put design to work for democracy.

A COMMUNICATIONS MAP OF THE AMERICAN VOTER'S EXPERIENCE

EDUCATION

LEARNING ABOUT VOTING RIGHTS AND DEMOCRACY

WORD-OF-MOUTH

Families are a primary source of civics education, but this method of voter education is inadequate.

HIGH SCHOOL CIVICS CLASSES

We learn about voting rights in high school civics classes, which are disappearing from U.S. education.

CITIZENSHIP CLASSES

Laborious self-study books are replacing the traditional citizenship classes required for naturalization.

DESIGN PROBLEM
DISAPPEARING CIVICS CLASSES

REGISTRATION

SIGNING UP TO BECOME A REGISTERED VOTER

PAPER REGISTRATION FORMS

Complicated, badly-printed voter registration forms are common in most states.

ONLINE REGISTRATION FORMS

Oregon Voter Registration
Download a Voter Registration Form now!

Many states are testing on-line registration systems. To minimize fraud, most states still print out a paper form.

MOTOR VOTER APPLICATIONS

Many states allow for voter registration on the driver's license application, but the check boxes can be hard to find.

VOTER ROLLS

Many voters are turned away from the polls because their registration is incomplete or inaccurate.

DESIGN PROBLEM
FORMS THAT ARE BARRIERS TO PARTICIPATION

PREPARATION

BECOMING INFORMED AND PREPARED TO VOTE

SAVE-THE-DATE CARD

VOTE

Everything you need to know is often lost on the poorly designed voting reminder postcard sent to every home.

VOTER REGISTRATION CARD

Each voter gets a registration card to tell them where to vote. Can they find it on election day? Maybe not!

PUBLIC SERVICE ANNOUNCEMENTS

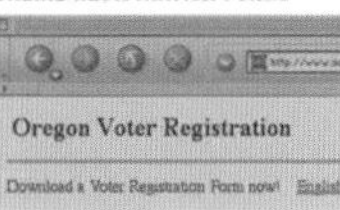

Non-profits produce get-out-the-vote campaigns during elections, but they need more money and design help.

PRE-ELECTION INFO PROGRAMS

Welcome to the 2001 Primary Election Voter Guide

Pre-election instruction packages come in the mail, but get lost or buried in piles of junk mail.

CAMPAIGN LITERATURE

Many voters rely on political campaign literature to prepare for elections. It is accessible, but is it objective?

SAMPLE BALLOTS

Sample ballots can help voters to rehearse and plan what they will do in the voting booth.

DESIGN PROBLEM
TOO MUCH OR TOO LITTLE INFORMATION

NAVIGATION

FINDING THE WAY TO THE VOTING BOOTH

EXTERIOR STREET SIGNS

ELDERLY AND HANDICAPPED VOTERS RESERVED PARKING
VOTER PARKING ONLY

Clear and legible temporary directional signs are needed to help voters find their way to the precinct door.

PRECINCT SIGNAGE

VOTE HERE

Temporary signs turn public buildings into precincts. They are often too small and poorly designed to be effective.

LINE AND BOOTH IDENTITY

42 ELECTION DISTRICT

How many voters waste time standing in the wrong line? Inadequate signage design and placement is often to blame.

PRECINCT WORKERS

Most voters depend on precinct staff to help them navigate the precinct.

CAMPAIGN WORKERS

Voters look for campaign workers as a signal that they are approaching the polls, but they are unreliable.

DESIGN PROBLEM
GETTING TO THE BOOTH ON TIME

VOTING

INDICATING A CHOICE IN AN ELECTION

HAND-COUNTED PAPER BALLOT

Paper ballots list all choices on a sheet of paper. They are easy for the voter to use, but hard to tabulate.

MACHINE-COUNTED PAPER BALLOT

DEMOCRATIC PARTY
WALTER F. MONDALE
GERALDINE A. FERRARO

This ballot is like a standardized test. It is designed for machine tabulation and not for voter ease-of-use.

MECHANICAL LEVER

These complicated machines make it easy to tabulate votes, but many voters find them difficult to use.

PUNCHCARD

The voter puts a ballot book over this punch card and pokes a selection. Sometimes it works, sometimes not.

DIRECT RECORD ELECTRONIC

New ATM-like voting machines are coming, but the interfaces need extra design attention to ensure ease-of-use.

VOTING INSTRUCTIONS

Pull the red voting handle from left
palanca grande de color rojo desde la izquierada

Even the best voting technology won't work if the user instructions are confusing or hard to figure out.

DESIGN PROBLEM
USER-UNFRIENDLY VOTING MACHINES

FEEDBACK

GIVING FEEDBACK ABOUT THE VOTING EXPERIENCE

CENSUS SURVEYS

12. Reason for not votin
01 Too busy
02 Illness or emergency
03 Not interested
04 Out of town
05 Didn't like candidates
06 Other reason

U.S. Census surveys are the best source of voter experience data. They track how, when and why people vote.

EXIT POLLS

Polls are a legal way to find out what is on the voter's mind. They rarely ask about the voting process.

VOTING EXPERIENCE SURVEYS

How long did it take you to get here from home?
How long did it take you to vote?
Did you get help with the equipment?
Who helped you?

Surveys of voters' experience are rarely done but are very much needed to design a better voting process.

DESIGN PROBLEM
FUTURE IMPROVEMENTS LACK VOTER INPUT

DESIGN TO THE RESCUE

ALL KINDS OF DESIGNERS CAN PARTICIPATE IN VOTER REFORM. HERE'S WHO SHOULD BE ON ANY VOTING DESIGN DREAM TEAM:

GRAPHIC DESIGNERS

ENVIRONMENTAL GRAPHIC DESIGNERS

INFORMATION DESIGNERS

ARCHITECTS

INDUSTRIAL DESIGNERS

EXPERIENCE DESIGNERS

HOW YOU CAN GET INVOLVED

THERE IS WORK TO BE DONE TO IMPROVE VOTING BY DESIGN, STARTING WITH YOUR OWN COMMUNITY. HERE ARE FIVE THINGS THAT ANY DESIGNER CAN DO, TO MAKE A DIFFERENCE BEFORE THE 2004 ELECTIONS:

1. BECOME A POLLWORKER

2. FORM A VOTING DESIGN COALITION

3. WORK WITH THE POLITICAL PARTY OF YOUR CHOICE

4. CALL YOUR CONGRESSPERSON ABOUT HR 3295

5. FORM A VOTING DESIGN ADVISORY TEAM

UNIVERSITY OF MINNESOTA

35.有机的网格

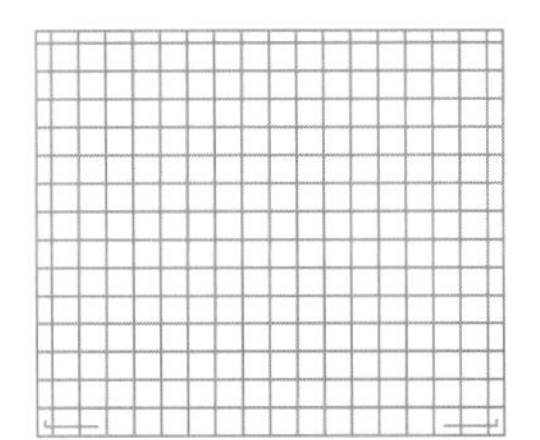

模块化网格的好处在于它不一定要方方正正。在一个连续的模块化项目中，可以改变网格的形状、大小和图案，以保持版式的秩序感和阅读时人的愉悦感。

REGINA, TO THE TRADE; NANCY CORZINE: 212-223-8340.

ELAN, FROM $84; PINE CONE HILL: 413-496-9700.

VINTAGE FABRIC, $95: ABC CARPET & HOME: 212-473-3000.

CHINESE EMPEROR, $165; BEDFORD & CO.: 212-772-7000.

DIYA OATMEAL, $265; JOHN ROBSHAW: 212-594-6006.

APPLE, $48; CARRERAS: 845-758-2200.

ETHNOA KUBA CLOTH, $265; CALYPSO HOME: 631-324-8146.

CAMINO, $195; RALPH LAUREN HOME: 888-475-7674.

19TH C. GOLD METALLIC APPLIQUE, $1350; B. VIZ: 318-766-4950.

TINA, $870; ARMANI CASA: 212-334-1271.

MODERN GOTHIC GRIFFIN, $198; MICHELE VARIAN: 212-343-0033.

DIYA AMBER, $80; JOHN ROBSHAW: 212-594-6006.

SEQUIN COVER, $83; THE CONRAN SHOP: 866-755-9079.

JACK RUSSELL, $98; L'AVENUE DES RÊVES: 212-396-9500.

ZARDOZEE, $90; ALPHA BY MILLI HOME: 212-643-8850.

STRIPES; $215; RANI ARABELLA: 561-802-9900.

112

限制颜色的变化以及为每一页选择一个色系，可以使页面给人一种平衡感

项目
《美丽之家》（*House Beautiful*）杂志

客户
《美丽之家》杂志社（*House Beautiful* magazine）

设计师
芭芭拉·德怀尔德

经过重新设计，一本杂志获得了新生

连续的、结构合理的排版是每个模块的基础，而那些均衡的、全大写的无衬线字体则像是一条条有质感的分隔线

SAN MARGHERITA; $245; RANI ARABELLA: 561-802-9900.

LATTICE, FROM $95; SEACLOTH: 203-422-6150.

CORAL ON WHITE LINEN, $185; HOMENATURE: 631-287-6277.

MARYANN CHATTERTON, $498; D. KRUSE: 949-673-1302.

SEABLOOM, FROM $110; OROMONO: 917-338-7568.

CHRYSANTHEMUM, $55; PINE CONE HILL: 413-496-9700.

TRANSYLVANIAN TULIP, FROM $83; AUTO: 212-229-2292.

SUZANI FLORAL, $212; MICHELE VARIAN: 212-343-0033.

IKAT, $500; D. KRUSE: 949-673-1302.

GREEK REVIVAL EMBROIDERY, $260; DRANSFIELD & ROSS: 212-741-7278.

PLAID, $135; ALPANA BAWA: 212-254-1249.

WEE LOOPY FELTED, $213; THE CONRAN SHOP: 866-755-9079.

VESUVIO, $395; DRANSFIELD & ROSS: 212-741-7278.

NIZAM, $83; JOHN DERIAN DRY GOODS: 212-677-8408.

CYLINER LINEN, $195; GH INTERIORS: 888-226-8844.

LINEN, $70; ALPHA BY MILLI HOME: 212-643-8850.

KAFFE FASSETT HIBISCUS, $68; PINE CONE HILL: 413-496-9700.

DAVID TURNER/STUDIO D

113

36.以全局的目光考虑图表

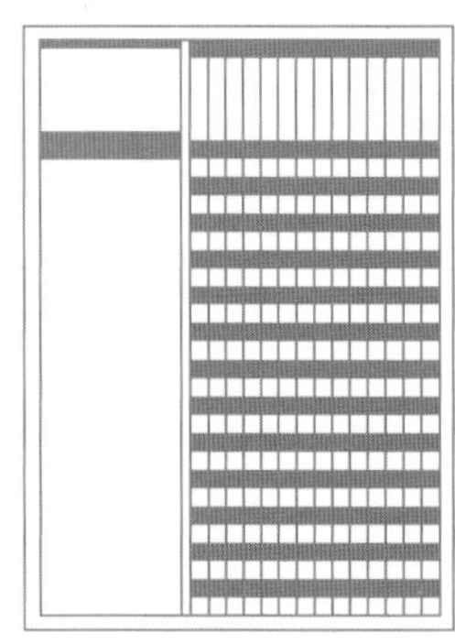

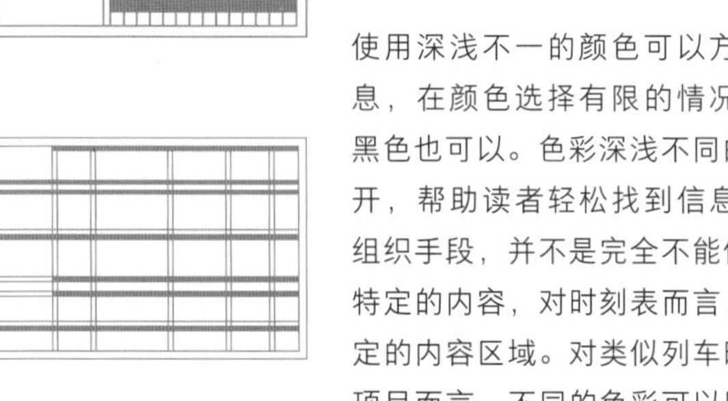

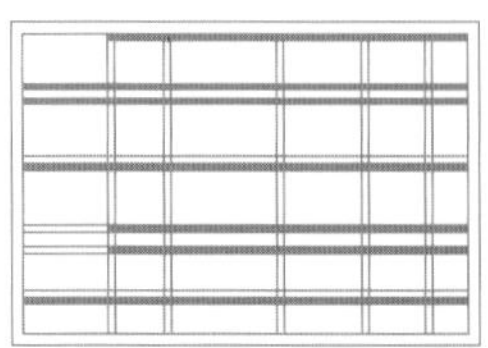

创建图表和时刻表是一种用数字信息完成的富有挑战性的任务。艾伦•卢普顿(Ellen Lupton)在她的著作《用字体思考》（*Thinking with Type*）中建议设计师不要滥用分隔线和文本框，这会形成她所谓的“数据监狱 (data prison) ”。按照卢普顿的建议，可以将图表、网格或时刻表看作一个整体，并考虑每一列、每一行或字段如何与整体设计相关联。

使用深浅不一的颜色可以方便读者浏览密集的信息，在颜色选择有限的情况下，即使只用白色和黑色也可以。色彩深浅不同的横条可以使各行区分开，帮助读者轻松找到信息。框架和分隔线作为组织手段，并不是完全不能使用。分隔线可以区分特定的内容，对时刻表而言，分隔线还可以定义特定的内容区域。对类似列车时刻表这样系统复杂的项目而言，不同的色彩可以区分不同的铁路或通勤线路。

如果没有展示内容，网格系统就毫无意义。在多栏网格中，干净的排版至关重要。对机场和火车站的指示牌来说，数据排版的方式会影响旅客出行的效率，要么轻松旅行，要么迷路。即使信息量巨大，也要确保足够的行间距，这样可以增加可读性，这是设计时刻表的第一准则。

交替使用不同的颜色条可以区分时刻表上的每一个车站。分隔线的使用并不多，却清楚地划分了主要信息和次要信息。垂直分隔线将站点与目的地分开，而水平分隔线则将主要地理区域区分开

时刻表的排版要点也适用于票价表。同样，也可以通过交替使用不同的颜色条表示不同的车站，利用水平分隔线和垂直分隔线标出标题，如单程及非高峰时刻往返的站点和票价

项目
新泽西州轨道交通时刻表

客户
新泽西轨道交通局（New Jersey Transit）

设计公司
212联合公司（Two Twelve Associates）

这些新泽西州的轨道交通时刻表表明，通过简化和流线化，设计师可以在没有太多分隔线、文本框的情况下组织素材。图标或箭头可能会显得老套，但使用常见的图标帮助出行的人浏览丰富的信息是一种相当有效的方式

在说明支付要求的段落中，标题使用了易于辨认的标志符号

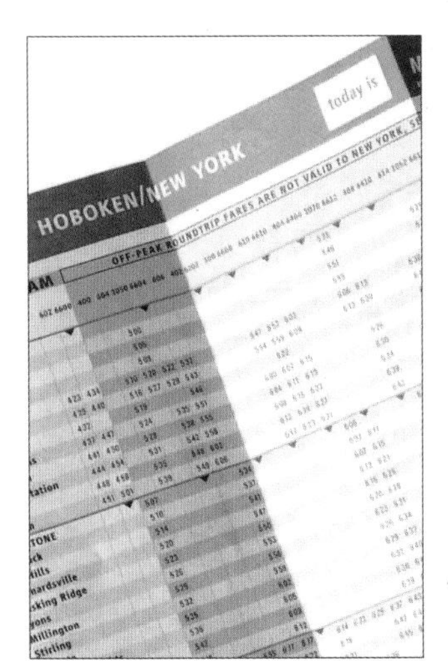

交通快线车站用箭头符号标记

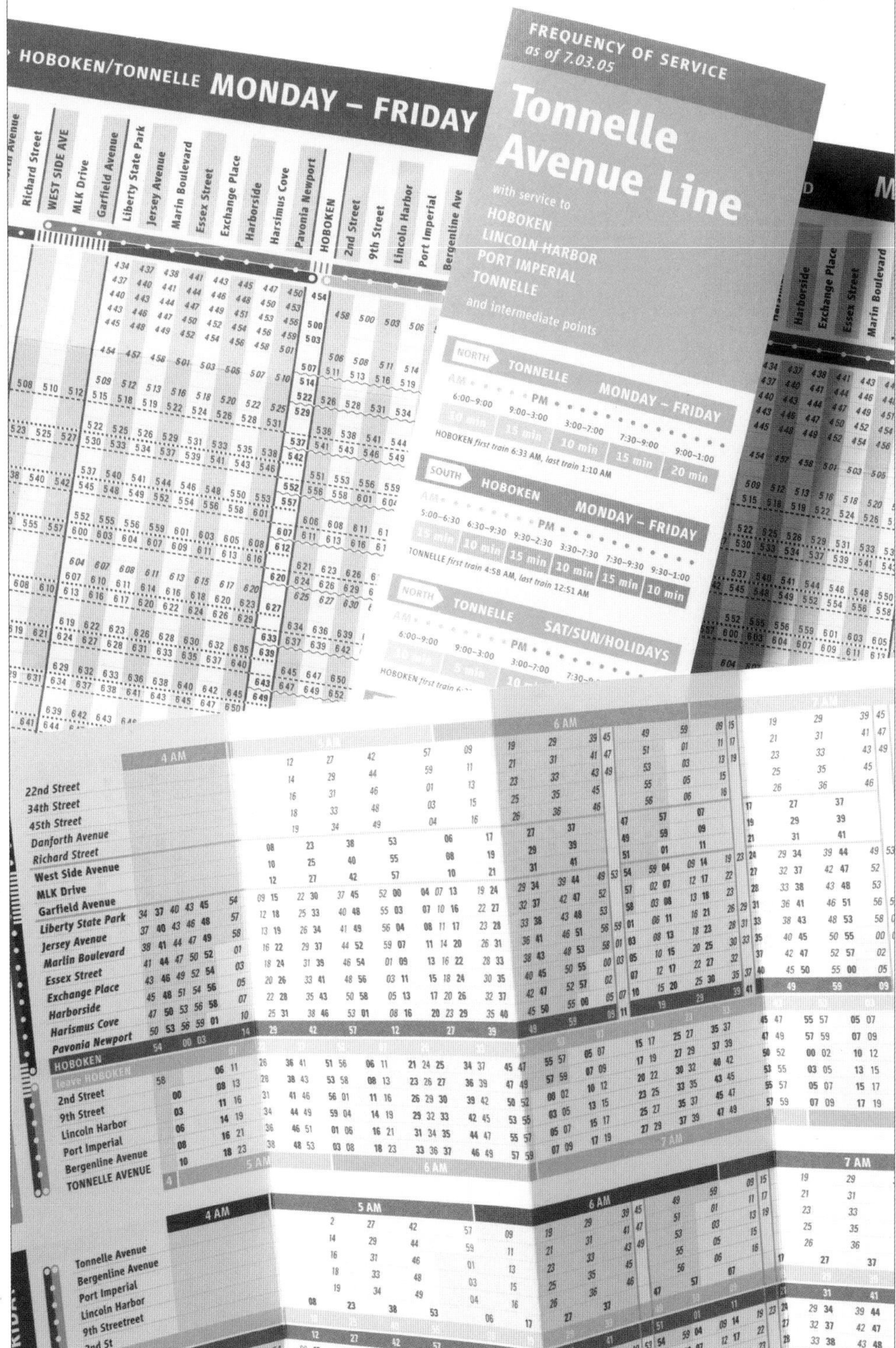

排版干净简洁。设计师在每一行和每一列周围都留出了足够的空间，使密集的信息显得宽松，易于阅读。破折号和波浪线使用得很少，但效果很好。白色箭头指明了火车前往的方向，蓝色的方框和下方附加的文字进一步说明了当天的车次安排

37.升华图表

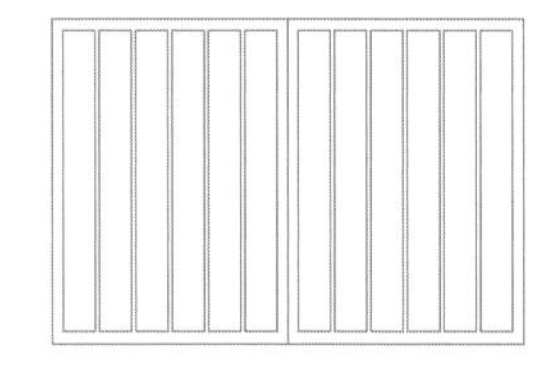

在这个项目中，必要的图表和表格包含了重要的信息，这些信息中，有些是固定条款，有些是用于说服股东或投资者的内容。这些信息通常是连续不断的单调的数字列表，但通过变换数字的大小和粗细可以构建出活泼的图形，形成清晰而生动的报告。即使只使用有限的色彩，也可以通过改变数字的尺寸和形状，让财务报表更有质感。

另见68～69页

精心设计的数字既生动又有规律，好像图形一样

项目
萨瓦德尔银行（Banc Sabadell）2017年年报与数字设计（印刷版）

客户
萨瓦德尔银行

设计公司
马里奥•埃斯肯纳兹工作室（Mario Eskenazi Studio）

设计师
马里奥•埃斯肯纳兹，杰玛•维尔加斯（Gemma Villegas）

萨瓦德尔银行年报中的数字设计非常成功，在很多宣传材料（圣诞节宣传册、礼物单等）中都曾重复使用

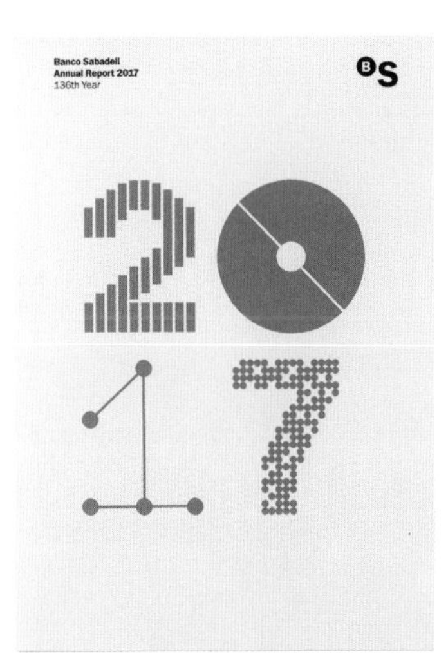

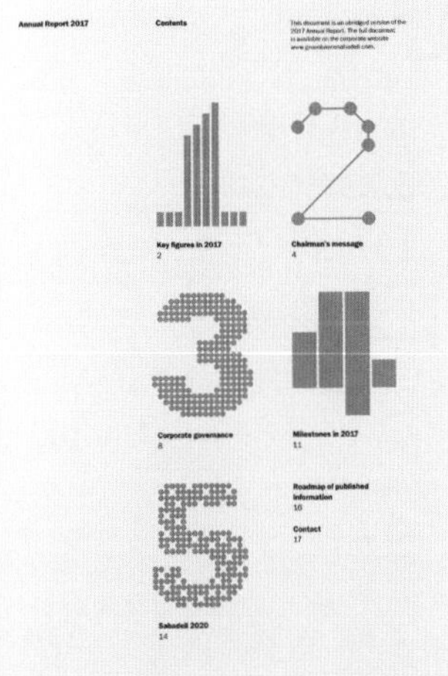

精心设计的数字为封面和内页增添了优雅的感觉，并且如项目序号一样会重复出现

醒目的统计数据在设计上使用了多样又整齐的网格，在垂直和水平方向上都层次分明，标题和副标题排列有序

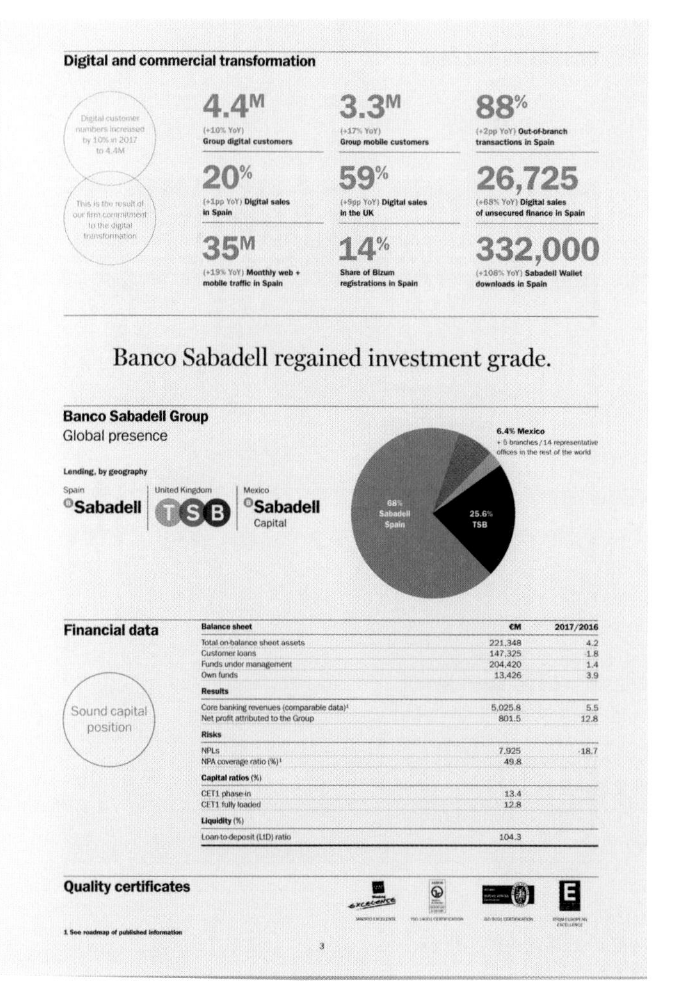

Balance sheet	€M	2017/2016
Total on-balance sheet assets	221,348	4.2
Customer loans	147,325	1.8
Funds under management	204,420	1.4
Own funds	13,426	3.9
Results		
Core banking revenues (comparable data)	5,025.8	5.5
Net profit attributed to the Group	801.5	12.8
Risks		
NPLs	7,925	-18.7
NPA coverage ratio (%)	49.8	
Capital ratios (%)		
CET1 phase-in	13.4	
CET1 fully loaded	12.8	
Liquidity (%)		
Loan-to-deposit (LtD) ratio	104.3	

● 38.增添趣味性

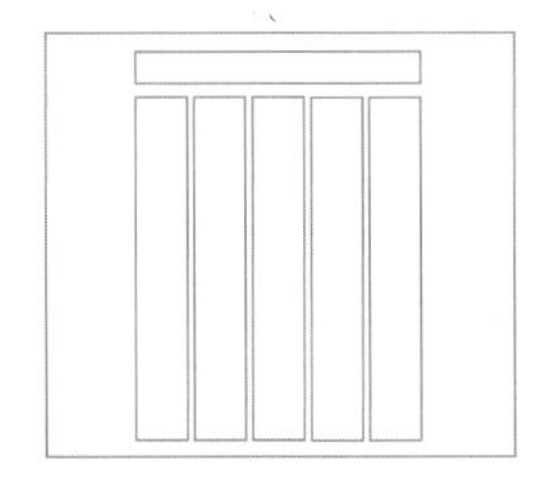

另见66～67页

自定义的数字在登录页面中给用户带来一种愉悦感，数字的设计在不同的载体上都保持了一致。地图的设计使用了与部分自定义数字（如“6”）相同的形式，可以很好地与条形图和统计数据搭配。

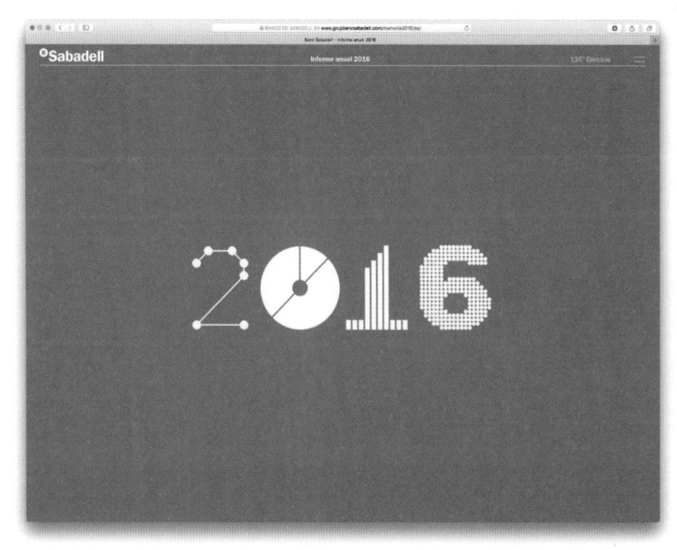

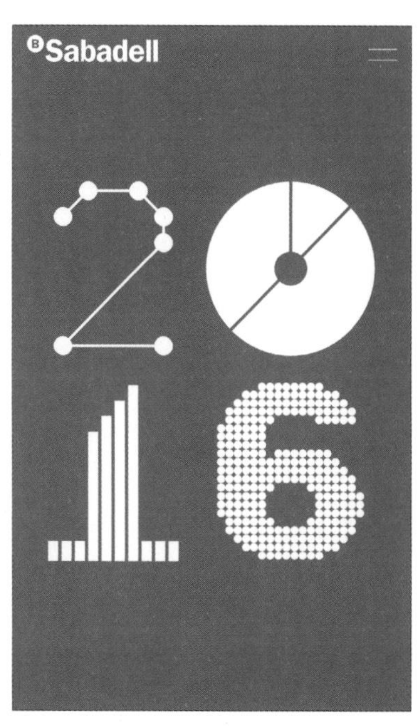

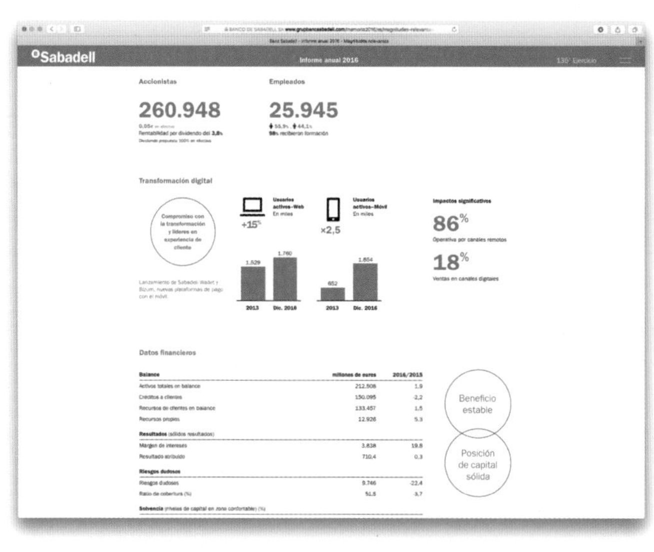

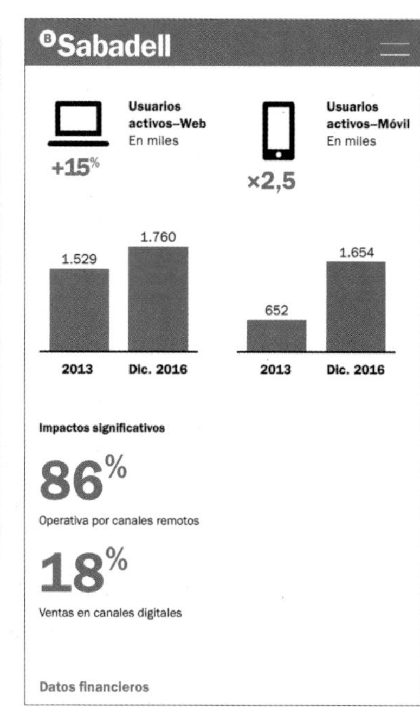

项目
萨瓦德尔银行2016年年报
（电脑及移动设备端）

客户
萨瓦德尔银行

设计公司
马里奥·埃斯肯纳兹工作室

设计师
马里奥·埃斯肯纳兹，杰玛·维尔加斯

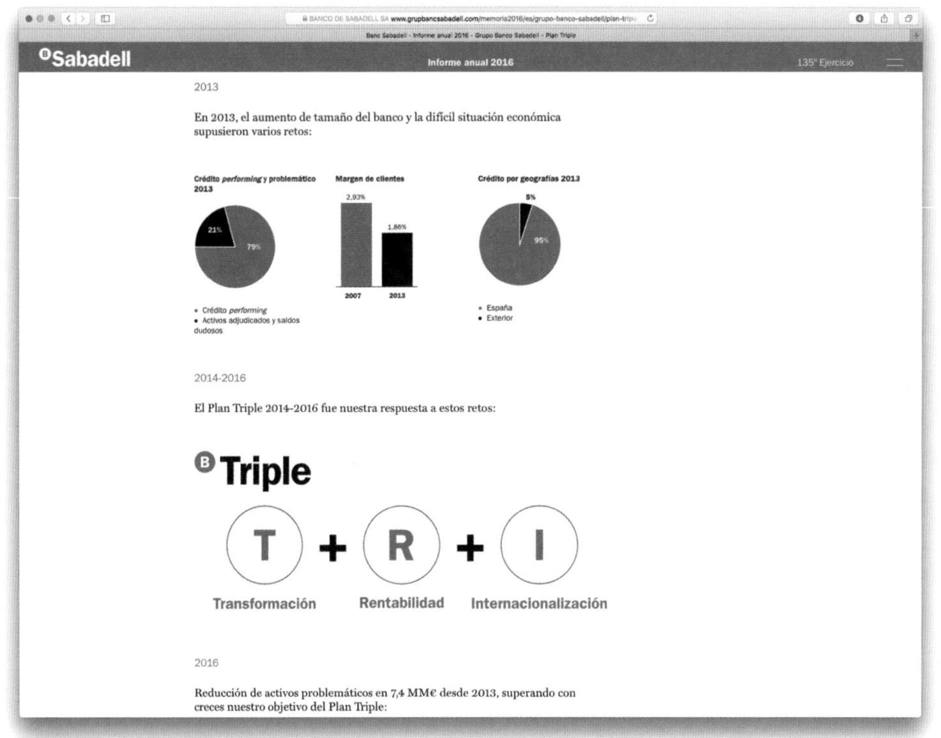
Sabadell
Informe anual 2016
135° Ejercicio
2013
En 2013, el aumento de tamaño del banco y la difícil situación económica supusieron varios retos:
2014-2016
El Plan Triple 2014-2016 fue nuestra respuesta a estos retos:
Triple
T + R + I
Transformación
Rentabilidad
Internacionalización
2016
Reducción de activos problemáticos en 7,4 MM€ desde 2013, superando con creces nuestro objetivo del Plan Triple:

Sabadell
El Plan Triple 2014-2016 fue nuestra respuesta a estos retos:
Triple
T + R + I
Transformación
Rentabilidad
Internacionalización
2016
Reducción de activos problemáticos en 7,4 MM€ desde 2013, superando con creces nuestro objetivo del Plan Triple:
Crédito performing y problemático 2016
12%
88%

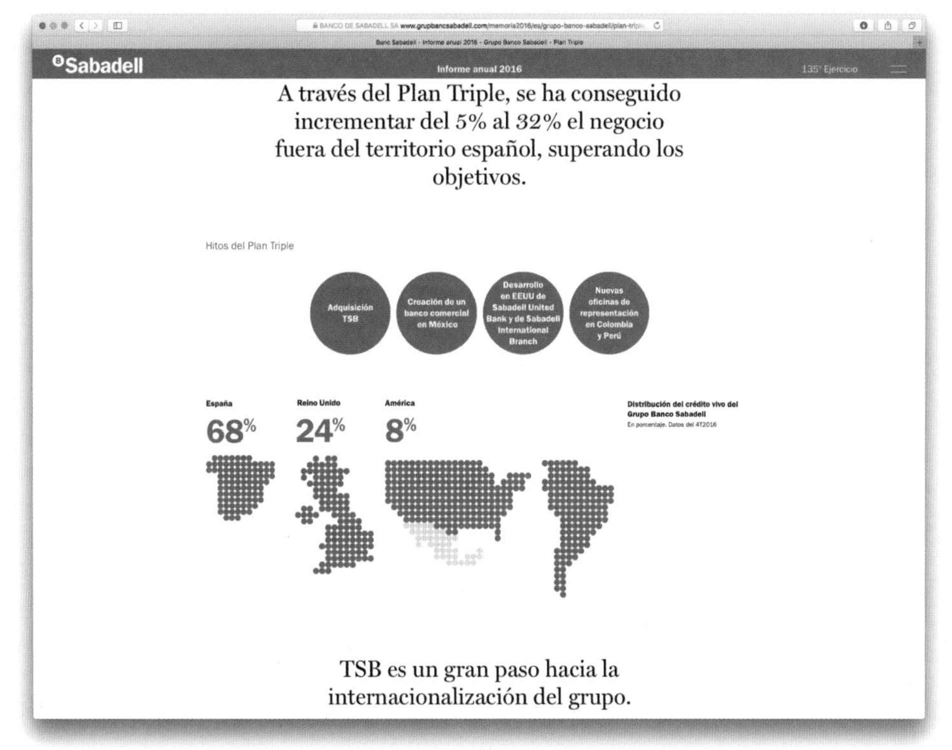
Sabadell
Informe anual 2016
135° Ejercicio
A través del Plan Triple, se ha conseguido incrementar del 5% al 32% el negocio fuera del territorio español, superando los objetivos.
Hitos del Plan Triple
Adquisición TSB
Creación de un banco comercial en México
Desarrollo en EEUU de Sabadell United Bank y de Sabadell International Branch
Nuevas oficinas de representación en Colombia y Perú
España
Reino Unido
América
68%
24%
8%
TSB es un gran paso hacia la internacionalización del grupo.

Sabadell
68%
24%
8%
Distribución del crédito vivo del Grupo Banco Sabadell
En porcentaje. Datos del 4T2016
TSB es un gran paso hacia la internacionalización del grupo.
TSB
6,4%
Cuota en cuentas corrientes

39.慎重地选择框架

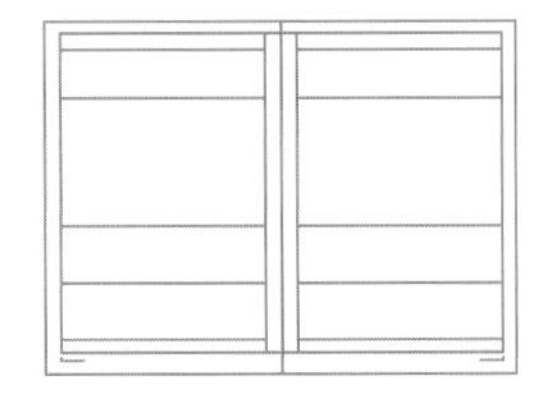

理想的情况下，使用表格可以避免形成混乱的框架。然而，有时信息中会涉及很多不集中的元素，因此，设计数据最简便的方法是为每个单元都加上框。

尽管设计订阅卡可以不受分隔线、框架和边框的约束，但在不同的区域使用不同粗细的分隔线和边框不仅可以呈现内容中的秩序，还可以创建一种有秩序感的布局。

郵 便 は が き

料金受取人払郵便

新宿北局承認

4121

差出有効期間
平成21年11月
23日まで
★切手不要★

1 6 9 — 8 7 9 0
133

東京都新宿区北新宿1-35-20

暮しの手帖社

4世紀31号アンケート係 行

ご住所 〒 —
電話 — —
お名前
メールアドレス @
年齢 [ ] 歳
性別 女 ／ 男
ご職業 [ ]
ご希望のプレゼントに○をつけて下さい。 □「日東紡のふきん」3枚箱入り □「花森安治の表紙絵ポストカード」5枚セット

いただいた個人情報は、誌面作り、当選プレゼントの発送、小社グループの商品案内等の送付に利用させていただき、厳重に管理、保管いたします。

＊ご回答は、184ページの記事一覧をご参照の上、番号でご記入下さい。

A．表紙の印象はいかがですか [ ]
ご意見：

B．面白かった記事を3つ、挙げて下さい [ ] [ ] [ ]

C．役に立った記事を3つ、挙げて下さい [ ] [ ] [ ]

D．興味がなかった、あるいは面白くなかった記事を3つ、挙げて下さい [ ] [ ] [ ]

E．今号を何でお知りになりましたか [ ]
その他：

F．小誌と併読している雑誌を教えて下さい

G．小誌を買った書店を教えて下さい [区市町村]

H．小誌へのご要望、ご意見などございましたらご記入下さい

◎ご協力、ありがとうございました。

本页和下页图：这些订阅卡中有不同粗细的分隔线。较粗的分隔线会突出某些类型的内容，让读者注意最重要的文本或标题。不同粗细的线既能让页面保持平衡，突出重点，也对附加材料进行了补充

项目
《每日笔记》杂志

客户
《每日笔记》杂志社

设计师
林修三，柳政明

这张订阅卡兼具了美观和实用性

● 【定期購読】【商品、雑誌・書籍】のお申込みは、こちらの払込取扱票に必要事項を必ず記入の上、最寄りの郵便局に代金を添えてお支払い下さい。
● 169項、183頁の注文方法をご覧下さい。
● 表示金額はすべて税込価格となっております。
● 注文内容を確認させていただく場合がございます。平日の日中に連絡のつく電話番号を、ＦＡＸ番号がございましたら払込取扱票にご記入ください。
● プレゼントの場合はご注文いただいたお客様のご住所、お名前でお送りします。

02 東京　払込取扱票　通常払込料金加入者負担

口座番号 00190-7-45321
金額 6300
加入者名 株式会社　暮しの手帖社
料金　特殊取扱

通信欄
※「暮しの手帖」の定期購読を
20＿＿年＿＿号より1年間（6冊）申し込みます
※プレゼントされる場合、送付先が異なる場合はご送付先を下欄へ記入下さい。
ご住所 〒□□□-□□□□
ご氏名　tel

払込人住所氏名 ※（郵便番号　）
（電話番号　－　－　）
（FAX　－　－　）

受付局日附印

裏面の注意事項をお読みください。（私製承認東第43990号）
これより下部には何も記入しないでください。

各票の※印欄は、払込人において記載してください。
切り取らないで郵便局にお出しください。

払込金受領証
口座番号 00190-7 通常払込料金加入者負担 45321
加入者名 株式会社　暮しの手帖社
金額 6300
払込人住所氏名 ※
料金　特殊取扱　受付局日附印

記載事項を訂正した場合は、その箇所に訂正印を押してください。

02 東京　払込取扱票　通常払込料金加入者負担

口座番号 00170-1-59128
金額 ※
加入者名 株式会社　グリーンショップ
料金　特殊取扱

通信欄
※プレゼントされる場合、送付先が異なる場合はご送付先を下欄へ記入下さい。
ご住所 〒□□□-□□□□
ご氏名　tel

払込人住所氏名 ※（郵便番号　）
（電話番号　－　－　）
（FAX　－　－　）

受付局日附印

裏面の注意事項をお読みください。（私製承認東第44327号）
これより下部には何も記入しないでください。

各票の※印欄は、払込人において記載してください。
切り取らないで郵便局にお出しください。

払込金受領証
口座番号 00170-1 通常払込料金加入者負担 59128
加入者名 株式会社　グリーンショップ
金額 ※
払込人住所氏名 ※
料金　特殊取扱　受付局日附印

記載事項を訂正した場合は、その箇所に訂正印を押してください。

40.跳出矩形外思考

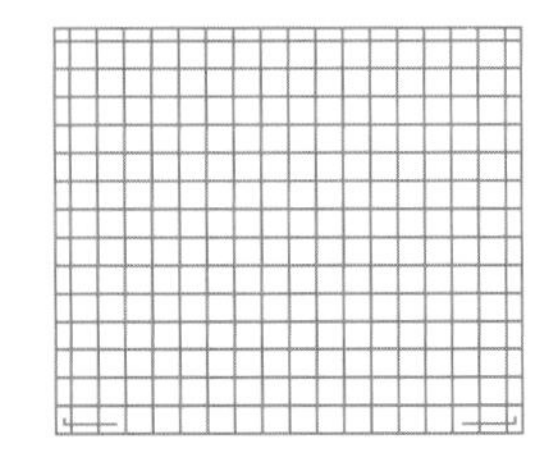

网格也可以用来组织非常规的形状，将空间分成不同的区域。例如，既可以把一个圆形水平和垂直地四等分，也可以切成多个扇形。

在卡片的一面，出血图片与文字之间形成对比。排版很简单，粗体的标题与标志相呼应，引起人们对标题和网址的注意。地铁车厢上的横线与文本区域的线条相呼应

NAME:

NEW YORK TRANSIT MUSEUM

Think About It....

When New York City's first subway opened on October 27, 1904, there were about 9 miles of track. Today the subway system has expanded to 26 times that size. About how many miles of track are there in today's system?

Most stations on the first subway line had tiles with a symbol, such as a ferry, lighthouse, or beaver. These tiles were nice decoration, but they also served an important purpose. Why do you think these symbols were helpful to subway passengers?

When subway service began in 1904, the fare was five cents per adult passenger. How much is the fare today? Over time, subway fare and the cost of a slice of pizza have been about the same. Is this true today?

WORLD'S FAIR
SPECIAL
←LOCAL-EXP→

Metropolitan Transportation Authority

Today's subway system uses a fleet of 6,200 passenger cars. The average length of each car is 62 feet. If all of those subway cars were put together as one super-long train, about how many miles long would that train be? (Hint: There are 5,280 feet in a mile.)

Redbird subway cars, which were first built for the 1964 World's Fair, were used in New York City until 2003. Then many of them were tipped into the Atlantic Ocean to create artificial reefs. A reef makes a good habitat for ocean life—and it is a good way to recycle old subway cars! Can you think of other ways that mass transit helps the environment?

To check your answers and learn more about New York City's subway system, visit our website: **www.transitmuseumeducation.org**. You'll also find special activities, fun games, and more!

The New York Transit Museum's programs are made possible, in part, with public funds from the New York State Council on the Arts, a state agency.
All photographs are from the New York Transit Museum Collection.

项目
圆形书籍教育工具

客户
纽约交通博物馆（New York Transit Museum）

项目开发商
勒内特•莫尔斯（Lynette Morse），维吉尔•塔里德（Virgil Talaid），美国教育部（Education Department）

设计公司
卡拉佩鲁齐设计公司

设计师
贾尼斯•卡拉佩鲁齐

这张旋转卡的内容包括了教育信息和各种活动，而且，它是活动的

在卡片的另一面，粗分隔线巧妙地包含了说明文字和标识区。蓝色和红色是纽约A号、C号、E号和1号、2号、3号地铁的颜色

41.用色彩吸引读者的注意力

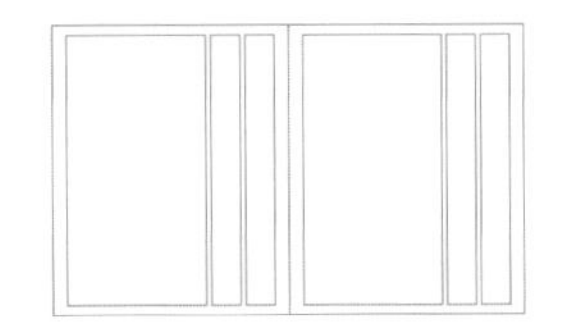

无论网格是固定的还是多样的，强烈的色彩都可以帮助出版物凸显某些章节、故事或文本。有彩色背景的页面与以白色或浅色为背景的页面之间的对比，可以改变阅读的节奏，引起读者的兴趣和注意。不同颜色的侧边栏或辅助文本框也可以用来放置不同信息，从而避免过度使用分隔线或边框。

由The Wing公司推出的这本杂志专门展示女性工作和社交的空间，杂志的第一、二期使用了大胆和多样的色彩，以此告诉读者：女性正在进步

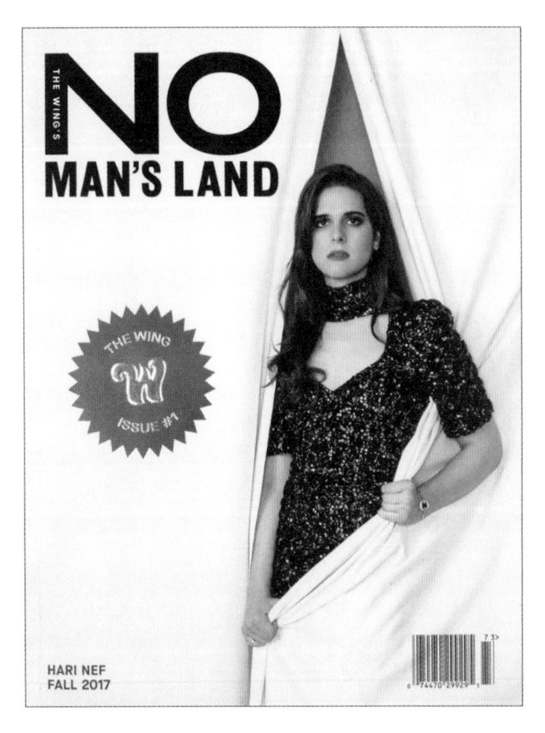

项目
《没有男性的国度》（*No Man's Land*）

客户
The Wing公司

设计公司
Pentagram公司

创意总监
艾米莉•欧本曼（Emily Oberman）

合伙人
艾米莉•欧本曼

高级设计师
克里斯汀娜•霍根（Christina Hogan）

设计师
伊丽莎白•古德斯皮德（Elizabeth Goodspeed），乔伊•崔罗（Joey Petrillo）

项目经理
安娜•梅克斯勒（Anna Meixler）

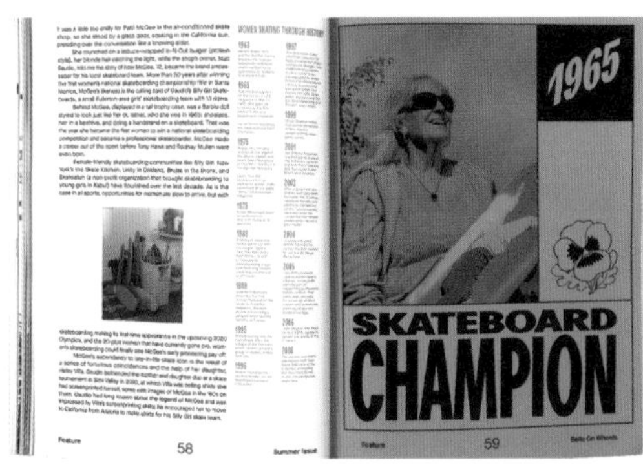

多彩的页面、背景和字体让这本立场鲜明的出版物发表了态度更加鲜明的宣言

上图：每个故事开头的色彩让页面有了多样化的节奏

左下图：彩色背景的对页将各种元素打散重组

右下图：表示时间轴的彩色文字通过不同宽度的分栏，与正文产生了鲜明的对比

McGee had seen and done a lot in her post-skate career. She worked on a fishing boat owned a topless bar got divorced raised her kids ran a trading post with her second husband. At 72, when life tends to slow down for most people McGee is ready for her next adventure

62 63

It was a little too chilly for Patti McGee in the air-conditioned skate shop, so she stood by a glass door, soaking in the California sun, presiding over the conversation like a knowing elder.

She munched on a lettuce-wrapped In-N-Out burger (protein style), her blonde hair catching the light, while the shop's owner, Matt Gaudio, told me the story of how McGee, 72, became the brand ambassador for his local skateboard team. Nearly 50 years after winning the first women's national skateboarding championship title in Santa Monica, McGee's likeness is the calling card of Gaudio's Silly Girls Skateboards, a small Fullerton-area girls' skateboarding team with 13 riders.

Behind McGee, displayed in a tall trophy case, was a Barbie doll styled to look just like her or, rather, who she was in 1965: shoeless, hair in a beehive, and doing a handstand on a skateboard. That was the year she became the first woman to win a national skateboarding competition and became a professional skateboarder. McGee made a career out of the sport before Tony Hawk and Rodney Mullen were

WOMEN SKATING THROUGH HISTORY

1963
Wendy Bearer Bull and her brother Danny become the first professionally endorsed skateboarders to be sponsored by Makaha Skateboard Club.

1965
Patti McGee appears on the cover of *LIFE* magazine in May 14, 1965. She goes on to become the first Women's National Skateboard champion.

Laurie Turner becomes the 1965 National Girls' Champion.

1975
Peggy Oki, the only woman on the legendary Zephyr skateboard

1997
The first issue of the *Villa Villa Cola* zine debuts, created by Tiffany and Nicole Morgan, two skateboarding sisters. It uses humor to encourage girls to skateboard and offers advice on how to overcome being intimidated by men in the field. Other zines, *Bruisers* and *50-50: Skateboarding and Gender*, soon follow.

1999
Elissa Steamer is the first woman to appear in Tony Hawk's skateboarding video game series.

2001
Jen O'Brien becomes the first girl to skate at

● 42.控制主色调

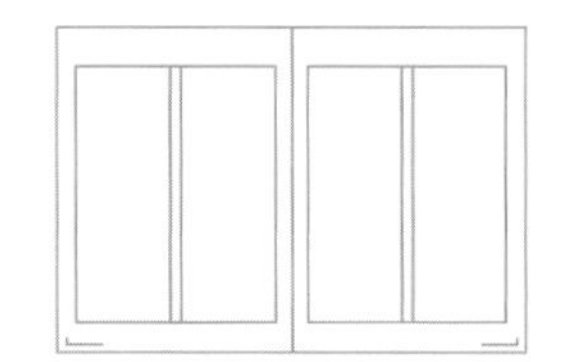

虽然很多颜色都能吸引人的注意力，但是过多的颜色会淹没信息。控制主色调可以让页面的主题保持连贯。

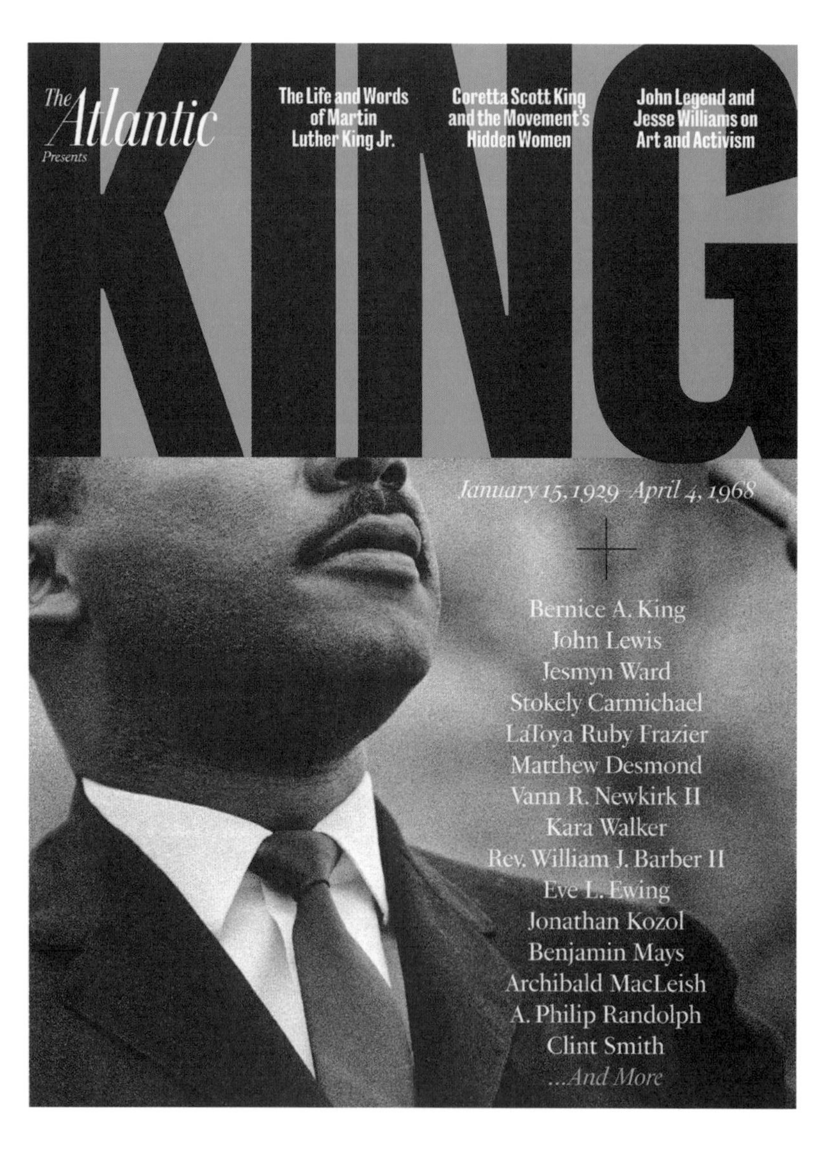

一个只有金色和黑色的封面把它与市场上其他同类产品区别开。经过慎重的剪裁，封面中的照片比完整的原图更耐人寻味，暗示着一个生命的猝然终止

项目
《金》

客户
《大西洋月刊》杂志社

创意总监
保罗•思培拉

艺术总监
戴维•萨默维尔

设计公司
OCD | 原创设计冠军

设计师
博比•马丁，詹妮弗•奇洛恩

黑白对比强烈的页面、与历史相呼应的版式、文本中的一点儿绿，这些设计都凸显了这部作品的戏剧性

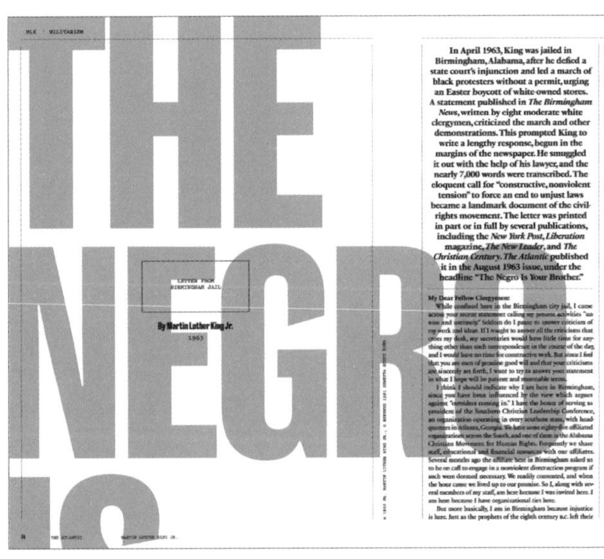

In April 1963, King was jailed in Birmingham, Alabama, after he defied a state court's injunction and led a march of black protesters without a permit, urging an Easter boycott of white-owned stores. A statement published in *The Birmingham News*, written by eight moderate white clergymen, criticized the march and other demonstrations. This prompted King to write a lengthy response, begun in the margins of the newspaper. He smuggled it out with the help of his lawyer, and the nearly 7,000 words were transcribed. The eloquent call for "constructive, nonviolent tension" to force an end to unjust laws became a landmark document of the civil-rights movement. The letter was printed in part or in full by several publications, including the *New York Post*, *Liberation* magazine, *The New Leader*, and *The Christian Century*. *The Atlantic* published it in the August 1963 issue, under the headline "The Negro Is Your Brother."

My Dear Fellow Clergymen:

While confined here in the Birmingham city jail, I came across your recent statement calling my present activities "unwise and untimely." Seldom do I pause to answer criticism of my work and ideas. If I sought to answer all the criticisms that cross my desk, my secretaries would have little time for anything other than such correspondence in the course of the day, and I would have no time for constructive work. But since I feel that you are men of genuine good will and that your criticisms are sincerely set forth, I want to try to answer your statement in what I hope will be patient and reasonable terms.

I think I should indicate why I am here in Birmingham, since you have been influenced by the view which argues against "outsiders coming in." I have the honor of serving as president of the Southern Christian Leadership Conference, an organization operating in every southern state, with headquarters in Atlanta, Georgia. We have some eighty-five affiliated organizations across the South, and one of them is the Alabama Christian Movement for Human Rights. Frequently we share staff, educational and financial resources with our affiliates. Several months ago the affiliate here in Birmingham asked us to be on call to engage in a nonviolent direct-action program if such were deemed necessary. We readily consented, and when the hour came we lived up to our promise. So I, along with several members of my staff, am here because I was invited here. I am here because I have organizational ties here.

But more basically, I am in Birmingham because injustice is here. Just as the prophets of the eighth century B.C. left their villages and carried their "thus saith the Lord" far beyond the boundaries of their home towns, and just as the Apostle Paul left his village of Tarsus and carried the gospel of Jesus Christ to the far corners of the Greco-Roman world, so am I compelled to carry the gospel of freedom beyond my own home town. Like Paul, I must constantly respond to the Macedonian call for aid.

Moreover, I am cognizant of the interrelatedness of all communities and states. I cannot sit idly by in Atlanta and not be concerned about what happens in Birmingham. Injustice anywhere is a threat to justice everywhere. We are caught in an inescapable network of mutuality, tied in a single garment of destiny. Whatever affects one directly, affects all indirectly. Never again can we afford to live with the narrow, provincial "outside agitator" idea. Anyone who lives inside the United States can never be considered an outsider anywhere within its bounds.

You deplore the demonstrations taking place in Birmingham. But your statement, I am sorry to say, fails to express a similar concern for the conditions that brought about the demonstrations. I am sure that none of you would want to rest content with the superficial kind of social analysis that deals merely with effects and does not grapple with underlying causes. It is unfortunate that demonstrations are taking place in Birmingham, but it is even more unfortunate that the city's white power structure left the Negro community with no alternative.

In any nonviolent campaign there are four basic steps: collection of the facts to determine whether injustices exist; negotiation; self-purification; and direct action. We have gone through all these steps in Birmingham. There can be no gainsaying the fact that racial injustice engulfs this community. Birmingham is probably the most thoroughly segregated city in the United States. Its ugly record of brutality is widely known. Negroes have experienced grossly unjust treatment in the courts. There have been more unsolved bombings of Negro homes and churches in Birmingham than in any other city in the nation. These are the hard, brutal facts of the case. On the basis of these conditions, Negro leaders sought to negotiate with the city fathers. But the latter consistently refused to engage in good-faith negotiation.

Then, last September, came the opportunity to talk with leaders of Birmingham's economic community. In the course of the negotiations, certain promises were made by the merchants—for example, to remove the stores' humiliating racial signs. On the basis of these promises, the Reverend Fred Shuttlesworth and the leaders of the Alabama Christian Movement for Human Rights agreed to a moratorium on all demonstrations. As the weeks and months went by, we realized that we were the victims of a broken promise. A few signs, briefly removed, returned; the others remained.

As in so many past experiences, our hopes had been blasted, and the shadow of deep disappointment settled upon us. We had no alternative except to prepare for direct action, whereby we would present our very bodies as a means of laying our case before the conscience of the local and the national community. Mindful of the difficulties involved, we decided to undertake a process of self-purification. We began a series of workshops on nonviolence, and we repeatedly asked ourselves: "Are you able to accept blows without retaliating?" "Are you able to endure the ordeal of jail?" We decided to schedule our direct-action program for the Easter season, realizing that except for Christmas,

设计师有时会使用“设计与电影的隐喻”这种方式。这些对页使用了“电影式”的方法展示出了标题——“这个黑人是你的兄弟（The Negro is Your Brother）”，这个标题是1963年《大西洋月刊》在介绍马丁•路德•金的文章《伯明翰监狱的来信》（*Letter from Birmingham Jail*）中使用的。绿色的标题相当于视觉上的画外音

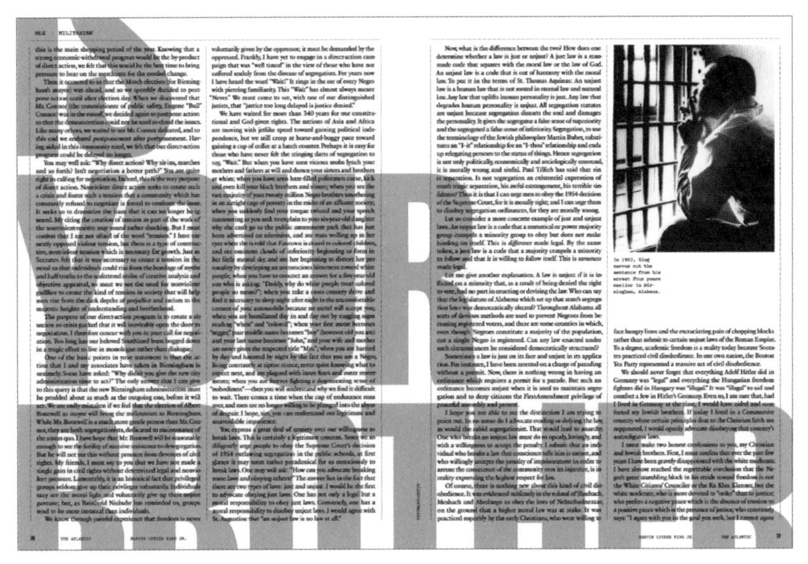

[illegible]

Now, what is the difference between the two? How does one determine whether a law is just or unjust? A just law is a man-made code that squares with the moral law or the law of God. An unjust law is a code that is out of harmony with the moral [illegible]

合理的间距、黑色的文本和绿色的标题确保整个页面作为一个整体吸引读者的注意，同时又突出了重要的观点。

尽管这个案例的设计强调了对色彩的控制，但这些页面也展示了许多网格使用的其他问题。要注意间距，尤其是在文章的开头。也要观察如何在不同的网格中放置标题，而不让标题横跨每篇文章。适当的间距使标题呼应了马丁•路德•金沉思的照片

[illegible]

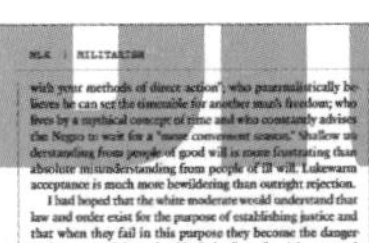

"Wait" has almost always meant "Never."

[illegible]

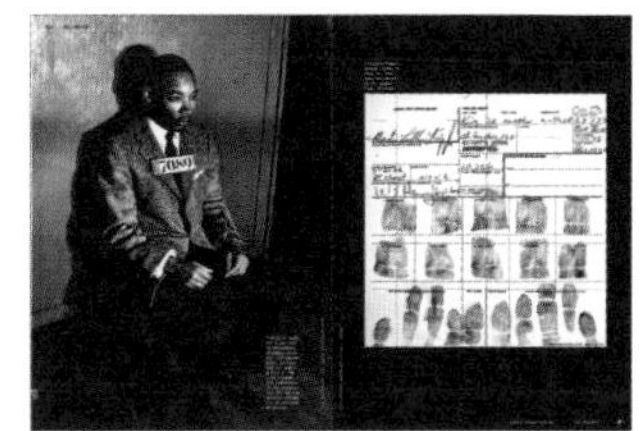

● 43.让色彩传达信息

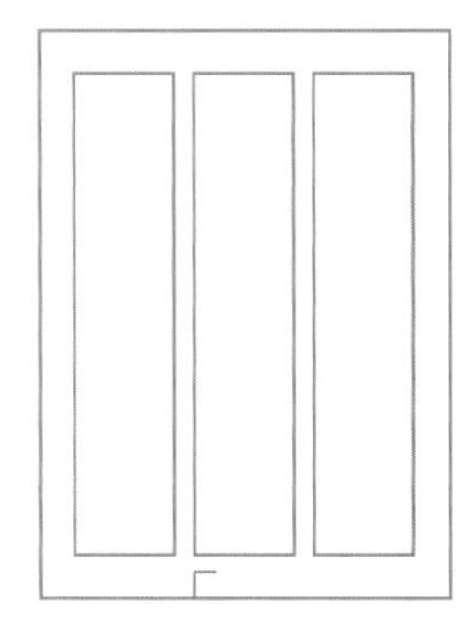

当页面已经有了一个确定的结构时（这种情况常见于杂志中），有时最好的办法就是放松一下，采用简单的排版，让色彩，尤其是一张绚丽的照片中的色彩成为页面的主角（有时是版式中的主角）。

本页及下页图： 尽管在一个全彩印刷的作品中，最大限度地使用色彩是一件很有诱惑力的事，但是使用一种简单的颜色（如黑色）可以反衬出高饱和度的图片，让读者把注意力集中到照片上。过多的视觉元素只会带来相反的效果

项目
《美丽之家》杂志

客户
《美丽之家》杂志社

设计师
芭芭拉•德怀尔德

一张色彩丰富、设计精妙的照片在排版中无需多余的装饰也能熠熠生辉

COLOR

Violet
GENTLE VIOLET (2071-20)
BENJAMIN MOORE COLOR PREVIEW

Peony
SWEET TAFFY (2086-60)
BENJAMIN MOORE COLOR PREVIEW

Magenta Anemone
FORWARD FUCHSIA
(SW 6842) SHERWIN-WILLIAMS

Hydrangea
TROOPER (26-14) PRATT & LAMBERT

Pink Rose
PEACHGLOW (90YR71/144) GLIDDEN

Water Lily
TULIPE VIOLET (30-14) PRATT & LAMBERT

Orchid
VESPER (70RB67/067) GLIDDEN

Hyacinth
ORIENTAL NIGHT (29-14) PRATT & LAMBERT

COOL SHADES A word about finishes: Light colors look darker in a flat finish. Dark colors look brighter in a gloss or semigloss. A flat finish will work well for the lighter shades here, but the deep purples and pinks will definitely look better with a sheen.

Pink Daisy
SMASHING PINK (1303)
BENJAMIN MOORE CLASSIC COLORS

FOR MORE DETAILS, SEE RESOURCES

28 HOUSE BEAUTIFUL

● 44.色彩与排版的完美结合

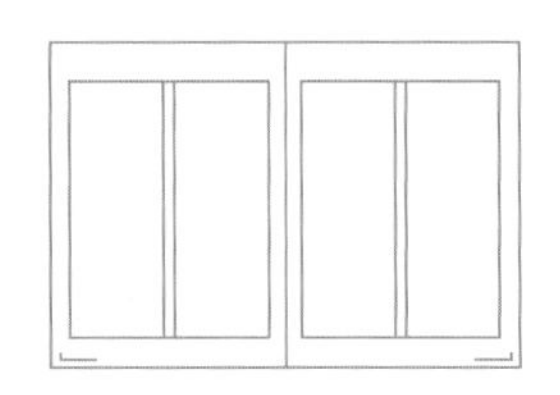

在一本彩色的指南类书籍中，要控制颜色的使用，确保指南的内容不会被页面上的其他元素喧宾夺主。但是，在一个既定的色调中，选择合适的色彩会让排版效果更好。

每个部分的开头都使用了覆盖整个页面的、色彩丰富的照片。大胆的排版与丰富的色彩相互碰撞

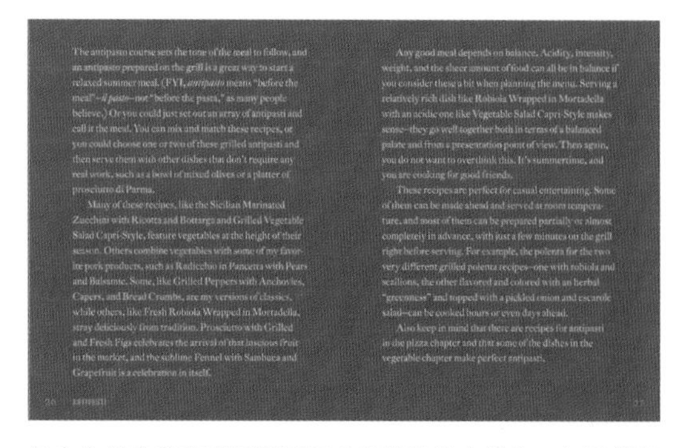

The antipasto course sets the tone of the meal to follow, and an antipasto prepared on the grill is a great way to start a relaxed summer meal. (FYI, *antipasto* means "before the meal"—*il pasto*—not "before the pasta," as many people believe.) Or you could just set out an array of antipasti and call it the meal. You can mix and match these recipes, or you could choose one or two of these grilled antipasti and then serve them with other dishes that don't require any real work, such as a bowl of mixed olives or a platter of prosciutto di Parma.

Many of these recipes, like the Sicilian Marinated Zucchini with Ricotta and Bottarga and Grilled Vegetable Salad Capri-Style, feature vegetables at the height of their season. Others combine vegetables with some of my favorite pork products, such as Radicchio in Pancetta with Pears and Balsamic. Some, like Grilled Peppers with Anchovies, Capers, and Bread Crumbs, are my versions of classics, while others, like Fresh Robiola Wrapped in Mortadella, stray deliciously from tradition. Prosciutto with Grilled and Fresh Figs celebrates the arrival of that luscious fruit in the market, and the sublime Fennel with Sambuca and Grapefruit is a celebration in itself.

Any good meal depends on balance. Acidity, intensity, weight, and the sheer amount of food can all be in balance if you consider these a bit when planning the menu. Serving a relatively rich dish like Robiola Wrapped in Mortadella with an acidic one like Vegetable Salad Capri-Style makes sense—they go well together both in terms of a balanced palate and from a presentation point of view. Then again, you do not want to overthink this. It's summertime, and you are cooking for good friends.

These recipes are perfect for casual entertaining. Some of them can be made ahead and served at room temperature, and most of them can be prepared partially or almost completely in advance, with just a few minutes on the grill right before serving. For example, the polenta for the two very different grilled polenta recipes—one with robiola and scallions, the other flavored and colored with an herbal "greenness" and topped with a pickled onion and escarole salad—can be cooked hours or even days ahead.

Also keep in mind that there are recipes for antipasti in the pizza chapter and that some of the dishes in the vegetable chapter make perfect antipasti.

包含介绍内容的对页紧随每个章节的出血照片。与开篇粗体的无衬线字体不同，介绍性的文字使用了衬线字体

项目
《意式烧烤》（*Italian Grill*）

客户
哈珀•柯林斯

设计公司
纽约麦莫制作公司（Memo Productions,NY）

艺术总监
丽莎•伊顿（Lisa Eaton），道格拉斯•里卡迪（Douglas Riccardi）

这本烹饪书以网格为版式基础，极具个人风格。书中使用了高饱和度、大胆的颜色，以及大胆的排版。每个章节的色彩在主色调的基础上都有细微变化，但是前后和谐一致

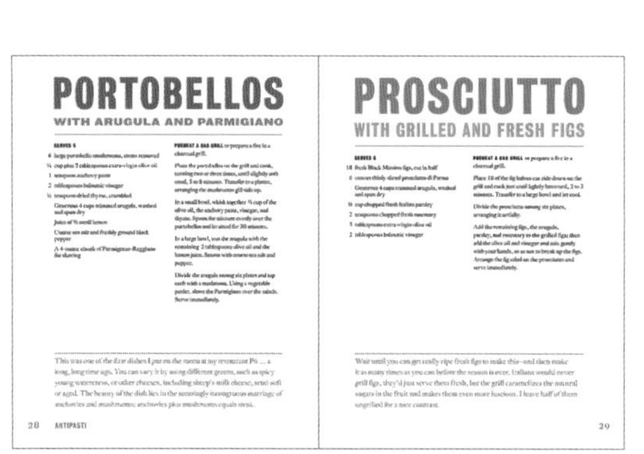

本页下面三张图和下页图：每个部分的颜色在主色调的基础上都有所不同，与彩色的照片互补

In Italy, cooking fish is all about freshness and simplicity—as I've said before, the philosophy of Italian fish cookery can be summed up in three words: *Leave it alone*. Complicated sauces and techniques are not part of the repertoire, and, in fact, Italians almost never serve any sauce at all with fish, other than an excellent olive oil. Lemon may sometimes appear, but even that is often considered beside the point. The one exception is *salsa verde*, the fragrant green herb sauce, which may sometimes accompany a fish with character enough to stand up to it, such as a whole grilled branzino (see page 126).

Few Italians would consider cooking anything other than local fish, whether from a mountain stream or the ocean, and I urge you to think in the same way: find a good fish market, and remember that what is freshest is best. If the specific fish called for in your recipe is not available—or doesn't look pristine and glistening—the fishmonger can help you choose another option (I include suggestions for substitutions in many of the recipes). If you are able to get fresh king mackerel for Mackerel "in Scapece" with Amalfi Lemon Salad, you will have the best mackerel dish you've ever tasted; if you can't find it, make the recipe with very fresh bluefish, or move on to another one. Most of the other fish recipes in this chapter, such as Monkfish in Prosciutto with Pesto Fregola and Swordfish Involtini Sicilian-Style, call for widely available varieties. But you'll want to be sure to get the best tuna available—sushi-quality, that is—for Tuna Like Fiorentina, and you really should use wild salmon for the Salmon in Cartoccio with Asparagus, Citrus, and Mint.

Cooking shellfish on the grill is easy, and the recipes in this chapter use several different techniques for achieving simple perfection. Clams in Cartoccio are wrapped in a foil package and allowed to steam in their fragrant juices. The shrimp in Shrimp Rosemary Spiedini alla Romagnola are threaded onto rosemary skewers, which impart their herbal fragrance and look sexy besides. I love cooking shellfish (and cephalopods) on a piastra, a flat griddle or stone placed on the hot grill (see page 000 for more on the subject), because it gives them a great sear and char, as in Sea Scallops alla Caprese or Marinated Calamari with Chickpeas, Olive Pesto, and Oranges.

Thinking globally while buying locally is especially important when you are buying fish. Some "trendy" fish have been overharvested to the point of extinction, and we now know that there can be problems with farmed fish as well, like salmon. The Monterey Bay Aquarium, at www.montereybayaquarium.com, maintains an up-to-date list of species that are being overfished in the United States and in the rest of the world. It's an invaluable resource, and I urge you to consult it when writing your shopping list, as I do both at home and at the restaurants.

86 FISH AND SHELLFISH

87

MARINATED CALAMARI

WITH CHICKPEAS, OLIVE PESTO, AND ORANGES

SERVES 6

CALAMARI

- 3 pounds cleaned calamari (tubes and tentacles)
- ¼ cup extra-virgin olive oil
- Grated zest and juice of 1 lemon
- 4 garlic cloves, thinly sliced
- 2 tablespoons chopped fresh mint
- 2 tablespoons hot red pepper flakes
- 2 tablespoons freshly ground black pepper

CHICKPEAS

- Two 15-ounce cans chickpeas, drained and rinsed, or 3½ cups cooked chickpeas
- ½ cup extra-virgin olive oil
- ¼ cup red wine vinegar
- 4 scallions, thinly sliced
- 4 garlic cloves, thinly sliced
- ¼ cup mustard seeds
- Kosher salt and freshly ground black pepper

OLIVE PESTO

- ¼ cup extra-virgin olive oil
- Grated zest and juice of 1 orange
- ½ cup black olive paste
- 4 jalapeños, finely chopped
- 12 fresh basil leaves, cut into chiffonade (thin slivers)

- 3 oranges
- 2 tablespoons chopped fresh mint

CUT THE CALAMARI BODIES crosswise in half if large. Split the groups of tentacles into 2 pieces each.

Combine the olive oil, lemon zest and juice, garlic, mint, red pepper flakes, and black pepper in a large bowl. Toss in the calamari and stir well to coat. Refrigerate for 30 minutes, or until everything else is ready.

Put the chickpeas in a medium bowl, add the oil, vinegar, scallions, garlic, and mustard seeds, and stir to mix well. Season with salt and pepper and set aside.

93

45.善用色彩

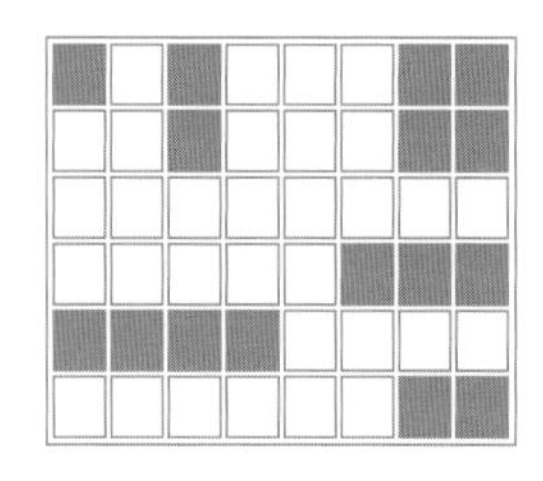

在一个整体网格内，大小一致的颜色模块的排列可以紧密又灵活，让文本和图片呈现出有趣的变化，如这个作品所示。方框和色彩可以形成一个整体的系统和框架，清晰地展示信息。当页面中需要列出许多详细的内容时，网格将一些颜色模块合并，使日期、信息与其他文本区别开，如网址链接、行动号召标语、主标题栏的广告横幅。

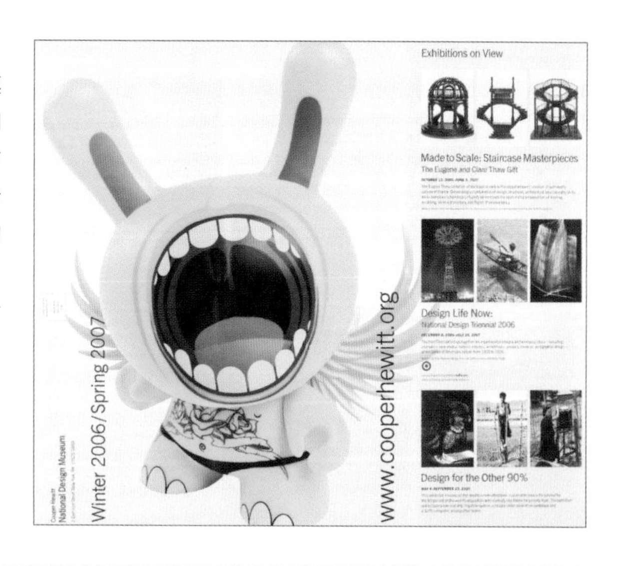

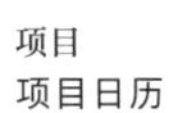

项目
项目日历

客户
史密森学会，美国国家设计博物馆库珀·休伊特设计博物馆

设计公司
曾·西摩设计工作室

设计总监
帕特里克·西摩

艺术总监
劳拉·霍威尔

网格系统为这个季节性的项目日历提供了统一的模式，但同时也允许色彩和图片有变化

主要展品的概要和它们的展出日期在项目日历的背面，与另一面的大幅图片互补，增加了页面的视觉张力和压缩感

不同尺寸的图片和特殊的轮廓图既遵循了彩色模块的框架风格，又有些跳脱

上页图：月份是用颜色编码的，以形成清晰的时间轴。每个事件都有自己的模块

首先，确定整块区域的大小，将其划分成同样大小的方块；然后，考虑整个外侧边距的大小。将方块当成独立的内容框使用，或者用两个甚至三个方块拼成一个水平或纵向的内容框。注意模块要包含的信息，可以按日期、月份、价格、事件或任何适合项目的内容用颜色编码。在设计大量的信息时，色彩应该起到传达和阐释信息的作用。

模块还可以用来放置照片和插图。与文本一样，图片可以放于一个模块、两个垂直模块、两个或四个水平模块或四个堆叠模块内。简而言之，有颜色的方块可支持多种变化，又能保证页面的秩序和整体性。为了增添更多的趣味，还可以在特殊图片上加轮廓，与内容框的网格形成对比，让生动的作品更有节奏感和视觉空间感。

由于有一系列色彩模式的结构性支撑，信息可以存在于各自的空间中。利用小字体、大标题和粗体信息，色彩模块可以让信息的层次性更强。不同大小和粗细的字体可以帮助读者迅速地找到日期、事件、时间和描述。多模块框中的大标题既增加了节奏感，也让相同层级的信息保持了一种一致性，例如，营销线路、客户或博物馆、行动号召标语和联系信息。

一个占据双面或者对页的项目，也可以利用模块化的方式，使用连续的模块（但也可以中断）清晰地划分区域。

● 46.通过色彩强调信息

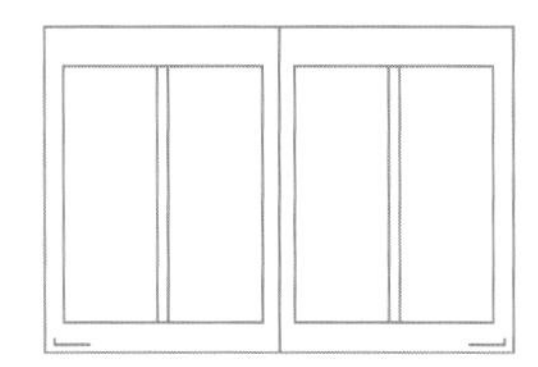

过多的颜色会让页面显得拥挤和混乱。但是，合理使用色彩可以帮助读者理清信息的先后顺序。如果辅以颜色，一个层次清晰的标题就更易于读者理解了。

白玉すいとん

あり合わせの根菜と一緒に白玉団子を煮込んだ手軽な汁料理。「主食もおかずも一度に食べられる。撮影など仕事の合間の昼食としても活躍した汁ものです。祖母もよく準備の手を休めて食べていました」。すいとんと言えばうどん粉が王道だが、阿部さんはより手軽な白玉粉を好んで使った。豆腐を練り込んだ白玉は、もちもちと柔らかな食感。

材料 4人分
大根6cm 人参1/3本 ごぼう10cm しめじ1/3房 まいたけ1/3房 油揚げ1枚 三つ葉8本 煮干し10本 薄口醤油大さじ2 豆腐約1/6丁 白玉粉約1/3カップ

作り方
1煮干しは頭と内臓を取り、鍋などで乾煎りしてから、水につけておく。(約6カップ、分量外)
2人参、大根は皮をむいて薄めのイチョウ切り、ごぼうは皮をたわしなどでよくこそいで洗い、薄く斜めに切って水にさらしておく。
3油揚げは熱湯をかけ、油抜きして食べやすい大きさに切り、しめじ、まいたけは小房に分けておく。
41に2を加え火にかけ、ひと煮立ちしたら3も加え、薄口醤油を半量入れてしばらく煮る。
5白玉粉に豆腐を混ぜ(写真)、みみたぶくらいの柔らかさにして形を整え、熱湯に入れて浮き上がってくるまで茹でたのち冷水に取る。
64に5を加え、ひと煮立ちしたら、残りの薄口醤油を加え味を調えて、ざく切りにした三つ葉を加える。

白玉すいとん

丸干しのチーズフライ

将标题中的第一个字放大并改变颜色，可以吸引读者对特定标题的注意

项目
《可颂》杂志

客户
《可颂》杂志社

设计师
马场诚子

插画师
小幡吉见（Yohimi Obata）

色彩巧妙地衬托了字体，让杂志版面更清晰，更有趣味性

じゃが芋団子

「おやつによく作ってくれたお団子。じゃがいもをすり下ろして、絞った汁から澱粉を取り出してつなぎにするんです。理科の実験みたいでしょう」。ひと手間をかけることで、たった数個のじゃがいもが、もっちりと食べ応えあるお団子になる。「甘いタレで食べるとおいしい。急に友達を連れて帰ったときにも、手早く作ってくれた記憶が」

じゃがいもひとつでできる、定番のおやつ。

材料 4人分
じゃがいも(小ぶりのもの)4個　黒練り胡麻大さじ6　醤油大さじ1と1/2　煮切りみりん小さじ1　ハチミツ大さじ1　昆布だし大さじ3　白煎り胡麻適量

作り方
1じゃがいもは皮をむき、おろし金ですり下ろす。
21をさらしなどで絞り(写真右)、絞り汁はコップなどに入れ(写真右)、汁以外はボウルなどに入れておく。
3絞り汁の底に白い澱粉が沈殿してきたら上澄みの汁を捨て、ボウルに入れた身に加えてよく混ぜ、団子状に丸める。
43を沸騰した湯の中に入れ、浮かんでくるまで茹でる。
5練り胡麻、醤油、みりん、ハチミツ、昆布だしをよく混ぜる。
64を水に取り、水気を切って皿に盛り、5をかけて、白煎り胡麻などをふる。

薩摩芋もち

じゃが芋団子と同じくらいの頻度で登場した甘味。「お餅に芋をつき混ぜて量を増やした、お腹に溜まるおやつでした。お餅だけだとすぐ堅くなってしまいますが、さつまいもが入っていると冷めても柔らかく、おいしく食べられる。甘くしたきな粉をまぶしてもいいですが、塩で食べるとさつまいもの自然な甘みが引き立ちます」

材料 4人分
さつまいも1/2本(約150グラム)　切り餅3切れ　きな粉適量　黒糖またはきび糖適量　塩適量

作り方
1さつまいもは洗って、皮付きのまま蒸気の上がった蒸し器に入れ、軟らかくなるまで20分以上蒸す。
21の、途中、残り6分くらいのところで切り餅も入れて蒸す。
31のいもの皮をむいてすり鉢などに入れてよく潰し、蒸して軟らかくなった餅とよくつき混ぜる。
4手に水を付けて3を適当な大きさにちぎり、きな粉、黒砂糖、塩などお好みのものをかける。

冷めてもおいしい、さつまいも入りのおもち。

茄子の胡麻煮

ごまの香ばしさが引き立つ、なすの煮物。とろりとしたなすの食感を出すために、皮はどうしても不要になる。「その皮を細く刻んで揚げて、切り昆布のようにトッピングに。祖母は何事も徹底した人で、皮を揚げるときも真っ黒になるまで揚げるのです。でも苦みが強くなるので、からっとしたら引き上げていいと思います」

皮も香ばしく揚げて、トッピングに使う。

材料 4人分
なす4本　だし1カップ　日本酒1/3カップ　みりん大さじ3　塩小さじ1　白すり胡麻大さじ4　薄口醤油適量　揚げ油適量

作り方
1なすは両端を切り落とし、縦に皮をむき十文字に深く包丁を入れて水に放し、ざるに上げる。皮は細切りにしてこれも水にさらし、ざるに(写真)。
2鍋にだし、酒、みりん、塩を入れてひと煮立ちさせ、水気を切ったなすを入れる。初めは中火、煮立ったら弱火にして軟らかくなるまで煮、すり胡麻を加えてしばらく煮て、味をみて薄口醤油を少々加える。
3なすの皮の水気をよく取り、170℃の油でカリッと揚げ、油をよく切る。
2を器に盛り、3を天盛りにする。

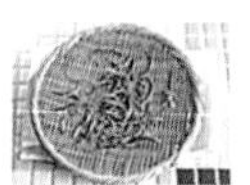

煮干しとごぼうの立田揚げ

「とくに祖母が気に入っていた一品。この料理専用のお皿が決めてあったほどです」。使うのは、ごぼうと煮干し。たったそれだけだが、酒と醤油に漬け、片栗粉をまぶしてカリカリに揚げるだけで、メインにも酒の肴にもなってしまう。魔法のようなレシピだ。「煮干しは比較的大きめのものを使うと、おいしく仕上がります」

材料 4人分
煮干し24本(1人前6本計算で)　ごぼう1/3本　醤油大さじ4　酒大さじ4　おろししょうが大さじ1　片栗粉適量　揚げ油適量

作り方
1煮干しは頭と内臓を取り、醤油大さじ2、酒大さじ2、しょうが大さじ1/2を合わせた漬け汁に漬けてしばらくおく(写真)。
2ごぼうは皮をたわしなどでこそいでよく洗い、薄く斜め切りにしてさっと水にさらし、水気を切って1と同量の漬け汁に漬けておく。
31と2の全体に片栗粉をまぶし、余分な粉をはたき落として170℃の油でからりと揚げて油をよく切る。

だしに使う煮干しも、立派なメインに。

在这个版式中，颜色用来区分不同的信息。对于指导性的材料来说，明确地区分信息非常重要。在这个烹饪书的对页中，小标题是彩色的。食谱中的数字是红色的，以便与其他文本区别开

お付き合い編　人付き合いを潤滑にする言葉づかい。

常識は知っておいてほしいですね。あとは「美意識」「センス」を磨いて。

Q お中元やお歳暮をお断りしたいのですが、相手に失礼にならない言い方はありますか？

Q グループのある人に会場の手配を頼みたい。上手にお願いする方法はありますか？

Q 近所の主婦の中傷合戦に巻き込まれ、私の名前も。どうすればいいですか？

Q 知人からセールスの勧誘を受けました。うまく断るにはどうすればいいですか？

Q 待ち合わせ場所に現れなかった友人に、ひとこと文句を言いたいのですが。

Q 親しい人との食事、きょうは自分が支払いたい。どう言えばいいですか？

问题的部分使用了不同的颜色，其字体的粗细、大小和阴影增加了页面的质感，也在视觉上引起了读者的注意

47.颜色让日历模块妙趣横生

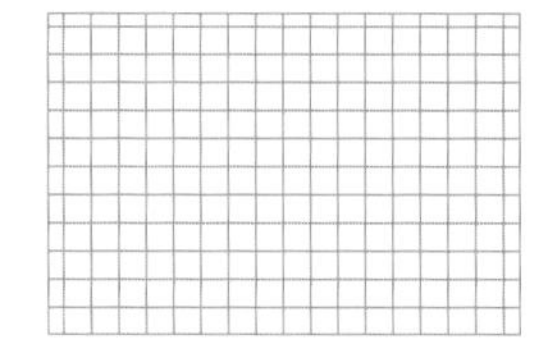

在日历中使用不同的颜色可以轻松地区分特定的元素，如星期几。这些信息既可以被凸显出来，也能与整个页面搭配。色彩应该与照片色调相辅相成。

对于需要强调的日期，应选用柔和的颜色搭配，不要让颜色喧宾夺主。低饱和度的颜色（包含较多灰色的颜色）更适合叠印文字，也就是在色块之上印刷文字。

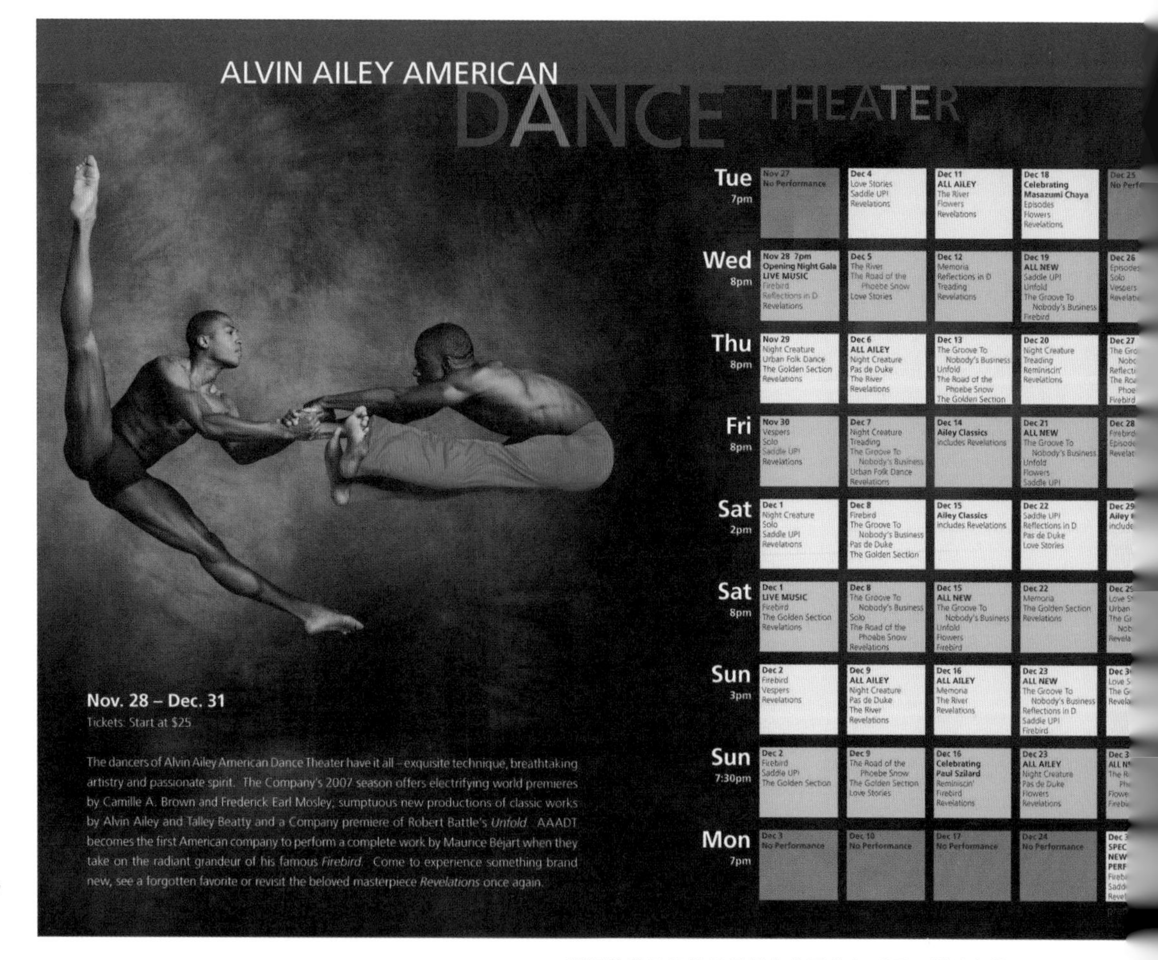

深沉的背景和极具震撼力的图片与采用了辅助色的日历形成对比

项目
大事件日历

客户
纽约城市中心（New York City Center）

设计师
安德鲁·杰拉贝克（Andrew Jerabek）

照片与一组色彩互相配合，形成了日历框的背景部分

右侧文本框的颜色温和但显眼，与精美的艺术照片和谐互补

这个对页使用了秋天的颜色，主色调是橙红色，极富戏剧性

● 48.利用色彩编码

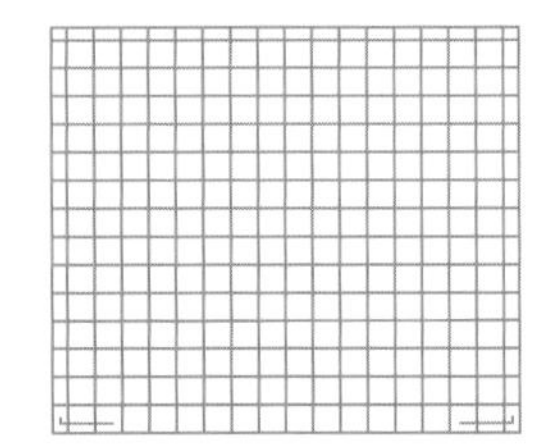

色彩编码可以帮助使用者迅速找到他们需要的信息。色彩结合图标，比单独的文字或色彩更能快速地传达更多信息。

色彩可以是中性的，也可以是明亮的，这取决于素材本身或客户的需求。高饱和度的色彩（包含少量灰色的颜色）可以迅速抓住读者的眼球。

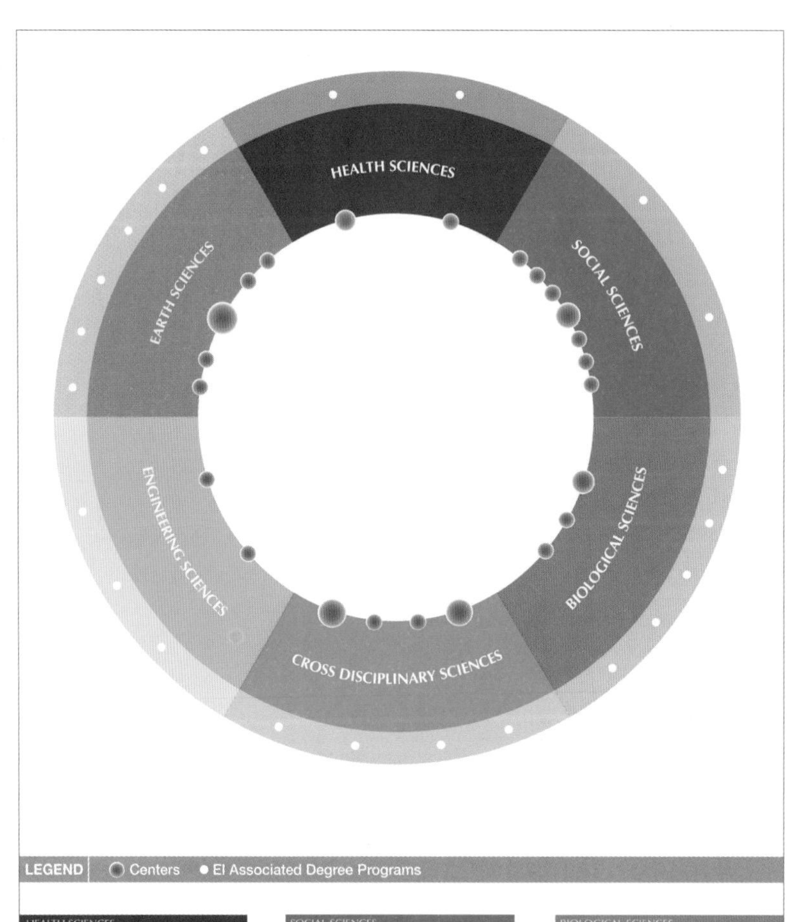

HEALTH SCIENCES

- Center for Global Health and Sustainable Development
- Center for National Health and Development in Ethiopia

SOCIAL SCIENCES

- Center on Globalization and Sustainable Development
- Laboratory of Populations
- Center for Capitalism and Society
- Center for Economy, Environment and Society
- Center for the Study of Science and Religion
- Center for Sustainable Urban Development
- Program on Science and Technology and Global Development

BIOLOGICAL SCIENCES

- Columbia University/UNESCO Joint Program on Biosphere and Society
- Center for Environmental Research and Conservation
- Black Rock Forest

CROSS DISCIPLINARY SCIENCES

- Center for Hazards & Risk Research
- Center for International Earth Science Information Network
- International Research Institute for Climate & Society
- Tropical Agriculture Program

ENGINEERING SCIENCES

- Earth Engineering Center
- Lenfest Center for Sustainable Energy

EARTH SCIENCES

- Lamont-Doherty Earth Observatory
- Center for Rivers and Estuaries
- Goddard Institute for Space Studies at Columbia University
- Center for Nonlinear Earth Systems
- Center for Climate Systems Research

通过设计，每一门学科都包含一组与学位课程相关的研究。每个学科都有特定的色彩系统

项目
标识项目

客户
哥伦比亚大学地球研究所（Earth Institute at Columbia University）

设计师
马克•英格利斯（Mark Inglis）

创意总监
马克•英格利斯

不同的颜色区分了哥伦比亚大学地球研究所的六门学科

图标也嵌入到了色彩系统内

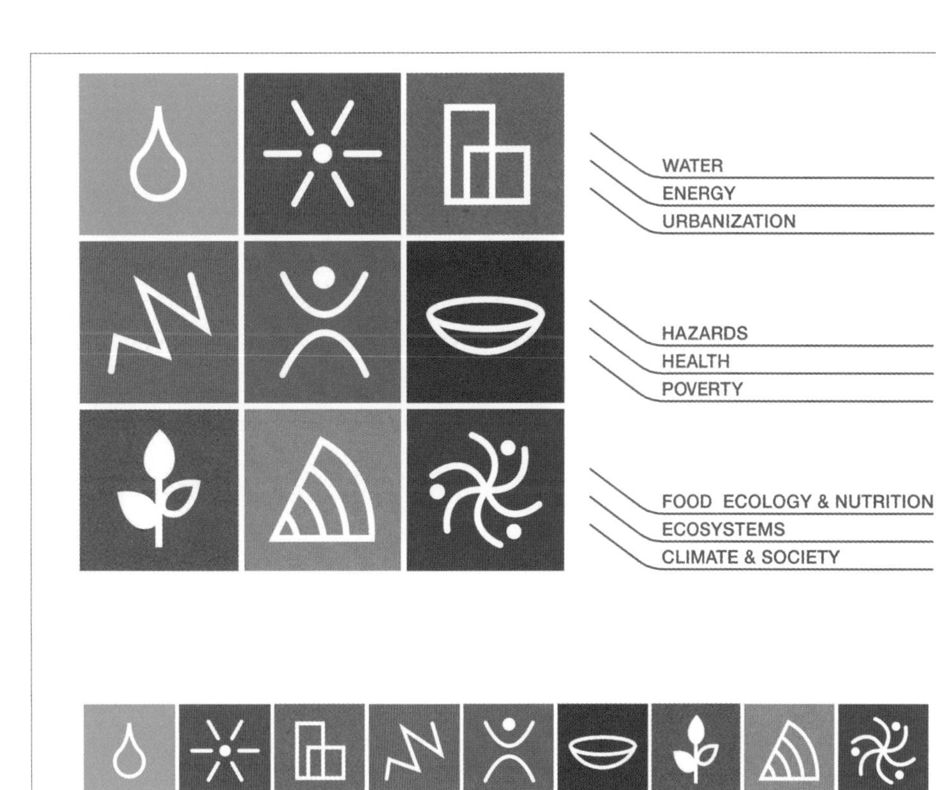

色彩与图标、色带或字体搭配

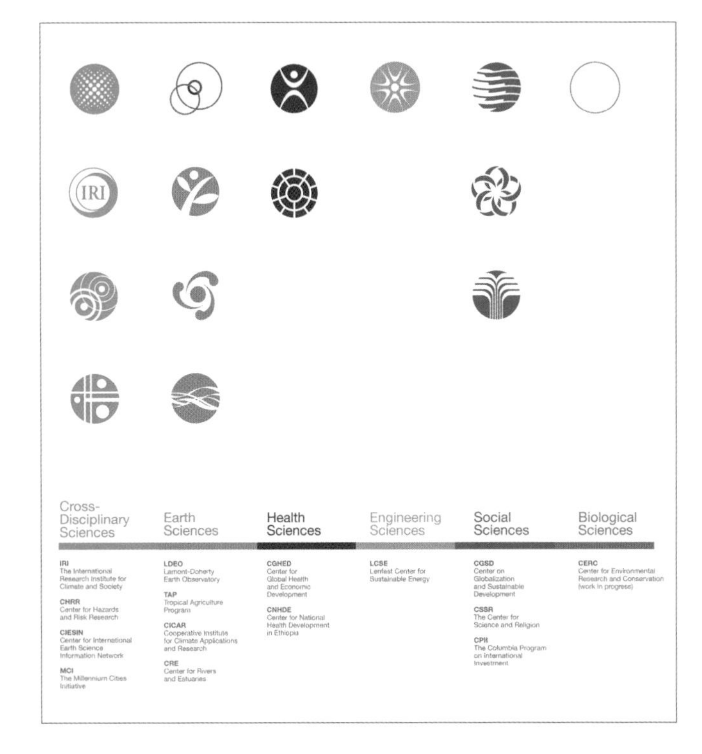

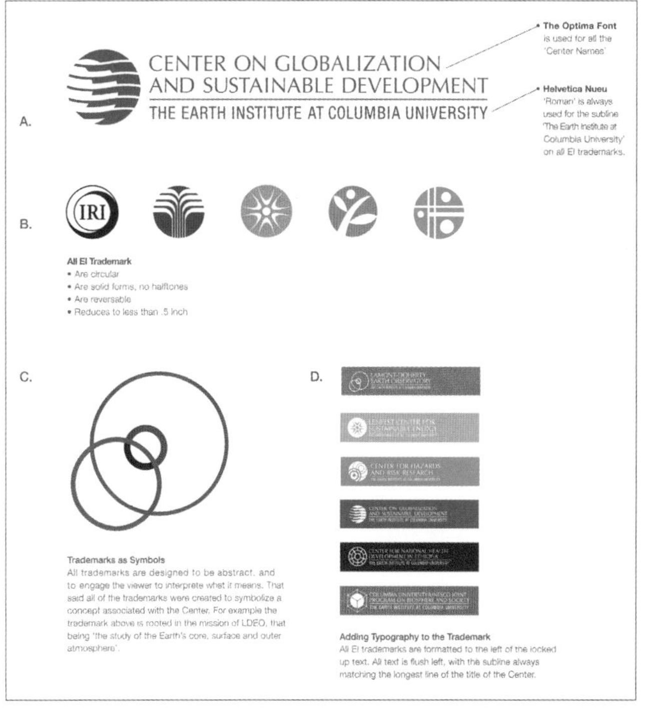

● 49.用色彩区别内容

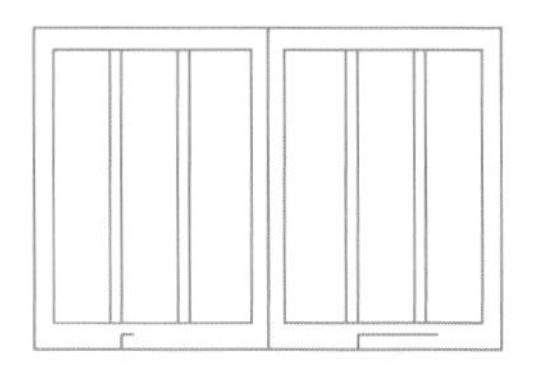

水平和垂直的分栏可以清晰合理地分离各部分元素，而不同灰度的色彩可以增加页面的活力和趣味性，并区分不同作者的观点。使用不同形式的文字效果更好，如反白字。

另见92～93页

凸显内容

反白字富有戏剧感。但是要注意，如果反白字量大且字号较小，可能会不够清晰，尤其是在使用衬线字体的时候，使用无衬线字体会更好一点儿。在屏幕上，黑色背景上的反白字可能会看不清楚，所以字体需要更大或更粗。

*AT SPOTCO, WE BEGIN WORK-*ing on a show by understanding its Event. I didn't invent this phrase—it was loaned to me by producer Margo Lion. But what I came to understand it as is quite simply the reason you see a show. Or even more simply, the reason you tell someone else to see a show. It can be straightforward,

上面这个例子展示了不同字体和字号的文字的对比。

项目
《百老汇：从出租到变革》（*On Broadway:From Rent to Revolution*）

客户
作者：德鲁·霍奇斯（Drew Hodges）；出版商：里佐利（Rizzoli）

创意总监
德鲁·霍奇斯

设计师
内奥米·米祖萨基（Naomi Mizusaki）

这本关于百老汇的回忆录使用了不同的颜色

不同的颜色代表本书的不同合作者。页面结尾的文字使用了黑色的文本框，与其他有关人物的内容不同，这部分是关于大事件的介绍，是体现百老汇理念的基本元素

1987 | ACT ONE

DREW HODGES
FOUNDER AND CHIEF EXECUTIVE OFFICER

WE OPEN ON A YOUNG design firm called Spot Design. It was named for a dog the landlord said we couldn't have. So I named the office as my pet.

After attending art school at School of Visual Arts in New York City, I had left working for my college mentor Paula Scher and began freelancing solo out of my apartment. I was working in the kitchen of my loft, across from the now-defunct flea markets on 26th Street and Sixth Avenue. This is the same kitchen where producers Barry and Fran Weissler came to see the early designs for *Chicago*—but I get ahead of myself. Ultimately, we were five designers and one part-time bookkeeper doing entertainment and rock 'n' roll work. We were young and laughed a lot. Ten years later, we had been privileged to work with Swatch Watch, MTV, Nickelodeon, the launch of Comedy Central and *The Daily Show*, as well as record work for Sony Music, Atlantic Records, and Geffen/DreamWorks records, where our most notable projects were album packages for downtown diva Lisa Loeb and iron-lunged Aerosmith. We grew adept at strategy, design, and collaboration with many downtown artists, illustrators, and photographers,—all people we would come to take full advantage of as we began our theater work. I went to the theater—it was a New York City joy for me. I had gone since early high school, riding the train to the city. But I never dreamed I would get any nearer than second-acting *Dreamgirls* from the mezzanine.

Two bold incidents happened to change that. First, Tom Viola and Rodger McFarlane were heroes of mine for the work they did through what was to become Broadway Cares/Equity Fights AIDS. A friend and art director from Sony Music named Mark Burdett

was assigned to work with Spot ad for the Grammy Awards prog the prime position of the back c was Martin Luther King Jr. Day clients were all away. So we moa doing yet another ad filled with minis of the labels and latest St release with hollow congratulat a waste of a great opportunity. V posited that Sony could be the fi record company to take a stand AIDS by making a donation an ing a red ribbon to the back of e issue of the program. And to pi idea, we called Rodger and Tom offices to help us fulfill it. The r was theirs after all. Remarkabl were working on the holiday ar answered the phone. They agree and the rest is history. I believe the first awards ceremony to pu script the concern over AIDS, a friendships were formed. Later year, Tom and Rodger called. T an ad due in three days for the show, *The Destiny of Me*, Larry sequel to *The Normal Heart*. W them a design based on a phot right hand—I guess we felt it s personal—and they loved it. T our first theater poster.

But it would have been a sh career without the second eve years later, we had just finishe the Aerosmith album for Geffe Sloane, David Geffen's star cre director, called and asked us t with the producers of *Rent*—G would be releasing the album. meeting with the ad agency in and got the assignment and a the hottest show in town a we had opened Off-Broadway. W year, we would have designed *Chicago*, and Jeffrey Seller sat in a mall in Miami to ask if w thought about starting an ad seemed a big risk—but it also like a world where you could meet the people doing the cre you were assigned to promote began to try and figure out ju ad agency worked anyway.

8

BRIAN BERK

O-FOUNDER AND CHIEF OPERATING
FFICER / CHIEF FINANCIAL OFFICER

THE SPRING OF 1997, designing the successful ad cam-
ıs for *Rent* and *Chicago*, we decided
empt to open a theatrical ad
y. The first question was: What
d we need to be able to pull this
For starters, we would need equip-
, office space, a staff, and most
rtantly, clients.
e equipment was easy. In order to
upfront costs down, we could
—a few computers and a fax
ine. From there, we could scrape
til we had some clients.
ice space: The design studio was
ıtly housed in Drew's apartment.
ew that for potential clients to
ler hiring us, we would need to be
theater district, and we would
o have a large conference room for
ekly ad meetings. I set out to look
e. One space was located in 1600
way. The building was fairly
wn (and we would later learn it
mouse problem), but it did have a
nd interesting history. It was built
very early 1900s as a Studebaker
and showroom. In the 1920s, it
nverted to offices and at one point
the original offices of Columbia
s, Universal Pictures, and Max
er Studios, creator of Betty Boop.
emed a fitting place for a theatri-
agency. By 1997, the building held
ination of offices and screening
(It has since been torn down.)
ace we looked at consisted of two
ffices, a big bullpen area for our
rs, and one large conference
ith the most amazing view of
Square. We actually found a
raph of a movie executive sitting
esk in the room that would
lly become our conference room.
e same wood paneling and
with the view. However, the
ı rug, which is seen on the floor
hoto, is long gone. The space had
r. We had to furnish it on the
Ve hired a set decorator friend to
office circa 1940s, so all the used
rniture we purchased would look like a very conscious design choice. We moved to the space in June 1997.

Staff: We already had a creative director (Drew), four graphic designers, an office manager, and me. I handled finance, administration, and facilities. We needed someone to head account services, an assistant account executive and a graphic production artist to produce all the ads. We'd hire a writer once we had some clients. For the production artist, we knew just who to hire: Mary Littell. She had worked for us before and was great. The person to head account services was harder to find. We needed someone who had worked at an ad agency before and understood media. From what Drew learned, one of the most respected account managers in the industry was Jim Edwards, or as was said by several producers, "He is the least hated." He had worked at two of the existing theatrical ad agencies. But would Jim join a startup? He was game and joined our team. Jim walked in the door on July 21, 1997. Mary was at her desk working on dot gain so she would be ready if we were ever hired to place an ad. Now, all we needed were some clients.

JIM EDWARDS

CO-FOUNDER AND FORMER CHIEF OPERATING OFFICER

OF DREW, BRIAN, MARY, BOB Guglielmo, Karen Hermelin, and Jesse Wann, I was the only one who had worked at an ad agency before. Little things like a copy machine that can make more than one copy every thirty seconds was not part of our infrastructure. I started on a Monday, and the pitch for *The Diary of Anne Frank* was that Friday. We didn't have any clients so that entire week was all about the pitch. Thursday night we were there late and inadvertently got locked in the office (how that is even possible still strikes me as odd). We couldn't reach anyone who had a key so we had to call the fire department to let us out. They did—and were adorable too.

Once we had a show, we became a legitimate advertising agency, which led to David Mamet's *The Old Neighborhood*, John Leguizamo's *Freak*, and Joanna Murray-Smith's *Honour* within months of being open for business.

Since SpotCo was a brand-new company, we had no credit with any of our vendors. The *New York Times* made us jump through so many hoops about establishing a relationship with them. I think we had to have a letter from the producers of *Anne Frank* saying they were hiring us as their ad agency. When it came time to reserve our first *Times* ad, about a month had passed since all those rules were handed down. Back then, you simply called the *Times* reservation desk and reserved the space. That's what I had been doing for years so I did it again, inadvertently forgetting that the ad needed to be paid for well in advance. I knew everyone there so they accepted my reservation without question. The ad ran. No one said anything. The bill came about a month later, and we paid it. About two months in, the *Times* noticed that we were sliding under their rules but since we were paying our bills and were current, they granted us credit. By that first Christmas, we had established credit everywhere, which was and is a big deal. Not many new companies can make that kind of claim.

In the first eighteen months of SpotCo, I never worked harder in my life. The hours were long (I gained a lot of weight during that time—do not put this in the book), and it wasn't easy, but we also saw direct results from all the hard work. The work was good, and people noticed what we were doing—and we were making money! The Christmas party of 1998 was particularly memorable. That day, we had just released the full-page ad for *The Blue Room*, which was pretty provocative because all we ran was the photo and a quote about how hot the ticket was. It was a big deal for us and kind of heady. The party was at some restaurant, and Brian had secured a private room. There were only three tables of ten, and we shared our Secret Santa gifts. Everyone was really into it and every time someone opened a gift, Amelia Heape would shout, "Feel the *love*, people! Feel the *love*!" Indeed.

TOM GREENWALD

CO-FOUNDER AND CHIEF STRATEGY OFFICER

WHAT AM I, NUTS? IT WAS early 1998. I had a good job, an amazing wife, three adorable kids under five years old, and a modest but nice house in Connecticut. In other words, I was settling in nicely to the 9 to 7, suburban commuting life. But I kept hearing about this guy Drew Hodges. First I heard about him through my friend Jeffrey Seller, who had worked with Drew on the designs for *Rent*. Then I heard about him through my friend David Stone, who was just about to start working with him on *The Diary of Anne Frank*. Then, I realized, they're talking about the tall guy who talked a lot and had barreled his way through meetings at Grey Advertising (where I worked at the time) while designing the artwork for *Chicago*. So when Jim Edwards called me and said, "Hey, I'm joining up with this guy Drew and we're turning his design shop into an ad agency and did you want to meet him?" I had to think about it. No way was I going to give up my job security, right? The odds of any theatrical ad agency surviving at all were miniscule, much less one with . . . wait, let me add them up . . . one client. And besides, I'd probably have to take a cut in pay. Only an insane person would consider it. "Sure," I told Jim. "Set it up."

So I went in to talk to Drew, and after about eighteen seconds, I'd made up my mind. When Drew asked if I had any questions, I had only one. "Where do I sign?" I joined SpotCo as the head (only) copywriter, head (only) broadcast director, and head (only) proofreader. On the downside, we were a very lean department. On the upside, I never had any problems with my staff.

Now here we are, almost eighteen years later. My wife is still amazing, my three kids are still adorable (although no longer under five), and my house is still modest but nice (although we redid the TV room). I never did settle in to that 9 to 7 job, though. Instead, I decided to take a chance, a flyer, and a crazy ride—and it's worked out pretty well. So yeah, I guess I was a little nuts. But it turns out craziness has its perks.

THE EVENT

AT SPOTCO, WE BEGIN WORK- ing on a show by understanding its Event. I didn't invent this phrase—it was loaned to me by producer Margo Lion. But what I came to understand it as is quite simply the reason you see a show. Or even more simply, the reason you tell someone else to see a show. It can be straightforward, or subtle. As much as I love advertising, branding and design and what it can do for a show, word of mouth is the number one reason a show becomes a hit. So the most effective branding hopes to invite, encourage, and curate that word of mouth. It has to be true to the experience of that show—otherwise it will fail miserably. If every great musical begins with an "I want" song, then every show's voice begins with "I want you to think of this show this way" tune.

What we as an agency have also learned is you choose your Event or it will be chosen for you, and you (and ticket buyers) may not like the sound of that choice. Lets call that the non-event.

At the beginning of each show presented here, we have listed the non-event first—the thing we feared the Broadway customer would default to when describing a show, knowing little else. This is followed by the Event (capital E!) that we wanted to communicate. Hopefully, this informs our successes and disappointments and, more to the point, why a certain idea was ultimately chosen. —D.H.

● 50.用元素的粗细、大小和形状代替颜色

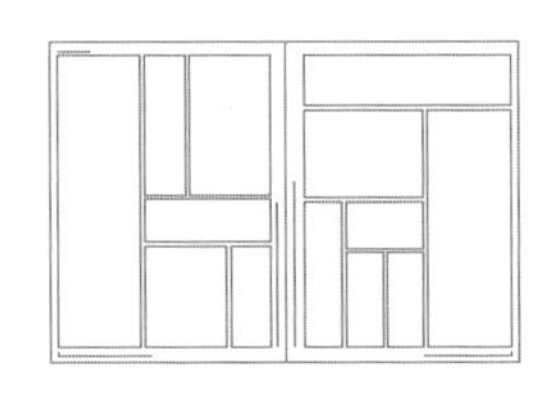

有时，为了减少不必要的元素，设计师会使用黑白两色，或尽可能少地使用颜色。在有些情况下，预算限制甚至要求设计师只能使用一种颜色。无论是使用有限的颜色还是只用一种颜色，都可以通过不同的字体、字号，以及字体和图片在网格中所占空间的比例来使页面达到有质感的效果。

一个通用的网格可以有多种宽度和大小，并有多种形式可以创造出变化。间距实际上也是一种图形元素，可以让页面结构更清晰，更有对比性。

另见90～91，104～105页

色调

如果客户要求或偏好使用一种或少数几种颜色，可以改变色调（颜色纯度的比例）。一个纯度为10%的出血黑色背景上可以很容易地叠印颜色纯度为100%的文字。背景越暗，100%纯度的字体可读性越差（这也与字号大小或纸张有关）。使用暗色背景时，可以选择使用反白字。

MLK | THE MAN

Eli Lee is an editorial fellow at *The Atlantic*.

The Arc of a Life

Martin Luther King Jr. was just 26 when he came to prominence, by leading a bus boycott in Montgomery, Alabama. He was only 39 when he was killed.

By Eli Lee

"I would turn to the Almighty, and say, 'If you allow me to live just a few years in the second half of the 20th century, I will be happy.'"

January 15, 1929: Michael King Jr. is born in Atlanta to Michael King Sr., a prominent local preacher and civil-rights leader, and Alberta King, a former schoolteacher. In 1934, his father changes both of their first names to Martin-by various accounts, to correct a birth-certificate mistake or to honor the theologian Martin Luther. His childhood is comfortable; he excels in school, skipping the ninth and 12th grades. He first becomes conscious of racism at age 6, when a white friend's father prohibits his son from playing with Martin.

June 8, 1948: King graduates from Morehouse College, in Atlanta, with a bachelor's degree in sociology. Later, he earns a bachelor-of-divinity degree from Crozer Theological Seminary, in Chester, Pennsylvania, and a doctorate in systematic theology from Boston University. (In 1991, a BU committee would determine that King had plagiarized passages of his dissertation, "A Comparison of the Conceptions of God in the Thinking of Paul Tillich and Henry Nelson Wieman," from other scholars' work.)

June 18, 1953: **King marries Coretta Scott, an activist and aspiring singer from Alabama studying at the New England Conservatory of Music. After his death, she would advance her husband's legacy by founding the Martin Luther King Jr. Center for Nonviolent Social Change.**

1955–56: Activists organize a boycott of the bus system in Montgomery, Alabama, after Rosa Parks, a black woman, is arrested for refusing to give up her seat to a white man. A former head of the state's NAACP calls on King, now the 26-year-old pastor of a local black church, to lead the boycott-because he's "young and intelligent with leadership ability" and has a "wonderful speaking voice." The protest lasts 381 days, ending in victory after the U.S. Supreme Court rules that segregation on public buses is unconstitutional.

January 10–11, 1957: Notables in the civil-rights movement form the Southern Christian Leadership Conference to coordinate nonviolent protest actions in the South. They soon elect King as the organization's president.

COURTESY OF MOREHOUSE COLLEGE; CHARLES MOORE; DON CRAVENS/THE LIFE IMAGES COLLECTION/GETTY

12 THE ATLANTIC MARTIN LUTHER KING JR.

February 1959: At the invitation of Prime Minister Jawaharlal Nehru, King visits India for a month, meeting with social reformers, government officials, and associates of the late Mahatma Gandhi, whose acts of civil disobedience to free the country from British rule inspired King's own approach to bringing about change. "I left India more convinced than ever before," King writes at the time, "that non-violent resistance is the most potent weapon available to oppressed people in their struggle for freedom."

August 28, 1963: At the March on Washington for Jobs and Freedom, then the largest protest in U.S. history, King addresses an estimated 250,000 people on the National Mall. His soaring call for racial justice comes to be known as the "I Have a Dream" speech, after his ad-libbed ending. Today, the speech is seen as both a rhetorical masterpiece and a defining moment of the civil-rights movement.

December 10, 1964: **King receives the Nobel Peace Prize. At age 35, he is its youngest recipient so far. He promises to donate the prize's $54,123 award to the civil-rights movement.**

March 1965: In response to the continued disenfranchisement of millions of black people across the South, the SCLC and other civil-rights groups demand voting rights in a 54-mile march from Selma, Alabama, to the state capital of Montgomery. Local police and white mobs react brutally to the nonviolent protest, beating many participants; two white demonstrators are murdered. The march hastens passage of the federal Voting Rights Act later in the year.

April 4, 1967: *In a speech titled "Beyond Vietnam," delivered at Riverside Church in Manhattan, King declares his opposition to the Vietnam War. He publicly criticized the war two months earlier, in a speech to the Nation Institute (see page 92). But the widely publicized Riverside Church speech upsets many of King's usual allies, who accuse him of hurting the cause of civil rights by alienating the American government and public. With this stand, King expands his calls for social justice at home into a broader, pacifist message.*

April 4, 1968: **While in Memphis to protest black sanitation workers' poor treatment by the city, King is shot as he stands on a balcony at the Lorraine Motel. He is declared dead about an hour later. The murderer, James Earl Ray, flees the country and is arrested two months later at Heathrow Airport, in London; he is known to be a racist, but his exact motive is never made clear. (Sentenced to 99 years in prison, he dies of natural causes in 1998.) The assassination sparks riots in more than 100 U.S. cities. President Lyndon B. Johnson declares a national day of mourning.**

November 2, 1983: After years of advocacy from labor unions and civil-rights groups, President Ronald Reagan signs legislation to commemorate the slain leader's birthday with a federal holiday on the third Monday in January. Martin Luther King Jr. Day is celebrated nationwide for the first time in 1986.

R. SATAKOPAN; FRANCIS MILLER/THE LIFE PICTURE COLLECTION/GETTY; PARAMONOV ALEXANDER; CHARLES KELLY/AP

MARTIN LUTHER KING JR. THE ATLANTIC 13

项目
《金》

客户
《大西洋月刊》杂志社

创意总监
保罗•思培拉

艺术总监
戴维•萨默维尔

设计公司
OCD | 原创设计冠军

设计师
博比•马丁，詹妮弗•奇洛恩

不同的字体采用了不同的字号和间距的设计成就了这本书“特刊”的地位

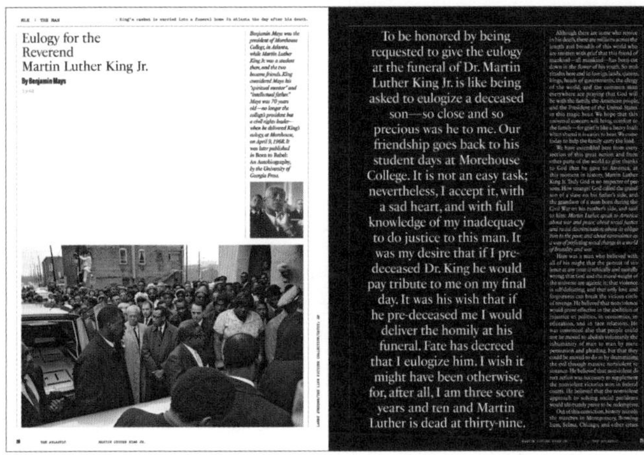
Eulogy for the Reverend Martin Luther King Jr.

By Benjamin Mays

To be honored by being requested to give the eulogy at the funeral of Dr. Martin Luther King Jr. is like being asked to eulogize a deceased son—so close and so precious was he to me. Our friendship goes back to his student days at Morehouse College. It is not an easy task; nevertheless, I accept it, with a sad heart, and with full knowledge of my inadequacy to do justice to this man. It was my desire that if I predeceased Dr. King he would pay tribute to me on my final day. It was his wish that if he pre-deceased me I would deliver the homily at his funeral. Fate has decreed that I eulogize him. I wish it might have been otherwise, for, after all, I am three score years and ten and Martin Luther is dead at thirty-nine.

"I make bold to assert that it took more courage for King to practice nonviolence than it took for his assassin to fire the fatal shot."

上页图：这本书展示了历史照片、名人语录和报告，它使用了通用的网格，凸显了页面的质感。三种不同的字体组合在一起，既唤起了人们对20世纪60年代的回忆，又为条理清晰的版面带来了一种清新的感觉。这些页面中各个部分的尺寸各不相同，但时间线是一致的

本页上图：四个章节开头虽然颜色上没有变化，但通过字号的变化产生了具有节奏感的效果

本页下图：悼词中充满了对马丁·路德·金的敬意。虽然设计上限制了颜色的使用，但通过设置更大、更宽或更窄的文本框，以及不同粗细的字体，突出了特定的文本。在网格中留出一定的空白，可以使留白像文章和图片一样动人

● 51.分解标志图形

设计标识极富挑战性，需要逻辑性、组织性和一致性。一个用于标识系统的图形网格系统（特别是在报刊亭、电话亭四周的装饰设计中）可以包含以下信息：

◎ 能按顺序搜索的信息层级，如选择1、选择2等；

◎ 二级信息（依然重要），如语言选择；

◎ 回答基本问题和需求的第三级信息，如机场的登机口信息、洗手间标志以及就餐地点等；

◎ 在根据指示操作的过程中，可能会产生的一系列复杂问题，例如，使用者突然意识到他要重复上面的步骤。

标志必须能让使用者看到，并轻松地阅读，即使在走路或开车时，也应该不受影响，所以，文字应该具有清晰的层次结构，颜色应该在不冲淡信息的情况下帮助标志引起使用者的注意。

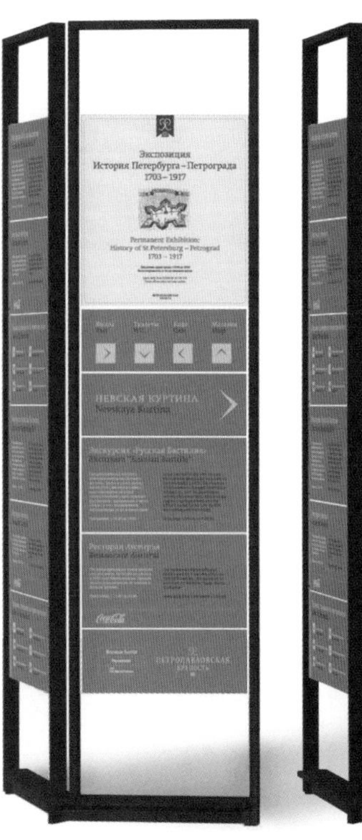

这个指路牌中，信息条由主要标志和图表牌构成

项目
指路牌和标志图形

客户
彼得和保罗城堡，圣彼得堡，俄罗斯（The Peter and Paul Fortress, St. Petersburg, Russia）

艺术总监
安东•金兹伯格

设计公司
RADIA工作室

这是为俄罗斯圣彼得堡的彼得和保罗城堡设计的指路牌，指路牌上有英语和俄语，帮助人们寻找方位。部分工程已完成

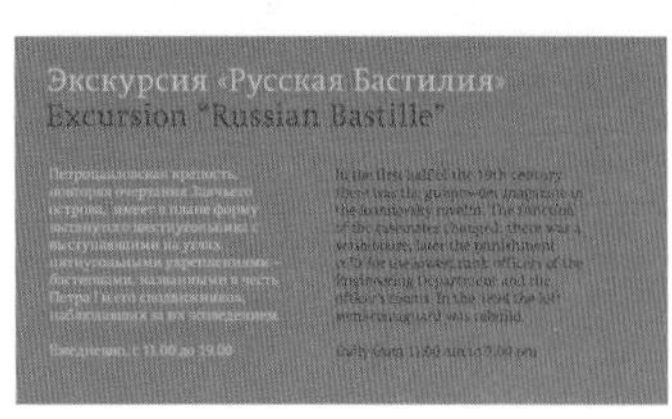

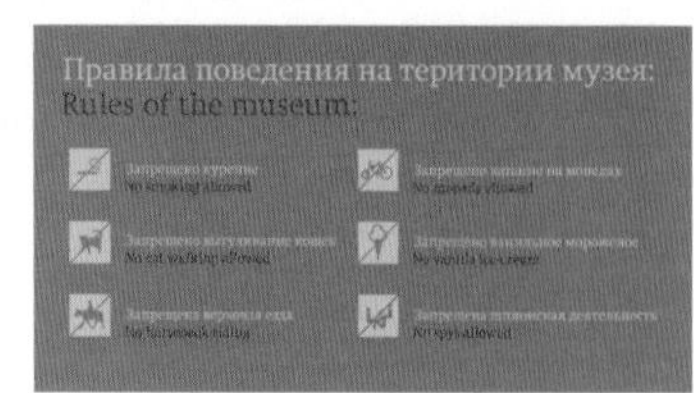

图表牌的细节显示了需要呈现和特别标明的信息

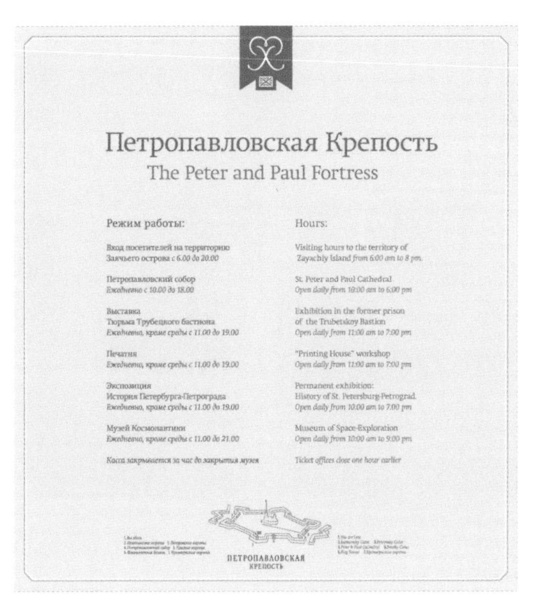

指路牌上的标志信息排版清晰，风格古典，与城市的历史相呼应

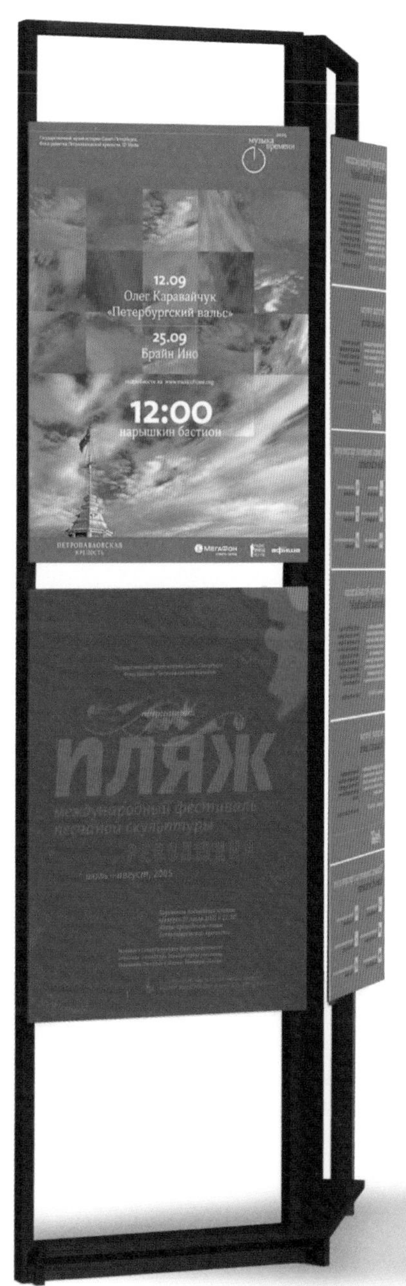

蓝色的面板是临时的通告栏，经过数字印刷后安装在指路牌上，用来发布临时事件，图中显示了海报的版式

52.系统化地使用条形

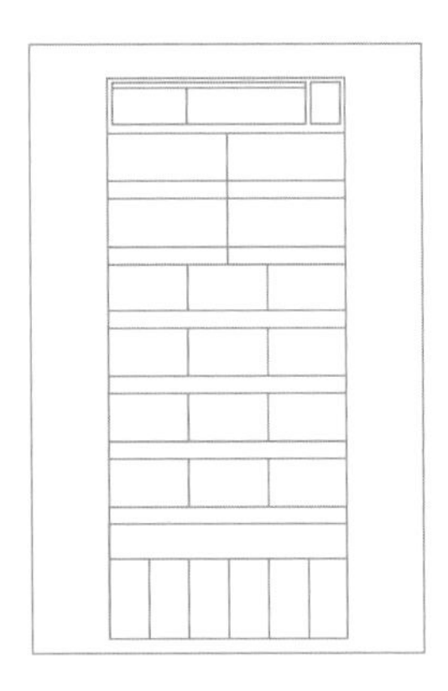

区分信息最简单的方式就是使用水平的层级。信息可以按条状排列，作为导航系统的一部分。这种层次结构不仅适用于网站，也适用于不同终端。

上图和下页左图： 横条提供了一个整体框架，上面是导航栏。导航栏中的选项可以拉出下一页的横向组织的信息

项目
犹太在线博物馆（Jewish Online Museum）网站

客户
犹太在线博物馆

品牌/前端开发
螺纹工作室（Threaded）

架构
Lushai设计机构

网页开发
Ghost Street工作室，Reactive工作室

犹太在线博物馆的网站是一个面向新西兰犹太社区的人群的网站。网站为访问者提供大量的教育资源

程序员的重要性

通常，前端设计师不是后端的程序员，这是众所周知的情况，但有的客户会认为设计师可以做完所有工作。犹太在线博物馆网站的设计师和负责架构的设计机构Lushai合作，开发出了满足犹太在线博物馆需求的动态架构，包括用户交互和对移动终端的响应式编译。

下页右图： 移动端的设计与网页保持一致，即使改变了横条的形状，也会使人联想到网站

JoM
JEWISH ONLINE MUSEUM
HISTORY
LIVES
ARTS
FAITH
HOLOCAUST
BROWSE BY
Te Jewry | An Online Exhibition
As Old as Auckland | An Online Exhibition
Portrait of a People | An Online Exhibition
Eduth | Soundscape
JEWISH LIVES IN NEW ZEALAND
A HISTORY
Eduth | Oral History
Jewish Lives in New Zealand
Jewish Lives | Anti-Semitism
Jewish Lives | Southern Lands
Karl Wolfskehl
Sir Julius Vogel
Dorothy Theomin
Emily Siedeberg
Karl Popper
CANTERBURY
Jewish Congregation.
Philip Aaron Philips
New Zealand Jewish Archives
Peter Munz
WORLD UPSIDE DOWN

JoM
MENU
TIMELINE
Jewish history all under one roof.

JoM
MENU
TIMELINE
HISTORY
BIOGRAPHY
ARTS
FAITH
HOLOCAUST
ABOUT
CONTACT
CLOSE X

JoM
MENU
TIMELINE
My birth coincided with the passing of the Nuremberg Laws

● 53.用间距界定水平区域

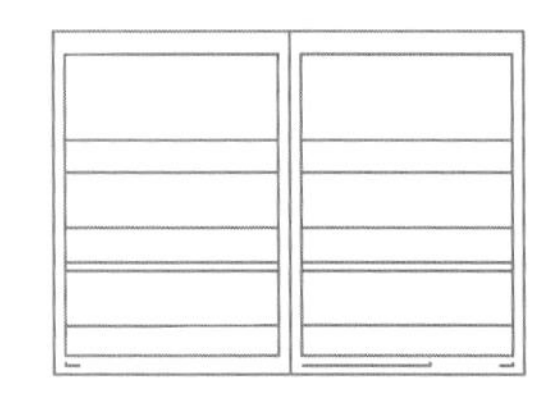

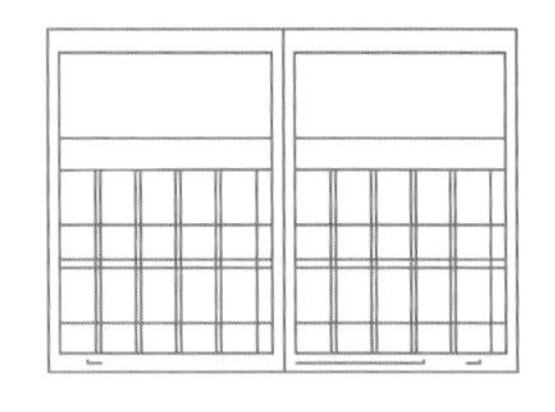

页面上适当的间距可以带来一种秩序感和平衡感。间距可以将介绍性素材（如标题和正文）与解释性素材（如图片说明和步骤说明）分开，分散的区块可以帮助读者有效地浏览页面信息。

正月

松飾りでお迎え

「明けましておめでとうございます」の挨拶で始まる元旦。新しい年の始まりを祝うとともに、お正月の神様を家に迎え、ともに過ごす特別なハレのひとときです。

お正月の神様は、歳神様とも呼ばれ、祖先の霊や豊穣の神と信じられてきました。歳神様を「こちらへどうぞ」と迎えるためのしるし、それが松飾りです。お正月になると、歳神様が空の向こうの神の国から、人の国へと飛来します。歳神様は、立てられた松を目印に、人々のもとへと降り立つのです。清らかに整えられた玄関に立つ青々とした松飾りは、お正月ならではの風景。一枝を飾れば、新年を迎えるにふさわしい、凛とした空気が生まれます。飾られる場所は家々の門口。地方によっては榊や竹など他の常緑樹を用いることもありますが、松が多く使われたため、門松とも呼ばれます。常緑樹である松の木は、神が宿る「依代」と考えられ、また、長寿の願いもこめられていたといいます。

お年玉のこと

子どもたちがお正月に一番楽しみにしているのは、何といってもお年玉。このお年玉は、古くはお餅でした。トシは歳神様のトシ、ダマは魂のタマ。歳神様の魂が宿った丸いお餅を子どもたちに分け与えたものが、もともとのお年玉でした。新年に贈られる、歳神様の魂がこもった丸餅は、生きる力、新たなるエネルギーをもたらすものでもありました。お年玉は、子どもたちの新しい一年の幸せを願う、今も昔もかわらぬ気持ちがこめられた、贈り物です。

年玉包み

お年玉を包む以外に、ちょっとしたお祝いやお礼、旅館などで渡す心づけを包むときにもいいでしょう。折り方を紹介している左側の包みは、赤い線が入った緣紅紙を使っています。白い紙でもできますが、緣紅紙を使うと、赤い線が、ちょうどアクセントになるデザインです。

59

新年の折形

指導　折形デザイン研究所

折形は、和紙を使った包み方の作法。日本のしきたりから生まれた、幸せを折る心のかたちです。門松を立て、お雑煮を囲んで新しい一年を迎える晴れやかなひとときに、折形を添えてみませんか。静かな気持ちで白い紙を広げ、清々しい一年を迎える準備を始めましょう。お正月はもうすぐです。

松飾り

大きな門松になると、家によっては、立てる場所などが難しいもの。一枝の松飾りなら、マンションなどの玄関にも自然になじみます。「木の花包み」といわれる折形を、松飾りに仕立てています。裏側にこよりなどで紐をつけ、ドアや壁に貼ったり、引っ掛けるとよいでしょう。

58

项目
《每日笔记》杂志

客户
《每日笔记》杂志社

设计师
林修三，柳政明

在有许多图片和信息的单页和对页中，合理的水平层级可以划分标题和步骤，给人一种有序和平和的感觉，使信息易于解析

间距清晰地将文本与图片分开，将信息分隔清楚

実際に和紙を折ってみましょう 4種類の作り方です

折形と日本のしきたり

折形は、室町時代に始まった、武家に伝わる礼法と伝えられます。折形をはじめとする日本のしきたりの数々は、本来の意味や由来は忘れられながらも、暮らしの中に生き続け、今に伝わったもの、と折形デザイン研究所の山口信博さんはおっしゃいます。

お正月にお雑煮をいただき、結婚のお祝いには水引をかけたご祝儀を贈る。生活に根付いたしきたりを、民俗学者の折口信夫は、「生活の古典」と呼びました。

「……私どもの生活は、功利の目的のついて廻らぬ、いわば、むだとも思われる様式の、由来不明なる「為来り」によって、純粋にせられることが多い。その多くは、家庭生活を優雅にし、しなやかな力を与える。門松を樹てた後の心持ちのやすらいを考えてみればよい。……」（『古代研究Ⅰ－祭りの発生』（古代生活の研究）中公クラシックスより）

松飾り

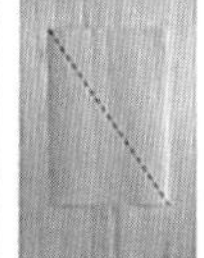

①檀紙を半紙の大きさに切って、左下の角を対角線で折り上げます。

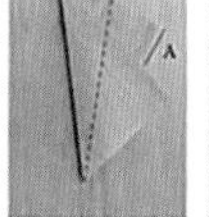

②Aを左端の辺に合わせて折ります。

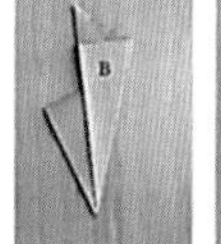

③Bを開き、その開いた折り目に合わせて、左端の辺を折ります。

④Bを元に戻し、下端を裏側に折り上げます。

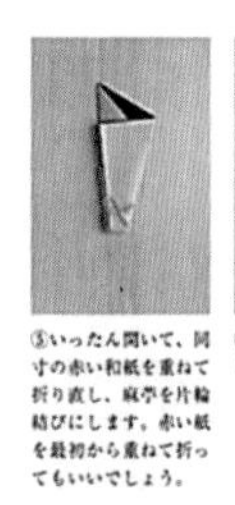

⑤いったん開いて、同寸の赤い和紙を重ねて折り直し、麻苧を片輪結びにします。赤い紙を最初から重ねて折ってもいいでしょう。

年玉包み

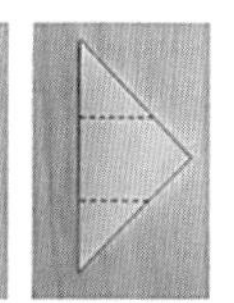

①18cm四方の縁紅紙を対角線で折り、左辺を三等分して、上の角、下の角の順に折ります。

②右端の角を左に折ります。

③上の角を上に引き上げて、右辺を中央に合わせて折ります。

④上の角の2枚の紙の間から三つ折りのお札や硬貨を入れ、上の角を、下の角の2枚の紙の間に差し込みます。

⑤出来上がりです。

------ は手前に折る折り目、------ は裏に折り返す折り目、—— は辺を示します。

お正月は、しきたりが特に身近になる時期。まずは小さな折形から、日本の豊かな心を感じてみてはいかがでしょう。

贈り物を包むことは紙を選ぶときから始まっています

折形には、和紙で出来た半紙を使います。和文具店などで手に入りますが、手漉きの和紙を使うと、やはり一味違うもの。今回は、折形デザイン研究所の美濃和紙、「折形半紙1/2」を使いました。半紙は多少サイズに幅がありますが、ここでは折形半紙の243×343㎜を目安にしています。今回の折形は、全て折形デザイン研究所のオリジナルです。

折形には真・行・草の格があり、紙と包み方の組み合わせを、贈り物や相手に合わせて選びます。包み方は同じでも、紙を変えれば格が異なってきます。松飾りと67頁右上の祝儀包みに使用した檀紙は、皺という凸凹のある格の高い紙。年玉包みに使用した縁紅紙は、縁に赤い線が入った正方形の和紙です。赤は、彩りのアクセントに使われ、少しのぞかせたり、内側に重ねた色を「におい」といいます。赤い「におい」は、今回は民芸紙を使いましたが、他の赤い和紙でも。67頁の青の紙幣包みは、片端に赤い線が入った折形半紙を使っています。絵の具で端に線を描いてもいいでしょう。

屠蘇散包み

①横半分に切った半紙を縦に置き、下辺を上辺に合わせて折り上げます。

②右下の角を上辺に合わせて折り、左辺を右に折り返します。

③下辺を左辺に合わせて折り上げます。

④上端の4枚の紙を2枚ずつに開いて屠蘇散を入れ、右上の角を、下の角に合わせて差し込みます。

⑤赤い紙を、下の三角の上端から少し出る大きさに切って、差し込みます。

箸包み

①半紙を横半分に切って図のように置き、左上の辺を、右上の辺と平行で、右に正方形ができる位置に折ります。

②上に、右と同じ正方形ができるように、右にある2つの角を、2枚一緒に左に折ります。

③左にある2つの角が右辺に接するように折ります。

④下端を裏側に折り上げます。

⑤いったん開いて、同寸の赤い和紙を重ねて折り直します。赤い紙を最初から重ねて折ってもいいでしょう。

一个完整的水平结构将介绍性素材分成不同区块。图片和图片说明横跨整个页面，让每组图片和图片说明都清晰易读，文章中步骤说明描述的内容清楚明了

54.用插图表现时间线

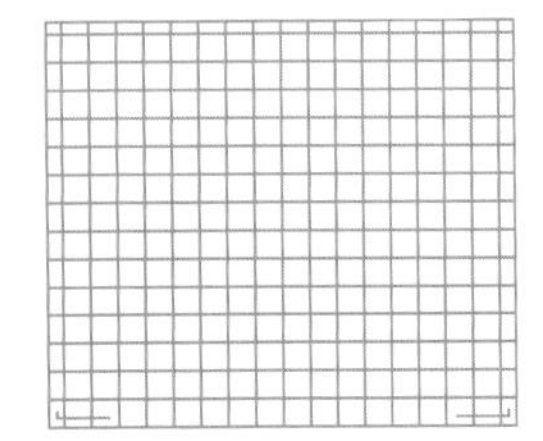

时间线不仅是一个功能性的信息，还可以代表一个人的一生或某个时期，所以设计要反映内容。

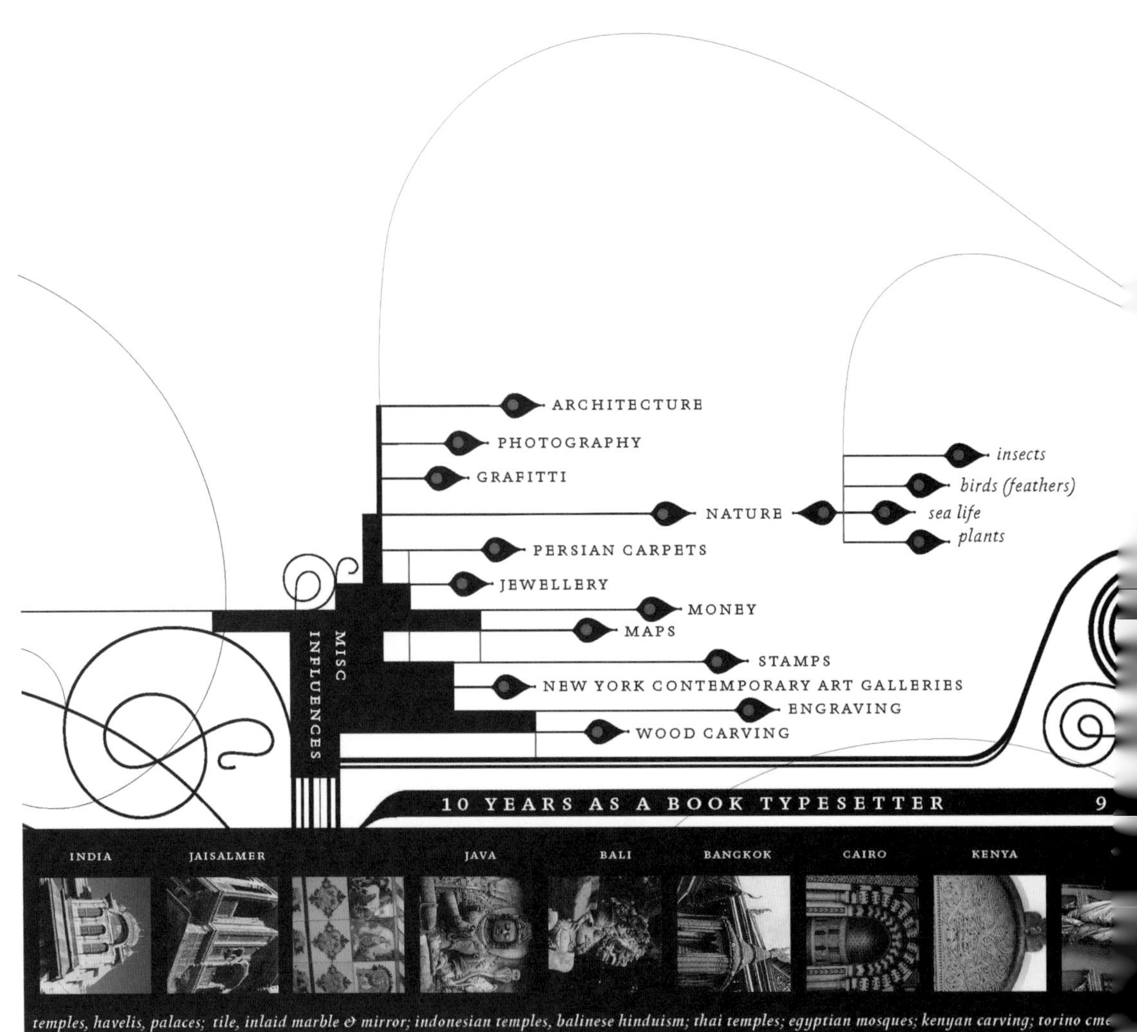

插图优雅的风格不仅源于其中的曲线，也源于粗细不同的分隔线。小标题的字母间距创造出了一种轻盈的质感，以连接符号“&”为灵感的曲线设计也很美。这个作品极富动感，精心设计的直线平衡了曲线的设计

项目
影响图

客户
玛丽安•班杰斯

设计师/插画师
玛丽安•班杰斯

在这幅玛丽安•班杰斯设计的关于影响力和艺术词汇的插图中，工艺和细节是最重要的。对艺术有影响的因素，如文化运动和流派都设计得非常明显。班杰斯从事书籍设计行业长达十年，在排版方面非常有天赋

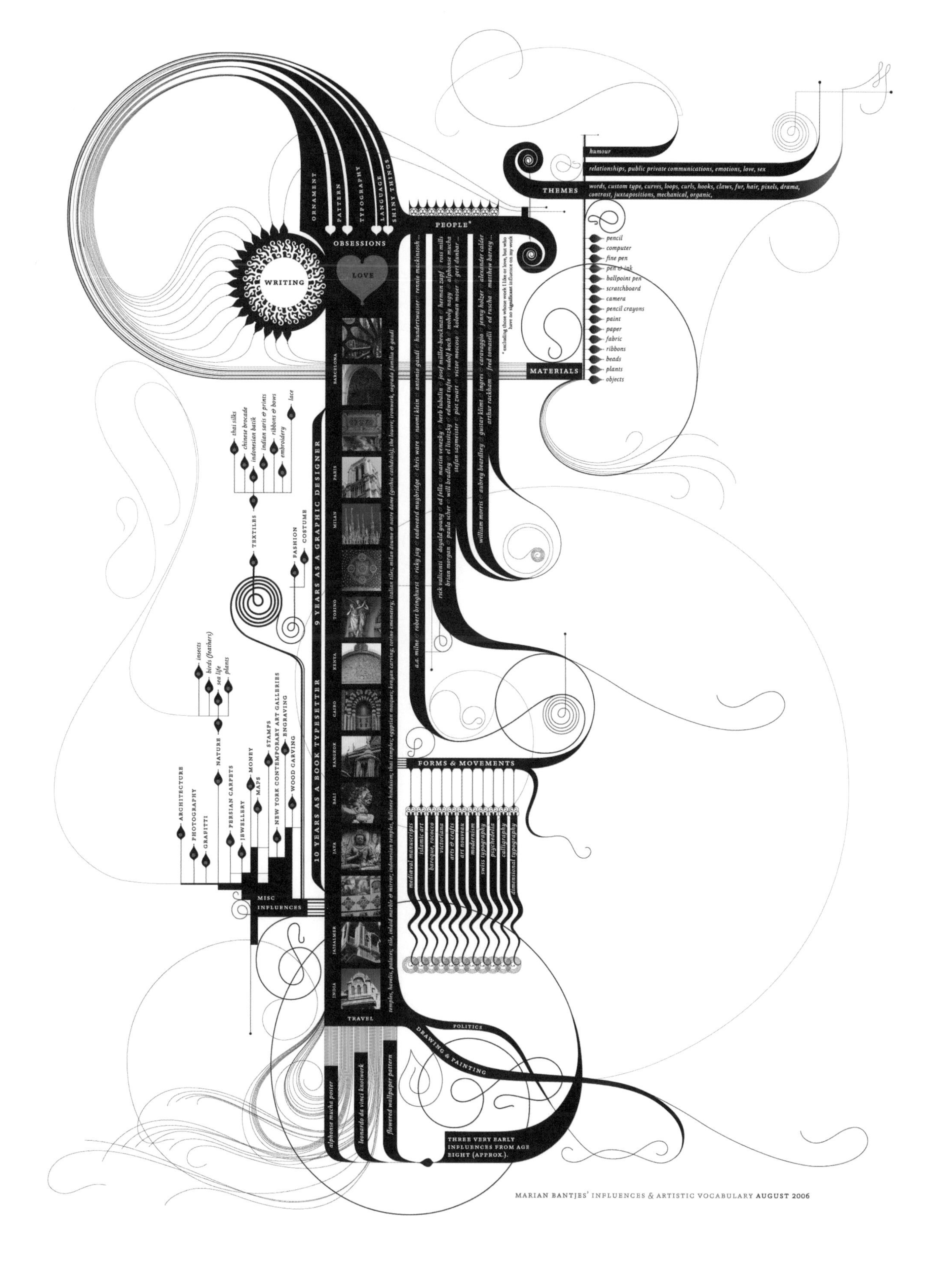

MARIAN BANTJES' INFLUENCES & ARTISTIC VOCABULARY **AUGUST 2006**

55.把导航栏当作标志

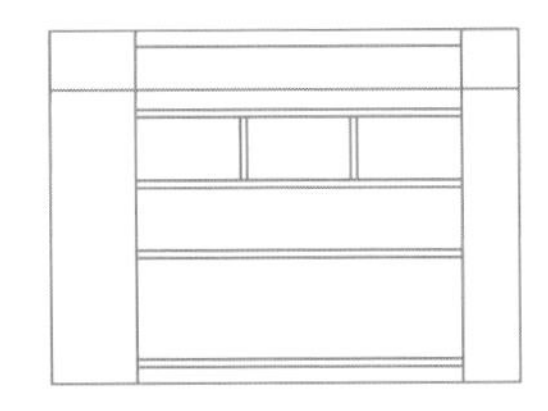

把不同项目分隔开最简单的做法就是把可利用的空间切分开。一个清晰的横栏可以作为一个标志，用来引起人们对头条内容或信息的注意。此外，在栏顶使用色彩填充可以让信息从标题中脱颖而出，创造一种具有张力的对比，阴与阳、浅色与深色、主导与从属。

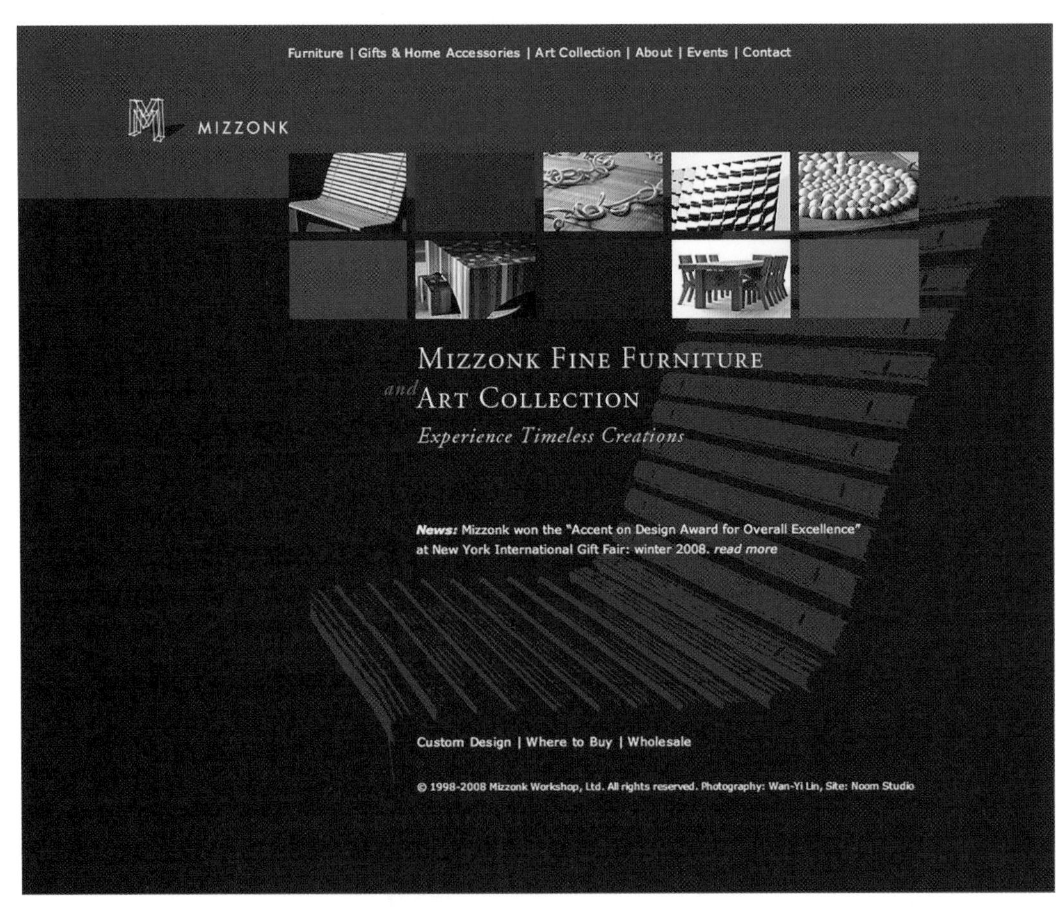

在水平结构里，读者可以从上到下浏览主页

项目
米佐克工作室（Mizzonk Workshop）网站

客户
米佐克工作室

设计
庞亚普•诺姆•其塔亚拉克（Punyapol "Noom" Kit-tayarak）

这是一个定制家具企业的网站，其特点是简洁、深沉的线条

在子页面上，导航栏仍然起到了强大的水平指向作用

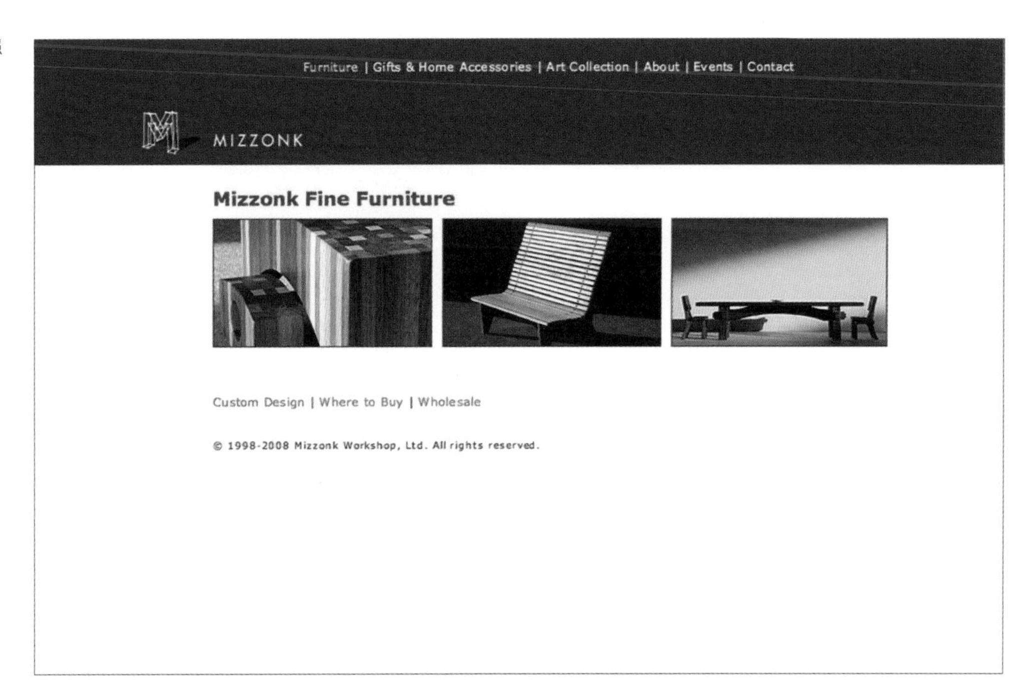

并不是所有元素的大小或与栏目底边的距离都相同。当文字下沉到栏目底部时，会产生一种流动感

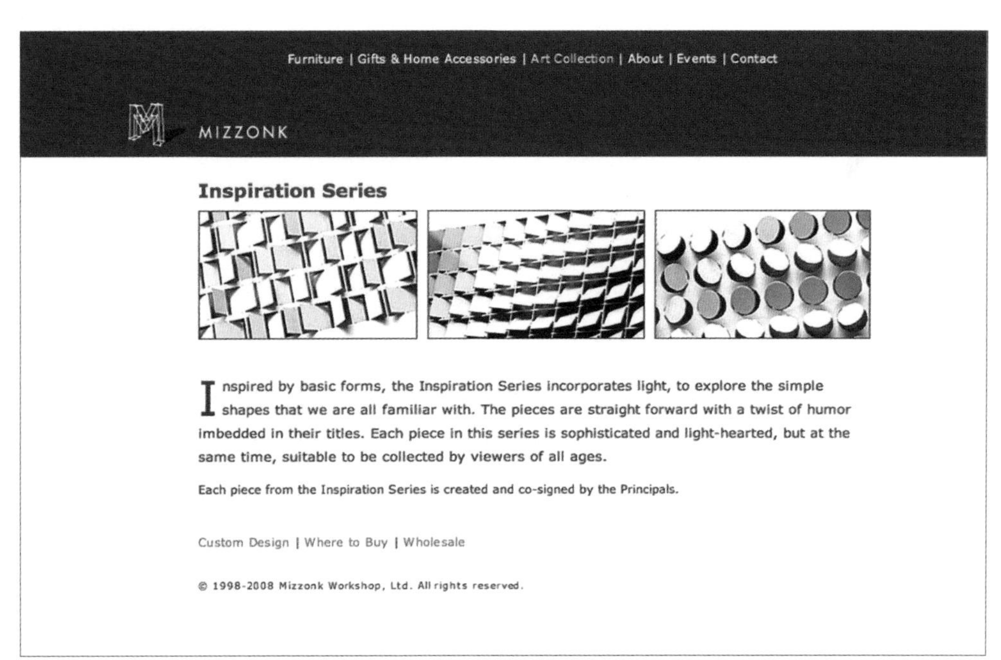

56.既要清楚又要有趣

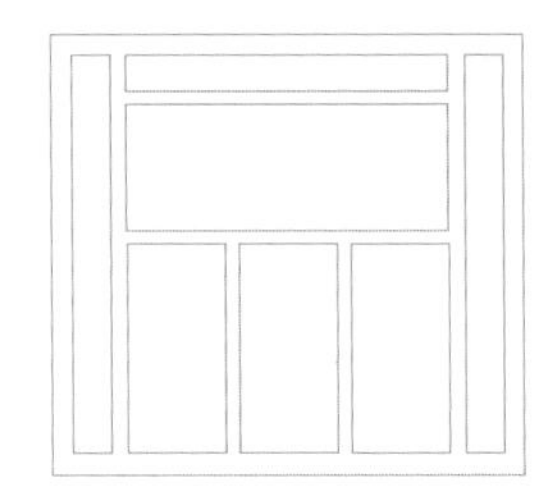

在这个案例中，标签上的信息不会让人产生歧义。两种语言和颜色都清清楚楚，并且所有读者需要知道的信息都经过精心设计，以系统的方式组织排列。这种网格的使用既轻松又有趣，也避免了把读者弄糊涂。

另见92～93页

项目
轻松达姆（Free Damm）无酒精啤酒的标志和包装

客户
达姆啤酒（Cervezas Damm）

设计公司
马里奥•埃斯肯纳兹工作室

设计师
马里奥•埃斯肯纳兹，马克•费雷尔•维维斯（Marc Ferrer Vives）

这个漂亮的包装有着严谨的分隔线和自由的风格

上页图：即使没有瓶子，标签也可以作为一张独立的海报。垫纸（上图）上使用了模块化的网格。这个标签只有两种颜色，但由于其字体大小、样式和显眼的白色字体的设计，依然具有很强烈的视觉冲击力

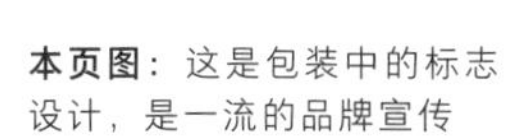
本页图：这是包装中的标志设计，是一流的品牌宣传

57.翻转一下

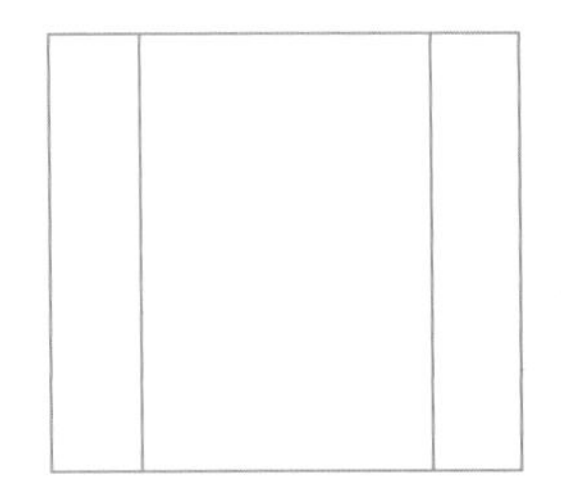

项目
《大鼻子情圣》（*Cyrano de Bergerac*）剧院海报

客户
首席制片人：苏珊•布里斯托（Susan Bristow）

设计公司
SpotCo

创意总监
盖尔•安德森（Gail Anderson）

设计师
弗兰克•加尔久洛（Frank Gargiulo）

插画师
埃德尔•罗德里格斯（Edel Rodriguez）

这则广告强调的是标题中最需要记住的部分，避免了因文字太多而使读者忽略某些部分的问题。名字选用了简洁有力的大号字体，而姓氏使用了较小的字号

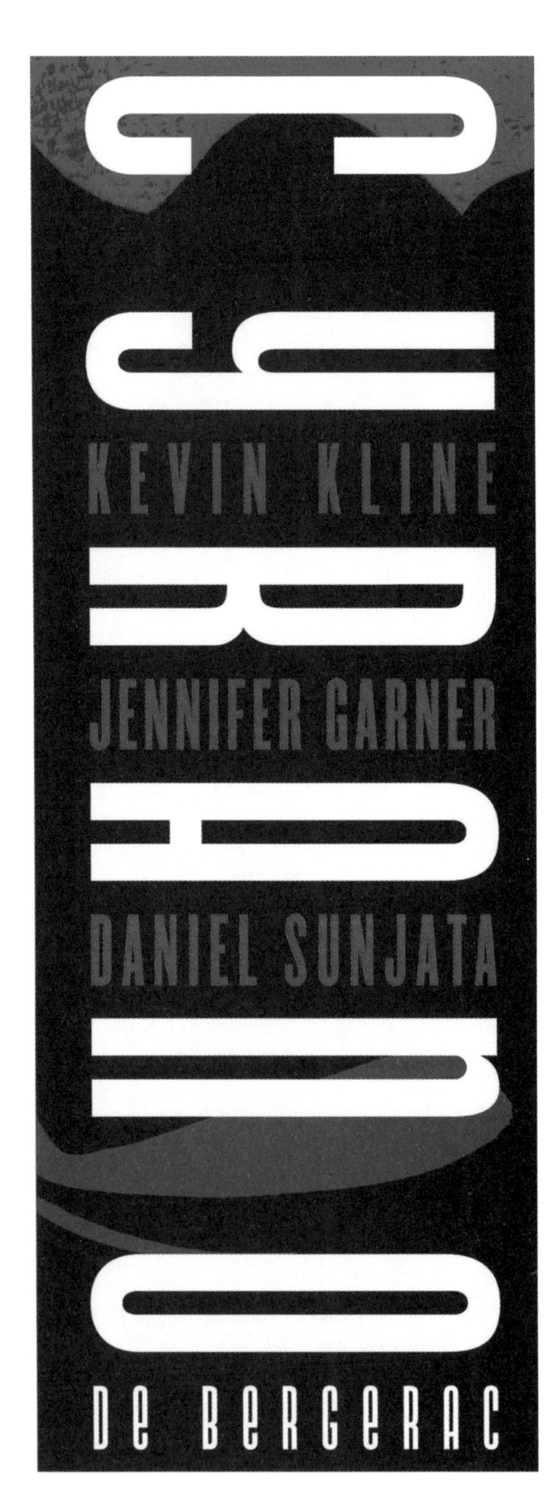

文字可以进行水平和纵向的翻转。大号字体就像一个容器，可以容纳素材中的其他信息。这个广告页里的每个名字的宽度都通过巧妙地调节字间距以及字体的大小和粗细来进行控制。

本页和下页图： 紧凑的排版和有限的色彩未必一定会造成作品呆板无趣。醒目的黑体字形成了一条中心信息栏。设计师将肖像插画和精美的版式结合，凸显了这名演出明星的特点

10 WEEKS ONLY

CYRANO DE BERGERAC

KEVIN KLINE

JENNIFER GARNER

DANIEL SUNJATA

BY EDMOND ROSTAND

TRANSLATED AND ADAPTED BY ANTHONY BURGESS

DIRECTED BY DAVID LEVEAUX

KEVIN KLINE JENNIFER GARNER DANIEL SUNJATA in CYRANO DE BERGERAC by EDMOND ROSTAND Translated and Adapted by ANTHONY BURGESS Also Starring MAX BAKER EUAN MORTON CHRIS SARANDON JOHN DOUGLAS THOMPSON CONCETTA TOMEI STEPHEN BALANTZIAN TOM BLOOM KEITH ERIC CHAPPELLE MACINTYRE DIXON DAVIS DUFFIELD AMEFIKA EL-AMIN PETER JAY FERNANDEZ KATE GUYTON GINIFER KING CARMAN LACIVITA PITER MAREK LUCAS PAPAELIAS FRED ROSE LEENYA RIDEOUT THOMAS SCHALL DANIEL STEWART SHERMAN ALEXANDER SOVRONSKY BAYLEN THOMAS NANCE WILLIAMSON Set Design by TOM PYE Costume Design by GREGORY GALE Lighting Design by DON HOLDER Sound Design by DAVID VAN TIEGHEM Hair Design by TOM WATSON Casting by JV MERCANTI Technical Supervision HUDSON THEATRICAL ASSOCIATES Press Representation BARLOW-HARTMAN Production Stage Manager MARYBETH ABEL General Management THE CHARLOTTE WILCOX COMPANY Directed by DAVID LEVEAUX

TICKETMASTER.COM or 212-307-4100/800-755-4000

GROUP SALES 212-840-3890 · RICHARD RODGERS THEATRE, 226 WEST 46TH STREET

58.保持干净

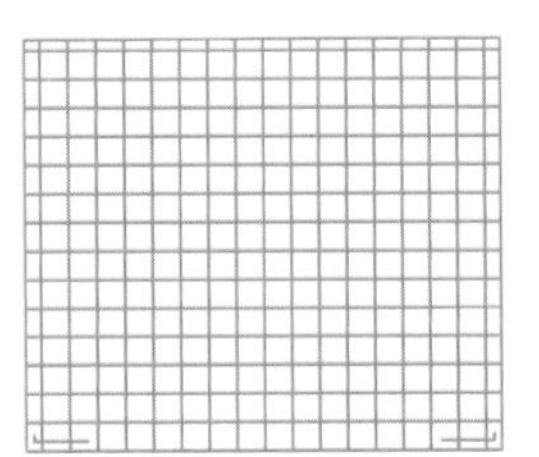

在网格中设计的字体和用于网格中的字体需要清新干净。但是，字母也可以环绕成曲线，或变成其他形式。

项目
维多利亚啤酒（Cerrezas Victoria）字体设计

客户
维多利亚啤酒公司

设计公司
马里奥•埃斯肯纳兹工作室

设计师
马里奥•埃斯肯纳兹，达尼•鲁比奥（Dani Rubio），马克•费雷尔•维维斯

这个字体表设计于2017到2018年间，在西班牙马拉加举办的庆祝维多利亚啤酒厂建立90周年纪念展览上用作瓷砖上的图案，也用于促销和广告宣传活动

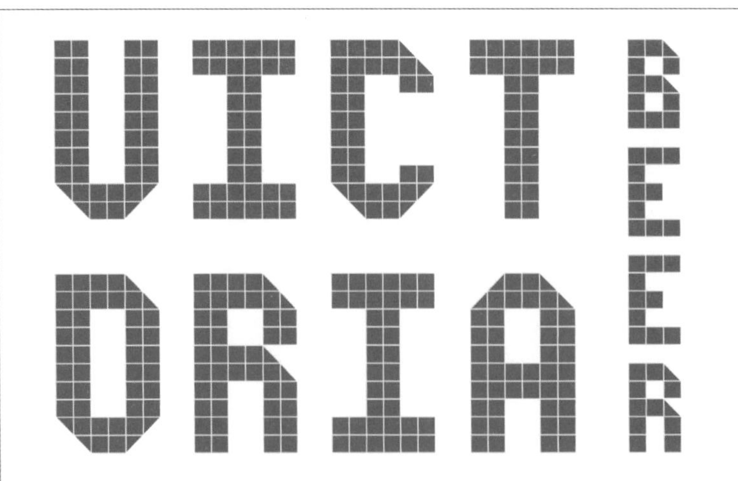

维多利亚啤酒的这个字体设计不仅是为了纪念啤酒厂建立90周年，也是为了庆祝该厂在离开西班牙马拉加市20年后又回到该地区。“Malagueña y exquisita”的意思是“马拉加精品”

MALAGUEÑA
Y EXQUISITA

MAL
AGA

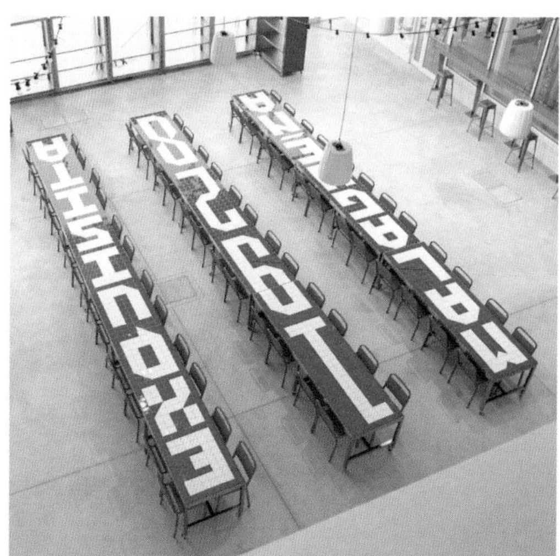

59.玩转网格

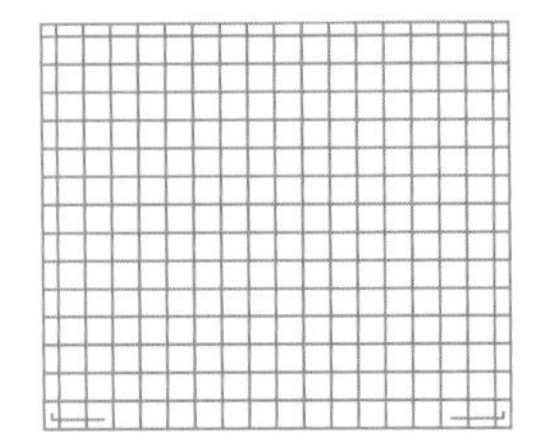

与爵士乐一样，版式也具有节奏。即使在一个紧凑的、精心设计的网格中，也可以通过改变字体的宽度、粗细和位置来“举办一场即兴爵士乐表演”。下一步就是看看当你把所有东西都翻转过来时会发生什么。

小号与大号无衬线字体相互依托，使这些文字具有很强的动感，叠印在背面的双层人物剪影上，文字就这样“摇摆”了起来

项目
林肯中心爵士乐团广告和宣传品

客户
林肯中心爵士乐团

设计公司
JALC设计部

设计师
博比•马丁

林肯中心爵士乐团宣传品的色彩明亮，井然有序，充满活力。广告的设计干净，具有瑞士风格，经过恰当的切分后显得非常酷

框中使用了不同大小和粗细的白色文字，与裁剪巧妙的图片形成强烈且富有韵律的对比

60.让读者参与其中

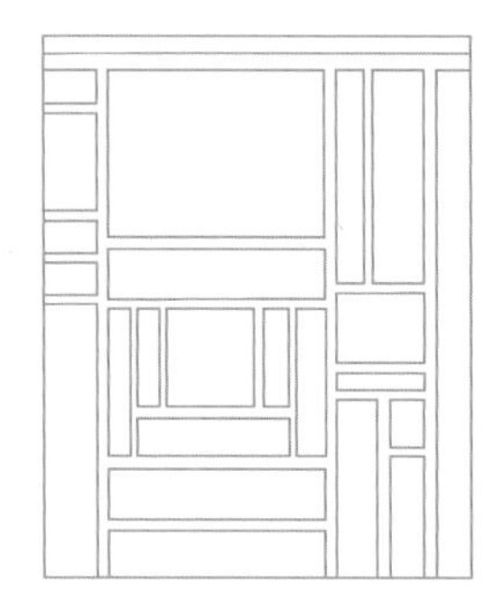

有时，网格需要超出它本身。字体的大小、形状以及粗细可以传递地域文化和全球文化，引起读者的兴趣，并呼吁人们行动起来。

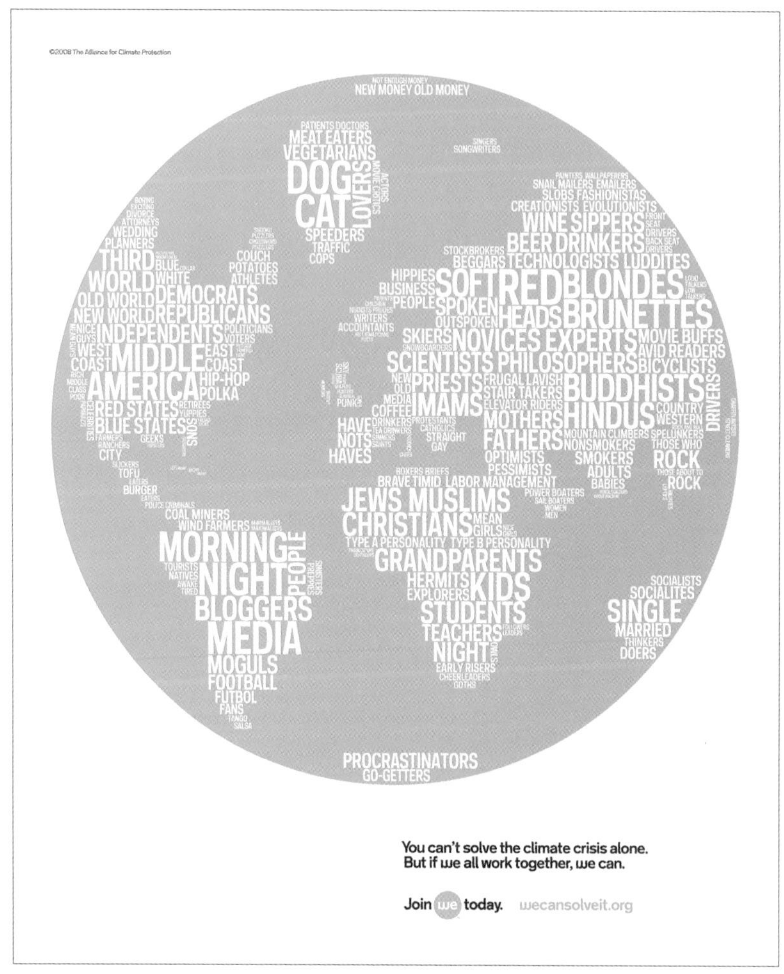

单词字号的大小是由特定的统计数据决定的。大一点儿的文字可以吸引眼球，而小一点儿和细一点儿的文字在视觉上可以起到黏合作用。对于呼吁环境保护的广告来说，明亮的绿色是最佳选择

项目
气候保护联盟广告

客户
WeCanDoSolveIt.org网站

设计公司
马丁公司（The Martin Agency），科林斯公司（Collins）

设计师
马丁公司：迈克•休斯（Mike Hughes），肖恩•莱利（Sean Riley），雷蒙德•麦金尼（Raymond Mckinney），泰•哈珀（Ty Harper）；科林斯公司：布莱恩•科林斯（Brian Collins），约翰•穆恩（John Moon），迈克尔•潘吉里南（Michael Pangilinan）

这则倡议环保的广告利用了大胆的版式来强调观点

NOT ENOUGH MONEY
NEW MONEY OLD MONEY
PATIENTS DOCTORS
MEAT EATERS
VEGETARIANS
DOG
CAT
LOVERS
MOVIE CRITICS
ACTORS
SPEEDERS
TRAFFIC
COPS
SINGERS
SONGWRITERS
PAINTERS WALLPAPERERS
SNAIL MAILERS EMAILERS
SLOBS FASHIONISTAS
CREATIONISTS EVOLUTIONISTS
WINE SIPPERS
FRONT SEAT DRIVERS
BEER DRINKERS
BACK SEAT DRIVERS
STOCKBROKERS
BEGGARS TECHNOLOGISTS LUDDITES
BORING
EXCITING
DIVORCE
ATTORNEYS
WEDDING
PLANNERS
THIRD
WORLD
BLUE
WHITE
COLLAR
SUDOKU PUZZLERS
CROSSWORD PUZZLERS
COUCH
POTATOES
ATHLETES
DEMOCRATS
REPUBLICANS
OLD WORLD
NEW WORLD
NICE GUYS
MEAN GUYS
INDEPENDENTS
POLITICIANS
VOTERS
RED SOX
YANKEES
FANS
WEST
COAST
MIDDLE
EAST
COAST
RICH
MIDDLE
CLASS
POOR
AMERICA
HIP-HOP
POLKA
CELEBRITIES
PAPARAZZI
RED STATES
RETIREES
YUPPIES
FREAKS
JOCKS
BLUE STATES
SONS
FARMERS
RANCHERS
GEEKS
HIPSTERS
CITY
SLICKERS
TOFU
EATERS
BURGER
EATERS
POLICE CRIMINALS
COAL MINERS
WIND FARMERS
MINIMALISTS
MAXIMALISTS
MORNING
NIGHT
PEOPLE
TOURISTS
NATIVES
AWAKE
TIRED
SINISTERS
PREPPIES
BLOGGERS
MEDIA
MOGULS
FOOTBALL
FUTBOL
FANS
TANGO
SALSA
HIPPIES
BUSINESS
PEOPLE
PARENTS
CHILDREN
NUDISTS PRUDES
WRITERS
ACCOUNTANTS
MATHEMATICIANS
POETS
SOFT
RED
BLONDES
SPOKEN
OUTSPOKEN
HEADS
BRUNETTES
LOUD TALKERS
LOW TALKERS
SKIERS
SNOWBOARDERS
NOVICES EXPERTS
MOVIE BUFFS
AVID READERS
SCIENTISTS PHILOSOPHERS
BICYCLISTS
SKY
SCUBA
GOLFERS
HUNTERS
CLASSICAL
PUNK
FANS
NEW
OLD
MEDIA
PRIESTS
FRUGAL LAVISH
STAIR TAKERS
ELEVATOR RIDERS
BUDDHISTS
DRIVERS
IMAMS
COFFEE
DRINKERS
TEA DRINKERS
PROTESTANTS
CATHOLICS
MOTHERS
HINDUS
COUNTRY
WESTERN
ROCK AND ROLL
HAVE
NOTS
SINNERS
SAINTS
STRAIGHT
GAY
HAVES
CHEFS
FATHERS
MOUNTAIN CLIMBERS
SPELUNKERS
NONSMOKERS
THOSE WHO
OPTIMISTS
SMOKERS
ROCK
PESSIMISTS
ADULTS
THOSE ABOUT TO
ROCK
BABIES
BOXERS BRIEFS
BRAVE TIMID LABOR MANAGEMENT
POWER BOATERS
SAIL BOATERS
WOMEN
MEN
JEWS MUSLIMS
CHRISTIANS
MEAN
GIRLS
NICE GIRLS
TYPE A PERSONALITY TYPE B PERSONALITY
PROSECUTORS
DEFENDERS
GRANDPARENTS
HERMITS
EXPLORERS
KIDS
STUDENTS
TEACHERS
FOLLOWERS
LEADERS
NIGHT
OWLS
EARLY RISERS
CHEERLEADERS
GOTHS
SOCIALISTS
SOCIALITES
SINGLE
MARRIED
THINKERS
DOERS
GRAFFITI ARTISTS
STREET CLEANERS
PROCRASTINATORS
GO-GETTERS

● 61.掌控紧凑的空间

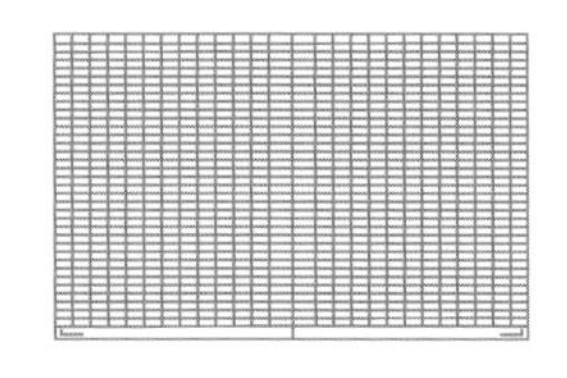

在一个构思精巧的网格内，窄边距也可以发挥大作用。当图片沿着明显的网格线对齐，并且留白与版式又得到了精心的设计时，窄的页边距可以成为别具匠心的点缀。在一个均衡的页面中，可以利用排版技巧和规则形成窄的页边距，为精心设计的页面带来不一样的效果。在开始设计时，便要留出调整的余地。无论是对初学者还是对经验丰富的从业者来说，边距都是不容易把握的。建立一个可以变化的网格系统涉及平衡与技巧，还需要反复练习与试错。大多数传统的胶版印刷商和出版商往往不愿意留过小的边距，因为小的页边距没有可供“跳跃”的空间（“跳跃”是指成卷的印刷纸在快速通过印刷机时产生的微小移动）。因此，出版公司的设计师通常会留出充足的页边距。

Typographie et Architecture

«que se passe-t-il quand le **texte sort de l'échelle habituelle** de l'imprimé ou de l'écran et va se confronter à notre corps, prendre **une importance physique**, quand il veut prendre une autre place et se développe dans la profondeur **comme une sculpture**?» **par presse papier** ■ ■ ■

«c'est quoi, finalement, la vraie forme d'un "M"?»

48 : 2.2008

Avec la "Façade aux mille lettres" (aussi appelée Moucharabieh typographique polyglotte) du musée des écritures du monde, à Figeac, et le "T" du tramway de Nice, Pierre di Sciullo vient de livrer deux nouveaux travaux qui font la part belle à la typographie dans l'espace public... L'occasion de revenir sur quelques-uns des travaux de ce graphiste, typographe, dessinateur de caractères, et de questionner avec lui les rapports entre typographie et architecture.Nous voici donc à la gare de Gretz-Armainvilliers, par un froid dimanche de décembre. Pierre di Sciullo, un béret noir – plutôt inattendu sous ces latitudes – vissé sur la tête, nous attend au bas de la passerelle métallique, pour nous conduire, tout près d'ici, à l'atelier qu'il anime avec Catherine di Sciullo et Juliette Cheval. Une discussion s'installe

Marie Bruneau (MB): Pierre, finalement, quel est ton métier? graphiste? typographe?
Pierre di Sciullo (PdS): J'ai reçu, à ce propos, une leçon dont je me souviendrai toute ma vie. Assez intimidé, j'avais pris mon élan pour appeler Adrian Frutiger: «Bonjour, je suis typographe, et j'aimerais bien venir vous voir.» Il me répond aussitôt, presque cassant: «Dans ce cas, jeune homme, vous n'êtes peut être pas avec la bonne personne: vous auriez dû rencontrer Emil Ruder; malheureusement, il n'est plus de ce monde. En ce qui me concerne, je ne suis que dessinateur de lettres.» J'ai vraiment compris ce jour-là la différence qui existe entre les dessinateurs de caractères qui créent des alphabets, les typographes qui mettent en forme le texte dans la page (avec généralement des caractères créés par d'autres), et les graphistes, qui sont souvent – mais pas toujours – typographes. En ce qui me concerne, je coche les trois cases... Ce qui explique qu'en tant que dessinateur de caractères, j'aie parfois tendance à être un peu expéditif! C'est un travail extrêmement minutieux, et long. Dans ce que j'appelle ma "boîte à outils", il y a de nombreux caractères qui ne sont pas encore complètement développés...
MB: Comment t'es-tu intéressé à la typographie?
PdS: Je situe le point de non-retour au numéro 5 de Qui? Résiste, le "Manuel de la femme", publié en mai 1985. C'est à ce moment-là que je me suis posé des questions du type: «C'est quoi, finalement, la vraie forme d'un "M"?». Rien de ce que je voyais ne me satisfaisait vraiment, et j'en suis venu à dessiner les titres à la main, avec une règle et un compas... et à composer le mot "femme" avec deux "M" (deux "aime"?) dissymétriques, et des "E" capitales à quatre barres horizontales, juste avant que The Face ne publie ceux de Neville Brody... (J'en suis assez fier même si je suis à peu près sûr qu'on trouve ça cinquante ans plus tôt, chez Kurt Schwitters!). Plus tard Adrian Frutiger m'a expliqué qu'il ne fallait surtout pas dessiner des lettres avec une règle et un compas; il m'a remis un petit livre bourré

2.2008 : 49

项目
《阶段》杂志

客户
金字塔/《阶段》杂志

设计师
安娜·图尼克

这本法国设计杂志有着清晰的网格系统，展现了一种秩序感。窄边距成为设计的一部分，可以让页面容纳更多信息

这个排列清晰，具有平衡感的页面充分展示了网格的灵活性。所有的元素都对齐排列，大号字体带来了动感。页面中的留白与外围的窄边距形成对比，效果极佳。排版设计均衡，不同粗细、大小和颜色的字体放在一起十分和谐

LA FAÇADE AUX 1000 LETTRES
Figeac, 2007
architecte
Moatti & Rivière
scénographie
Pascal Payeur
aussi appelée moucharabieh typographique polyglotte pour le musée des écritures du monde.

在这个对页中，所有元素都整齐排列，窄的边距与图片之间的空间是相互呼应的

programme d'identité visuelle, ni la mise en forme des supports de communication?
PdS: Non. Ce qui m'a valu une belle déception! Je savais que cela ne faisait pas partie de la commande initiale, mais j'ai bien vite vu que mon travail risquait de remettre en question l'identité du CND. Au moment de l'inauguration du bâtiment, une consultation pour le graphisme a été lancée... Mais personne n'a pris la peine de me prévenir. Cela m'a d'autant plus surpris que le directeur du CND, s'était révélé un maître d'ouvrage remarquable.
MB: À Figeac, on est dans le même cas de figure?
PdS: Non, l'histoire est très différente... mais on retrouve, de la part de la maîtrise d'ouvrage, la même incohérence flagrante, la même incompréhension des enjeux... Au bout du compte, ici aussi, les supports de communication ont été traités dans une logique qui ignore l'esprit du travail mené sur le bâtiment.
BG: Voilà qui pose clairement la question du statut du graphisme: quand tu interviens dans le cadre du 1 %, ou de la commande publique artistique, on te situe dans le champ de l'art, on te reconnaît en tant qu'auteur et créateur de formes. Mais par contre, quand il est question d'identité graphique, les commanditaires pensent encore que c'est du ressort de "la communication"...
PdS: ...Et on change de camp! Si on prête bien attention au sens des mots employés dans les différents contrats qu'on te fait signer, c'est très clair. Je prends l'exemple du musée Champollion, à Figeac. Le concours portait exclusivement sur l'architecture et la scénographie. Pascal Payeur, le scénographe, a convaincu facilement Alain Moatti, l'architecte, qu'il n'était pas idiot, pour un musée de l'écriture, de faire équipe avec un graphiste passionné par l'écriture. Voilà comment nous nous sommes retrouvés à trois sur ce concours. Dans notre projet, la "Façade aux mille lettres" est l'un des éléments qui a intéressé le jury. Après que nous ayons été choisis, j'ai signé une convention avec la ville de Figeac pour la façade, dans laquelle je suis désigné comme "l'artiste". Par contre, pour traiter le graphisme à l'intérieur du musée, je suis sous-traitant du scénographe, et l'attitude de la maîtrise d'ouvrage a été différente.
BG: Ton intervention dans l'exposition permanente était-elle prévue dès le stade du concours?
PdS: Oui, puisque la première question à résoudre était "comment intégrer le programme dans le futur bâtiment, quelle visite proposer"? Les deux domaines sont devenus plus tard autonomes: j'ai travaillé la façade avec l'architecte, et l'intérieur avec le scénographe. Naturellement, nous avions régulièrement des réunions de mise en commun.
BG: Si on revient sur ces trois projets, après "Santé", qui est un gros mot accroché sur une façade, "Danse" qui joue dans le registre d'une enseigne sur un bâtiment, on est, à Figeac, devant une façade qui ne nomme rien, mais qui exprime pourtant la nature du musée qui est derrière. Voilà une autre façon d'intégrer de l'écriture dans l'architecture...
PsS. Oui, et même dans la "chair" du bâtiment. Et j'aurais d'ailleurs envie d'aller encore plus loin dans ce sens. J'aimerai travailler avec des architectes, vraiment dès le début de la conception d'un bâtiment.
MB: C'est souvent une question de rencontre, et de désir réciproque...
PdS: Il y a "l'homme lettres", de Geoffroy Tory, j'aimerais faire "la maison lettres", ou plutôt "la maison texte".
BG: Les mots sont plus importants que les lettres?
PdS: Le problème, quand on veut inscrire des textes sur une architecture, c'est qu'on est parfois limité par des questions économiques: il faut des textes très courts. "Santé", c'est un texte de cinq lettres... Les (pas tout à fait) mille lettres de la façade du musée Champollion ne font pas un texte: c'est un texte potentiel, même si, en réalité, il y a des

网格（12栏）的架构允许某些栏目没有填充信息，以平衡窄边距，并为内容丰富的对页留出呼吸的空间

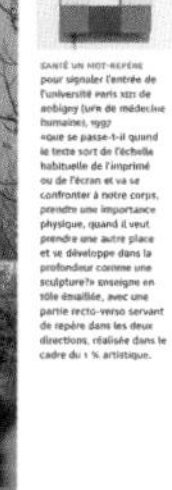

SANTÉ UN MOT-REPÈRE
pour signaler l'entrée de l'université Paris XIII de Bobigny (UFR de médecine humaine), 1997
«Que se passe-t-il quand le texte sort de l'échelle habituelle de l'imprimé ou de l'écran et va se confronter à notre corps, prendre une importance physique, quand il veut prendre une autre place et se développe dans la profondeur comme une sculpture?» enseigne en tôle émaillée, avec une partie recto-verso servant de repère dans les deux directions, réalisée dans le cadre du 1 % artistique.

Quand on suit le tracé du tram, le bâti change complètement d'échelle: parfois ça s'élargit et se grandit, ailleurs ça se rapproche et se tasse... Tout d'un coup, la ville s'écarte, le paysage se révèle: les collines et les quartiers nord apparaissent.
MB: Tu as donc choisi de travailler un "T", en volume...
PdS: Comme il s'agit d'une série de vingt enseignes identiques, je voulais éviter la monotonie. Le "T", c'est un objet toujours différent, suivant qu'on le découvre sur sa face ocre rouge, sa face bleue, ou de profil... Et c'est un signe dont la lecture est simple, et en même temps un peu mystérieuse. Si on arrive par une rue en pente – ou depuis la fenêtre d'un bâtiment – on comprend que c'est un "T" de profil: trois pas plus loin, ça disparaît... Je voulais aussi que le paysage traverse l'enseigne. J'ai donc choisi ces deux couleurs: l'ocre rouge – Nice est une ville italienne – et un bleu très dense, pour permettre justement de lire le bleu du ciel.
BG: Tu as pensé aux signes du métro?
PdS: Évidemment! Celui de Paris, avec notamment la très belle enseigne extérieure de la station "Sentier", et celui de Londres. Mais l'enjeu n'est évidemment pas le même: dans le cas du métro, il faut faire savoir que quelque chose se passe là, sous le trottoir; alors que dans le cas du tram, il faudrait être vraiment dans la lune pour ne pas voir les rails, ou la station! C'est ce qui m'excitait dans ce projet: voilà une intervention dégagée de toute fonction signalétique, qui offrait la possibilité de construire un signe poétique – ce qui ne veut pas dire qu'on ne puisse pas faire de la poésie en signalétique...
BG: Oui, mais tu as quand même fait un "T"!
PdS: Oui. J'ai fait un "T". D'ailleurs, j'aimerais bien qu'une autre ville me propose encore de faire une enseigne pour son tramway: je serais obligé de faire un autre "T"... ou autre chose qu'un "T"!
BG: Voilà qui illustre peut-être la relation entre typographie et architecture: depuis l'Antiquité – et notamment à Rome –, on grave des inscriptions, en capitales monumentales, sur les monuments, souvent en accompagnement d'un programme de sculpture. Et on retouve des inscriptions et des allégories tout au long de l'histoire... Au début du 20e siècle, l'Art Déco utilise encore largement la typographie dans son vocabulaire formel. Mais petit à petit, et surtout à partir du Mouvement moderne, on évacue tout recours à la décoration, et on en vient même à éliminer toute inscription sur les bâtiments: à partir des années 1980, plus personne n'écrit "Liberté, Égalité, Fraternité" sur les façades des mairies... J'ai le sentiment que les choses sont aujourd'hui en train de changer: un dialogue est en train de se renouer entre graphistes et architectes,

DANSE POUR LE CENTRE NATIONAL DE LA DANSE
Pantin, 2004,
réalisé dans le cadre de la commande publique et du 1 % artistique.
Typographie:
Minimum plancher (retravaillé)
10,00 m x 5,50 m,
aluminium, néons

de ses conseils: j'en ai suivi certains, j'ai soigneusement évité d'en appliquer d'autres...
Bertrand Genier (BG): Le dessin des lettres est donc, à un certain moment, devenu le centre de ton travail?
PdS: Oui, dans les années 1988-89. Une époque très fertile: j'avais l'impression d'avoir ouvert une porte, et de me trouver dans un paysage immense, infini. Chaque jour me venait une nouvelle idée. Je ne savais rien sur les lettres. L'une de mes amies, Maria Arnold, m'a conseillé (ordonné?) de lire le Typographie d'Emil Ruder. J'ai été à la fois frappé par la beauté de l'ouvrage, touché par certains éléments de son contenu, et choqué par plusieurs de ses affirmations péremptoires, dogmatiques même...
BG: Les représentants du Mouvement moderne avaient des certitudes – que nous avons perdues...
PdS: Bien sûr! Ils ont pensé: «Nous allons rendre la lecture plus facile, développer une esthétique qui pourra améliorer la vie des gens, mettre de l'ordre dans le monde...» Ce n'est probablement pas pour rien si Adrian Frutiger nomme un de ses caractères l'Univers: une façon d'affirmer, sans rire, avoir défini une fois pour toute le bon rapport de hauteur entre les majuscules et les minuscules, en tenant compte de la fréquence d'utilisation des capitales pour les langues européennes...
BG: Attitude typiquement moderne!
PdS: Pour vérifier le dessin de l'Univers, il m'a raconté avoir choisi un caractère important par siècle (Baskerville, Garamond, etc.), et superposé leurs dessins, au moyen de hachures. D'après vous, quelle est la résultante de ces croisements de formes, de ces fantômes d'intersections? L'Univers, naturellement! L'Univers, comme aboutissement – point final? – de toute la tradition occidentale de dessin de l'alphabet. Cqfd!
BG: On retrouve la même chose en architecture: pour Le Corbusier, par exemple, c'est le Modulor qui est censé résoudre tous les problèmes... Essayons justement, après ces considérations typographiques, de nous tourner un peu vers l'architecture... Peux-tu nous dire quelle est la première commande qui t'a donné l'occasion d'intervenir sur un bâtiment?
PdS: C'était à Bobigny. Mais la première fois où j'ai pu installer des lettres "en grand" – des "gros mots" à l'échelle du corps – c'était en 1995, à Maastricht, quand j'ai reçu le prix Charles Nypels1. J'y ai réalisé une installation avec les lettres du mot "Secret". C'est une chose que j'avais envie de faire depuis longtemps et j'ai pu vérifier alors mon pressentiment: il y a bien une émotion particulière à se trouver face à une écriture monumentale, dans un contact très rapproché. Je doutais fortement du dogme, très Mouvement moderne, qui institue la neutralité du signe et des caractères: «L'écriture est impalpable, invisible... elle appartient au domaine de l'abstraction, parce que son humble et noble mission, c'est de relayer la pensée.» Tout cela vient sans doute de la tradition humaniste: les dessinateurs de lettres étaient d'abord des libraires, des éditeurs, et surtout des amoureux du texte. Pour eux, il ne s'agissait surtout pas "d'interférer", mais de créer des caractères adaptés à la transmission des œuvres des penseurs de leur époque. Idée notamment affirmée avec force par Emil Ruder. Cette histoire de "neutralité" m'a tracassé: la neutralité a évidemment un sens politique... Les caractères suisses sont aussi "neutres" que l'était la Suisse durant la dernière guerre mondiale!
MB: Écrire "en grand" surprend, devient donc un événement en soi...
PdS: Ça fait aussi penser à de l'enseigne... J'ai à ce propos une anecdote qui illustre bien le trouble que fait parfois naître mon travail: nous attendions l'autorisation administrative de poser le mot "Danse" sur le toit du Centre national de la danse, à Pantin. Les mois passaient... à tel point que j'ai vu le moment où l'on allait inaugurer le bâtiment sans mon travail. Pourquoi? Parce que l'on ne parvenait pas à s'accorder sur sa dénomination exacte: "sculpture typographique" ou "enseigne"? Du point de vue administratif, il s'agissait dans le premier cas d'un objet culturel, alors que dans le second, il avait des connotations commerciales. À mon avis, les deux mots sont justes: aucun risque si c'est une enseigne, qu'on la confonde avec celle de Carrefour!
MB: Ce projet entrait dans le cadre du 1 % artistique?

j'ai vu le moment où l'on allait inaugurer le bâtiment sans mon travail. Pourquoi? Parce que l'on ne parvenait pas à s'accorder sur sa dénomination exacte: "sculpture typographique" ou "enseigne"?

PdS: Oui, comme le mot "Santé", à Bobigny, et la "Façade aux mille lettres", du musée des écritures du monde, à Figeac...
MB: Dans les trois cas, tu as inscrit des mots dans l'architecture; et dans les trois cas, ce qui est défini comme relevant du domaine de l'artistique assume finalement une certaine fonctionnalité...
PdS: Le décret qui définit le cadre d'application du 1 % artistique précise bien que ce type de commande peut être passé à des graphistes, des paysagistes, etc. Ce qui a donné lieu à certaines polémiques: est-il normal, par exemple, de réaliser, dans ce cadre, la signalétique d'une préfecture? En ce qui concerne le mot "Santé", à Bobigny, il n'était nullement demandé d'écrire quelque chose: le cahier des charges signifiait simplement qu'il fallait trouver un moyen d'indiquer l'entrée du bâtiment.
MB: Oui, mais tu as proposé ce mot...
PdS: Le bâtiment en question est situé juste en face de l'hôpital Avicenne; il abrite une "UFR de médecine humaine". J'ai proposé d'accrocher le mot "santé" en débord sur la façade, comme un repère. Ensuite, j'ai changé la plaque d'information placée au-dessus de la porte d'entrée: elle était trois fois trop grande, et comme on n'avait pas le recul nécessaire pour la lire, les gens ne trouvaient pas leur chemin!
BG: En ce qui concerne le Centre national de la danse, tu as également traité le programme de signalétique?
PdS: Oui, mais les deux études sont séparées: c'est un autre programme et un autre financement.
MB: Par contre, on ne t'a pas confié le

LE CENTRE NATIONAL DE LA DANSE
s'est installé à Pantin, dans le bâtiment de l'ancienne cité administrative construit par Jacques Kalisz en 1972, et reconverti par les architectes Antoinette Robain et Claire Guieysse en 2004. «Écrire le mot danse parce que c'est ce qui se passe à l'intérieur du bâtiment, parce qu'il n'y a pas de raison de se priver de la simplicité et de la clarté de ce mot, parce qu'à cette échelle et sur ce bâtiment, cela résonne aussi pour les passants comme une injonction à l'impératif présent, décalée des interpellations publicitaires ou institutionnelles.»

62.借助终端陈述观点

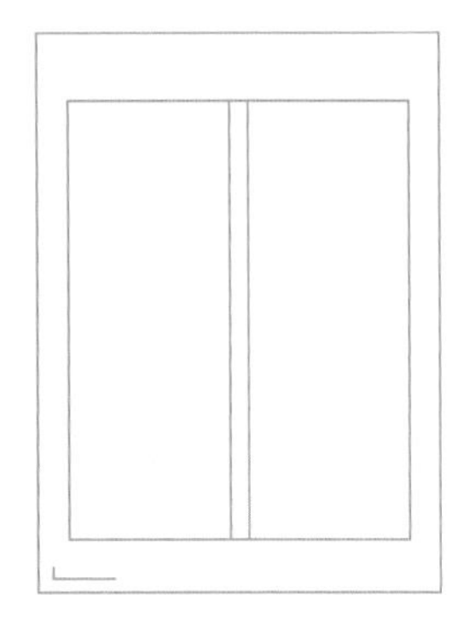

有些内容主题涉及很多细节和复杂的问题。当需要将大量丰富的信息装入有限的空间时，需要借助一些技巧来突出主要观点。比如，醒目的版头、有色条的侧边栏、项目符号、引人注意的图标，以及彩色的标题和文本。

每页的版头都展示了一个完整的标志系统，与本页话题相关的标志突出显示，成为每个段落的标记

项目
公共事件的材料和展示

客户
哥伦比亚大学地球研究所

创意总监
马克•英格利斯

设计师
金圣熙（Sunghee Kim）

这些复杂而详细的教育展览材料综合运用了完整的图标和色彩系统来表示每个章节要讨论的问题。各种各样的设计元素，如图标、标题、文本、图片和图形区分了各个章节，使信息更易于查找，也实现了既定的目标——空间布局得当、结构清晰、色彩鲜明、组织有序、黑白对比恰当、文字易读。页面中使用了很多有教育意义的图表，简洁利落的对齐方式可以让使用者浏览时更便利，不会产生混乱

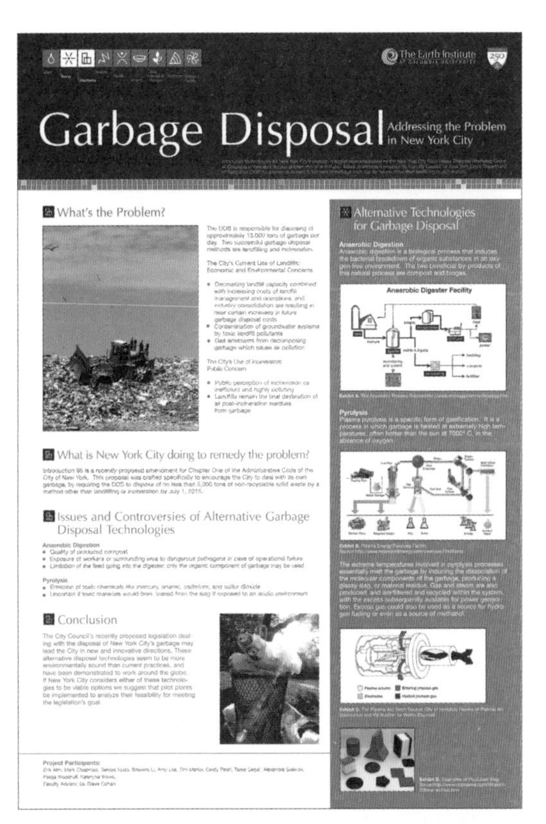

整个版式经过精心设计，使用连贯的黑色带作为版头的底色贯穿所有页面。版头包含的信息有图标系统、哥伦比亚大学地球研究所的徽标、标题和副标题等

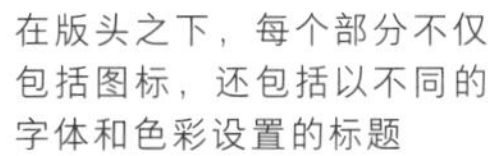

在版头之下，每个部分不仅包括图标，还包括以不同的字体和色彩设置的标题

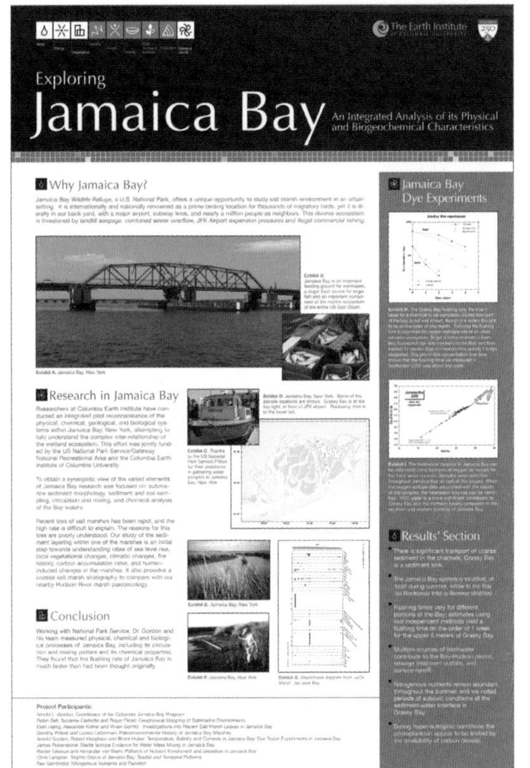

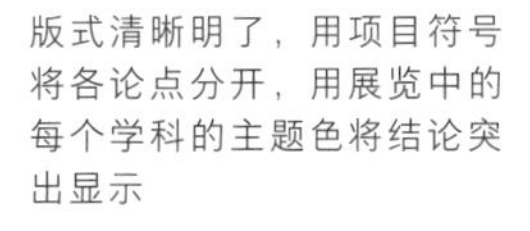

版式清晰明了，用项目符号将各论点分开，用展览中的每个学科的主题色将结论突出显示

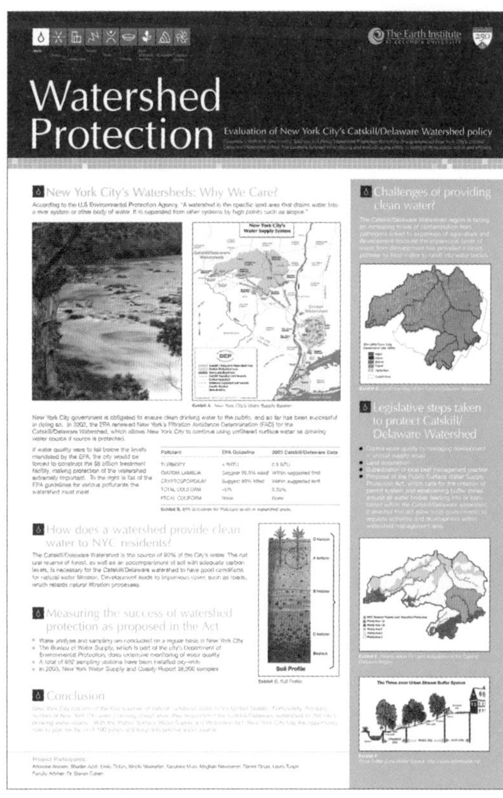

每一个系统的侧边栏也都进行了色彩编码，包含了实验和研究等类别的信息

● 63.打破网格

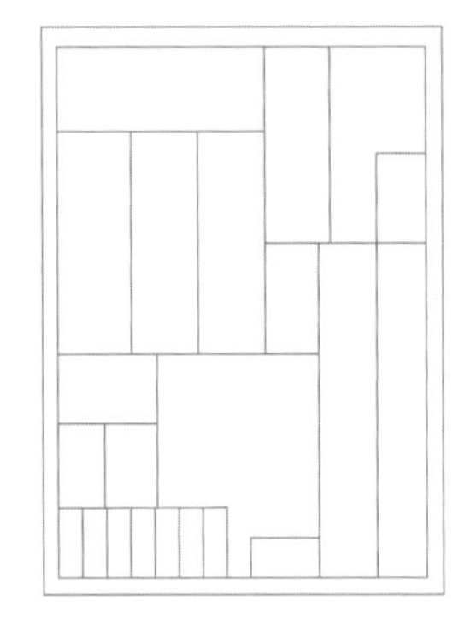

如何处理大量密集的文字、图片、表格、卡通画、统计数据等信息？将所有内容放进一个松散的框架，然后打破这个框架。

下页图： 这份报纸的创始人和设计师曾指出，他们的报纸没有使用网格。这份报纸的设计初衷是让城市居民在拥挤的城市中有一种归属感。即使如此，这份报纸的设计师也肯定了解且使用了网格系统

New York's Stories & Secrets & Supermarkets & Sex & Metadata & Terrorists & Herbalists & Scie... Lovers

EXCLUSIVE TELFAR POSTER

Civilization

A FUTURE HISTORY OF NEW YORK　PROTECT THE CHILDREN ● ISSUE 1 / VOL 1

Father, Mother, God, Bathe Me In Light

Darcie Wilder

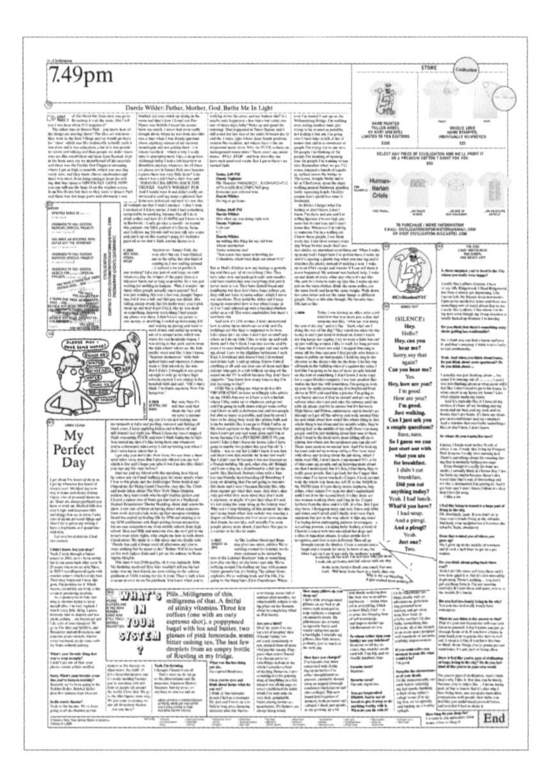

7.49pm

Darcie Wilder: Father, Mother, God, Bathe Me In Light

STORE

My Perfect Day

WHAT'S IN YOUR SYSTEM ?

End

项目

《文明》（*Civilization*）

客户

理查德•图利（Richard Turley）

编辑/设计师/作者

理查德•图利，卢卡斯•马斯卡泰洛（Lucas Mascatello），米娅•克林（Mia Kerin）等

这是一份反映纽约生活的报纸。报纸的内容就像纽约这个大都市一样，充满了沉思、真相和虚幻。在2018年开始出版的时候，创始人特意选择了维多利亚时期的排版风格，将新闻传递给这个被数字迷惑的世界

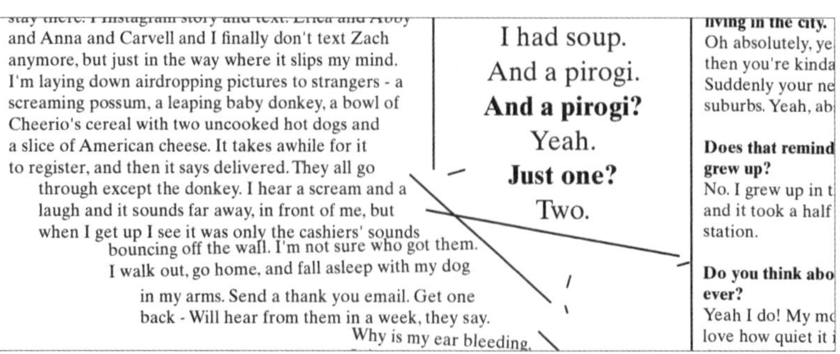

and Anna and Carvell and I finally don't text Zach anymore, but just in the way where it slips my mind. I'm laying down airdropping pictures to strangers - a screaming possum, a leaping baby donkey, a bowl of Cheerio's cereal with two uncooked hot dogs and a slice of American cheese. It takes awhile for it to register, and then it says delivered. They all go through except the donkey. I hear a scream and a laugh and it sounds far away, in front of me, but when I get up I see it was only the cashiers' sounds bouncing off the wall. I'm not sure who got them. I walk out, go home, and fall asleep with my dog in my arms. Send a thank you email. Get one back - Will hear from them in a week, they say. Why is my ear bleeding.

I had soup.
And a pirogi.
And a pirogi?
Yeah.
Just one?
Two.

living in the city.
Oh absolutely, ye
then you're kinda
Suddenly your ne
suburbs. Yeah, ab

Does that remind
grew up?
No. I grew up in t
and it took a half
station.

Do you think abo
ever?
Yeah I do! My mo
love how quiet it i

瑞士的网格系统与具象诗歌相遇，如图中的细节所示

3.1

Darcie Wilder

Cover story: We used to pass this park all the time on the bus, Dead Dog Park. There was a sign and everything. They were trying to up the number of Manhattan parks so they walked looking for squares of grass that could technically be parks. They ran out of names by the time they got to 168th and saw a dead dog and now it's Dead Dog Park, there's a sign and everything.

Father, Mother, God, Bathe Me In Light

○ 2018
HELL'S KITCHEN
1PM
15 NOTIFICATIONS

So I wake up in bed underneath a 25 pound weighted blanket but I just wanted the heaviest thing to put me out and the first thing I touch and look at is this blazing screen, and my eyes are still opening and they're making shapes adjusting to the light, I'm scrolling and looking for my dog and just scrolling and putting on the same podcasts as every morning before and trying to go back to sleep and failing. And then I'm using the new body wash that smells like pears and it's still so cold out, and I just found out my shower curtain was in the wrong place, years of making daily puddles on the floor, except who showers daily, more of a weekly puddle. I just got a new toothbrush that zaps me every thirty seconds, all I do it make it touch my teeth, moving around to a different tooth. My teeth were so soft they were stinging me when I brushed, a shooting pain up my gums. I'm trying to figure out what. I'm trying to. Every morning I can't fall back asleep but can't stay awake doing the things I want to do.

It's not that it's especially bad but that it just keeps never getting better, and I wake up around 1pm today so the kids at the high school next door are on their way home. Two recognizable character actors are talking on a bench, I'm throwing a tennis ball with a chihuahua mix and some puppy named Lemon comes up trying to steal it and does for a second. It's the kind of weather that could be fall but is March and taking it's time and even when it finally gets good enough to walk it'll be too hot, too much, too heavy, and I'll be wearing sleeves even when I shouldn't.

App goes off PUSH notification - assault 5 blocks away. Last night, suspicious package. Tonight a car'll jump the curb and a manhole a few miles away will blow off, like I used to imagine happening when I stood on them. The helicopter in the East River, I was a few blocks away. Repeated footage over and over, New York 1 in doctor's office, this is footage of people dying. This is the last moments of five peoples' lives.

Taking her off-leash even though she could die. But everyone, all the time. Aren't we all just trying to keep ourselves alive? It's crazy, they should teach kids about aneurysms earlier. But she's too scared of the curb, too scared of cars, too scared of the brushing whizzing noises to ever go near. Sometimes I worry she'll start chasing a rat, kill a rat with her bare paws and I'll have her, covered in blood, to clean up. My friend's dog killed a rat before they had time to stop her. Rats on 43rd run next to the parking lot bricks, behind the trees, burrowing deep. I used to sit on the wooden benches with Chiquita till my neighbor Kathy told me. Also the homeless sit there sometimes and you never know about bedbugs. My stomach itches. Peeling, flaking.

Why do croissants always stain the bag? So many greasy patches and they just soak through and I ate a croissant every day, I don't even like them. I guess I crave them when I go without but those bagels from the carts, you know they're either stale or not boiled, they're just straight up bread cut in a circle, they've got to be. Once I saw a delivery guy wipe out, spill all of them on Amsterdam Ave, scattered. Straight up on the pavement. Cart guy picked them up, dusted them maybe a little bit, sold them all day. Like when this girl in high school bought cart coffee, milk and sugar - I only did sugar then, and now I do nothing - and it started splitting out the seams of the cup. In her hand, driblets and then you know it was just all going to pour out. So then the guy's like, doing that hand motion like c'mere, come on, and he takes it from her and says like, they had gotten wet and he let them dry and tried to use them. Soggy cups.

A few weeks ago the G wasn't running and I was stranded at that PS 1 station. I ended up walking over that bridge to Greenpoint thinking I was going to be late because I was the only one that had to deal with it and I walked into the room and no one was there. It was one of those moments where the sun was setting, sky was some color maybe pink or even red, the water was making it colder and I was listening to that Jimmy Eat World song that sounds like a stomach ache when you're on the subway home with too much energy and not enough, and you're just turning everything, everything is about to bloom, it's a fade out that just fades into something else, I was the only one on the bridge taking a video and we were all going against the wind from the exhaust of the cars.

Coco comes when I call and I force the slimy tennis ball out her mouth. She's stripped it bald, tearing out neon green fuzz. I gnawed my teeth down to bits over the years and now they fit like puzzle pieces. This week I couldn't find her paperwork and had one of those spiraling meltdowns, started praying to anyone or no one or something or whatever, and then thought that I had to convince myself I'd never find it because you only get what you want when you don't want it anymore. Found it in a pile of loose papers in the closet.

○ 4PM
TIMES SQUARE

I walk on 42nd by my old office, the corporate office in a huge tower I used to walk there every morning during the Polar Vortex. When I was only wearing skirts and the same torn tights every day, and had to buy new tights and change before any of the bosses saw me. The wind would whip against my legs and my thighs got this stinging feeling, I wore Doc Marten dress shoes that gave me blisters and read the internet all day. Sometimes I had busy days, making coffee, ordering from Staples, time sheets. I remember phases by the lunches, brief stint doing halal but then the weeks I only craved guacamole, mushrooms, chicken strips. I would order Lenny's egg wraps and dunk them in a tiny tub of Tabasco hot sauce just to feel anything, just so bored and craving anything, and the salt would rush into my, my eyes watered, my tongue hung out of my mouth. I used to drink as much water as I could as fast as possible, Poland Spring after Poland Spring, out of boredom. Then I'd google "water poisoning."

She came from LA, surrendered 10/26/2016, up for adoption for nine months until I found her. The time I spent in LA is a blur, I can't remember if I worked in retail for 3 weeks, 6 weeks, or 3 months. They checked my bag going in and out so I left it in my car and had to pay $20 for a parking spot. Parked next to a matte'd out black sedan and dissociated after work circling the aisles of Target till I drove home.

They tore down the gas station on 44th and now it's a pit. Nearly empty, construction equipment moves around piles of dirt and drills in the middle of the day when I'm checking DMs. It says "Seen." But he hasn't written back and the real read receipts don't have periods but should, to imply that it's the end of the conversation. When did we stop using periods? I don't. The drill goes on and on so I got the blackout curtains and wake up in the single digit afternoon hours, after I wake up around 5 or 6 and look at my phone to avoid evil spirits. A ghost can't get you if you're looking at a screen, so I spent hours and hours staring at my phone in my bedroom, which used to be grandma's bedroom, where I cried and wept next to her bed when they said I should say goodbye to her for the last time. She laid there, and dead bodies look different in every stage, from dying to dead to been dead to on display.

So I light a lot of sage and whisper to myself and say the same non-denominational prayer my uncle told me when I was sleeping on the floor in LA, too scared to sleep in my own room alone, I closed my eyes and a thing, I don't know, something, gripped my throat and cut off my breath and I couldn't speak, couldn't breathe, couldn't, and so I reflexively repeat "Father Mother God Bathe Me In Light" and ignored the God part for years, and forgot I was saying it for brief stretches of time, the times when I was doing molly on weeknights and coming down when the sun was coming up and sitting in a rocking chair waiting while they met the guy and got more, doing more and feeling dead or dying and scared of all of it suddenly. I forgot the words but they came back, and now when I close my eyes to say FUCK I accidentally find myself with them, repeating them over and over like a lotion that absorbed all it can and now it's just swimming on top of pores, it's just too much, like everything I want is always over the surface trying to get in but there's no room, there's no space, we're all just waiting for it to combine, to infuse, to germinate and it never does anything but slick the surface.

○ 2017
9AM
MARCH, APRIL, MAY, JUNE

It takes me awhile to get out of bed and I finally make coffee, Cafe Bustelo, from the grinds too small, too many of them, and the french press mesh, pushing down the plunger it always flies up, it splashes every day but today burns my fingers, grazes my hand and the counter is filled with puddles of grime and grinds that stay there for weeks, maybe just short of a month, because I feel guilty buying paper towels but it's more that I feel better avoiding tasks, evading errands.

I'm trying to come up with a way where I do what I need. I walk down the West Side Highway, I miss when it was April and May and June, the exhaust would mix with the humidity for something so unpleasant and rank but in that way where you destroy something, you keep on flossing too hard and keep touching a cut, you keep stinging yourself. Walk all the way down listening to headphones so much the soles of your foot hitting the pavement reverb up, bouncing up into your ears against the headphones. Taking selfies on West Street, whatever it's called. Walk until you reach the cobblestones. Up until a certain point and on weekends it's so crowded and I can't leave my house. In high school I rode my bike down Sixth Ave. on the fourth of July listening to the cracks of fireworks and avoiding it all, street so clear. Bike rides every month, Critical Mass, used to bike up the highway outnumbering the cars. I gave up when there were too many white bikes, ghost bikes, memorials of hits and dead people. I got hit once or twice.

Remember that guy I got in the cyber fight with? The twitter feud shit thing a few weeks ago? He was biking in Central Park and his handlebars shook, apparently, and he wiped out lost control and his face slid across the pavement.

There was this time right before I kept stealing stuff - but only from Whole Foods and Barnes and Noble - where, for a whole year, I wore these fake glasses. And I never talk about it but a whole year, and no one ever said anything when I stopped wearing them except Carla, but they were $8 glasses from St. Marks

I like the hot when the diner booths stick to the backs of your thighs and it hurts to move. It stings to stand up. Sometimes they don't even have AC, or maybe they all got AC now. Last time I was at Waverly Diner this guy flashed a blade, maybe just a scissor but still a metal edge, between his cotton glove and the glass window, maybe it's plastic? Flashed a blade while we ate omelets and we told the waitress and she said, "What do you want me to do about it?" He crossed the street, approaching the NYPD van, seemed like he was trying to get arrested. The van revved up, engine on, and pulled out into traffic, ignoring him. And I thought I knew rejection.

We used to come here in the mornings before we both got fired. I asked Dad to tell me about the scar again. Leaving Waverly Diner, where I cried in front of Steve that time in 2010. He tells me about Stuytown again, growing up on 15th and 1st, and climbing the metal fence in the 60s. A group of adults watched him throw himself over and catch his skin on the spike, sliced open, huge scar running down. He tells me about the crying girl in the amputation ward, who just lost her arm. I've never heard this part of the story but each time I find something else new out it's like, yeah.

○ 2014
MALT LIQUOR, COCAINE, RED WINE

I started to worry about myself when the grocery store employee, the guy stocking the milk, asked if I was OK. I thought I could just keep circling, no phone service just listening to music, and that no one would bother me but sometimes it's like, the city buffer gives out, it wears down like a rubber sneaker sole, and I end up touching the floor, the cement ground, the bottom, end up having a conversation with someone.

○ 2007
ALEX
QUEENS

There was that time [redacted] ran out of the bookstore because I casually mentioned I made out with Alex - who I think moved to Portland and had a kid? And writes books on anarchist theory? He was living in Queens and no one else was really there, and his Greek landlady would give him liquor and he just stood in his doorway and maybe, looking back ten years, I should've left? Was that a social cue? It's like, weird with social cues because I have enough of my own brain telling me I'm a piece of shit and that I should just go, that like, I'm fighting that enough that I think normal things are social cues to leave and then I end up missing the real social cues because I'm talking myself down from the imagined stuff, you know?

○ 2014
IPHONE 5
RED WINE, WHISKEY, VODKA, WEED

There was that cat, on my way home when I was in Sunset Park, the year before my bag was locked up. When I was on my way home I saw a orange cat, infested and dirty, these huge black splotches of grime, and I kept petting it, talking to it, calling it a rat. My little orange rat. I thought it'd bite me eventually and I didn't want to deal with the paperwork so I kept walking. An hour and a half home on a yellow line train but no one calls them that but that's what they are. On the way home a guy told me, he was like, 40s maybe? That I should be a teacher. That I'm a "nice girl." It was the same subway station where I called that ambulance the last day of high school. She was nodding out on a bench, poisoned, and the cop asked her why her birth year was scratched out. Ended up going to Methodist and getting an IV and then I took a $60 cab ride home. The air is different at the top

CONT. PAGE 16

3.2

HERE:
11.09AM, CLOUDY
COULDN'T GET UP
DAY RATE
WELCOME TO ASANA

YOU HAVE 2 TASKS DUE: 2 INFOGRAPHIC... AND NEW LAYOUT * ... - [redacted] CREATIVE

YOU HAVE 2 TASKS DUE: 2 INFOGRAPHIC... AND NEW LAYOUT * ... - [redacted] CREATIVE

YOU HAVE AN OVERDUE TASK: 2 INFOGRAPHICS - NOT YET STARTED - [redacted] CREATIVE

REMINDER: [redacted] INVITED YOU TO JOIN [redacted] CREATIVE

YOU HAVE AN OVERDUE TASK: 2 INFOGRAPHICS - NOT YET STARTED - [redacted] CREATIVE

SPECIAL PROJECT: [redacted] [redacted BRAND CUE]

YOU HAVE AN OVERDUE TASK: 2 INFOGRAPHICS - NOT YET STARTED - [redacted] CREATIVE

DESIGN: [redacted] [SPECIAL PROJECT: [redacted]]

ASSIGNED TO YOU: DESIGN: [redacted] [SPECIAL PROJECT: [redacted]]

YOU HAVE AN OVERDUE TASK: DESIGN: [redacted] - BUFFY CREATIVE

WELCOME TO ASANA

REVIEW DECK AND MAKE NOTES [#[redacted] FACE]

ASSIGNED TO YOU: CHECK OUT THE [redacted]?

HOLD: [redacted] CARD ON [redacted] PAPER [[redacted]]

YOU HAVE A TASK DUE TODAY: CHECK OUT THE [redacted]? - [redacted] CREATIVE

SET LAUNCH DATE [#[redacted]]

YOU HAVE AN OVERDUE TASK: CHECK OUT THE STICKERS? - [redacted] CREATIVE

YOU HAVE AN OVERDUE TASK: CHECK OUT THE STICKERS? - [redacted] CREATIVE

BRIEF SIDE PROJECT ON COMMERCIAL SCRIPT [SPECIAL PROJECT: [redacted]]

UNBOXING UPDATE [[redacted] BRAND CUE]

OUTLINE DECK: PRODUCT FOCUS CONTENT [[redacted] BRAND CUE]

YOU HAVE AN OVERDUE TASK: CHECK OUT THE STICKERS? - [redacted] CREATIVE

BRAINSTORM ALTERNATIVES TO THE [redacted]: [SPECIAL PROJECT [redacted]]

YOU HAVE AN OVERDUE TASK: CHECK OUT THE STICKERS? - [redacted] CREATIVE

IDEAS FOR [redacted] VIDEO? [SPECIAL PROJECT: [redacted]]

ASSIGNED TO YOU: EDITOR INVITE IMAGE [SPECIAL PROJECT:

[CONT 16.2]

3.3

EMILY SEGAL

CLOSE YOUR EYES. THINK ABOUT HOME. DESCRIBE WHAT YOU SEE

"The living room of my mom and step dad's apartment on 91st Street. Oriental rug from my great grandpa, kitschy drippy orange and green glass vases from the 60s, a colossal plant that holds the spirit of my step dad's late mother Evelyn, heavy wrought iron coffee table my parents had made from a bank grate in Philadelphia when they were still in their 20s. Squeaking and squawking cars and sirens from Amsterdam Avenue, coming in from the street."

How many pillows do you sleep on?
Two.

Are you a thief?
Unfortunately I am no longer a thief, at least as far as physical objects are concerned. As a teen I was a consistent, if not terribly ambitious, shoplifter – bent on amassing the sum total of makeup available at a constellation of local CVS's and Duane Reades. I got caught taking a straightening iron (particularly unnecessary, because I have straight hair) one afternoon when I was fourteen.

I remember planning the heist during English class that afternoon. The manager told me she would call my mother, and I looked withering…"my mother is dead". Then, once they threatened to call my school (something I found humiliating beyond measure), I recanted and revealed my mother was still alive. Once she arrived to pick me up they told her I'd said she was dead, which she said she thought was pretty ingenious, given the circumstances. I was in huge trouble.

For years later I limited my shoplifting to the costume jewelry at Top Shop – it was kind of my stealing methadone. Eventually I gave that up, too.

To whom are you indebted?
Literally all of my friends (spiritually, not financially)

Changing the names to protect the guilty, what's the worst thing anyone's ever done to you?
Besides a couple gnarly heartbreaks and more-or-less typical divorced-family trauma – the worst things have been creative or collaborative betrayals…like when former collaborators had gone behind my back to talk trash or change an artwork, or bosses made announcements about me to large groups without talking to me first. I'm a Scorpio, obsessed with my work and horrified by loyalty transgressions!

Favorite soup?
Vichyssoise

Favorite scent?
Hinoki; Musty Basements

How do you deal with guilt?
I FaceTime my best friends Alexa and Rachel repeatedly until one of them tells me it's all right, or I write about it, or I psychically bury it!

If you could relive one moment in your life what would it be?
Toss up between a particular lecture and a particular fuck – both experiences of "flow" (blech)

Where do you sit on the many-worlds interpretation of quantum mechanics and the probability of infinite variants of every single moment in time?
Staunchly "pro"

Starsign?
Scorpio, Aries Rising, Leo Moon (<– Profoundly 'unchill' combination)

How long do you sleep for?
Eight hours. Sleep is the greatest determining factor in my sometimes disturbingly volatile mood landscape.

Tell me about the last time someone held your hand?
My long lost friend Elizabeth, dragging me through the streets of Hollywood to pluck an orange off the tree in someone's backyard!

Have you got your dues?
Not even close ○

— EMILY SEGAL IS A STRATEGIST & FOUNDER OF K-HOLE/NEMESIS

3.4

CONTAMINANTS DETECTED IN NEW YORK CITY DRINKING WATER

For the latest quarter assessed by the US Environmental Protection Agency (July 2017 - September 2017)

CONTAMINANTS DETECTED **ABOVE** HEALTH GUIDELINES:

- BROMODICHLOROMETHANE
  Cancer forming
- CHLOROFORM
  Cancer forming
- CHROMIUM (HEXAVALENT)
  Cancer forming
- DICHLOROACETIC ACID
  Cancer forming
- TOTAL TRIHALOMETHANES (TTHMS)
  Cancer forming
- TRICHLOROACETIC ACID
  Cancer forming
- OTHER DETECECTED CONTAMINANTS
  - Chlorate
  - Chromium (total)
  - Haloacetic acids (HAA5)
  - Monochloroacetic acid
  - Nitrate
  - Nitrate and nitrite
  - Strontium

From October 2014 to September 2017 this water utility was in **VIOLATION** of health-based drinking water standards

Source: ewg.org

3.5

SNEAKERS SEEN AT HOME DEPOT IN BED STUY.

NIKE	AIR JORDAN 1	RED/BLACK
ROCHE	RUN	RED
NEW	BALANCE	BLACK
NIKE	AIRMAX 180	WHITE
REDBOX	CLASSICS	WHITE
NEW BALANCE	(UNKNOWN)	GREY
ADIDAS	(UNKNOWN)	ORANGE
REEBOK'S	(UNKNOWN)	BLACK
NEW BALANCE	(UNKNOWN)	BLACK
CONVERSE	HI	ORANGE
NIKE	PEGASUS	GREY
SAUCONY	(UNKNOWN)	GREY
NIKE	AIR MAX LOW	BLACK
PUMA HI	(UNKNOWN)	BLUE
NIKE	HUARACHE	ORANGE
CHAMPION	(UNKNOWN)	BLACK

3.6

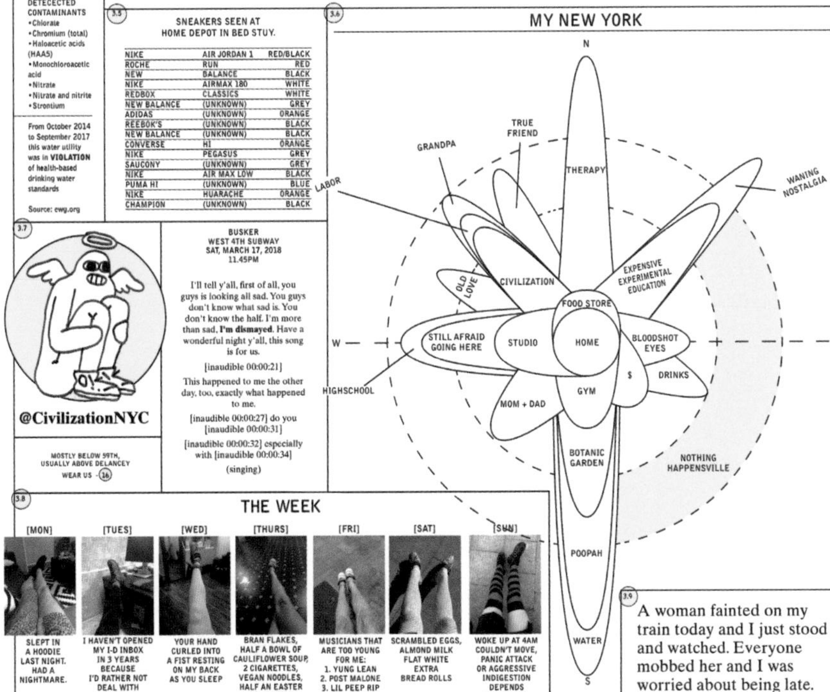

3.7

@CivilizationNYC

MOSTLY BELOW 59TH,
USUALLY ABOVE DELANCEY
WEAR US - 16

BUSKER
WEST 4TH SUBWAY
SAT, MARCH 17, 2018
11.45PM

I'll tell y'all, first of all, you guys is looking all sad. You guys don't know what sad is. You don't know the half. I'm more than sad, **I'm dismayed**. Have a wonderful night y'all, this song is for us.

[inaudible 00:00:21]

This happened to me the other day, too, exactly what happened to me.

[inaudible 00:00:27] do you [inaudible 00:00:31]

[inaudible 00:00:32] especially with [inaudible 00:00:34]

(singing)

3.8

THE WEEK

[MON] SLEPT IN A HOODIE LAST NIGHT. HAD A NIGHTMARE.

[TUES] I HAVEN'T OPENED MY I-D INBOX IN 3 YEARS BECAUSE I'D RATHER NOT DEAL WITH THESPAM

[WED] YOUR HAND CURLED INTO A FIST RESTING ON MY BACK AS YOU SLEEP

[THURS] BRAN FLAKES, HALF A BOWL OF CAULIFLOWER SOUP, 2 CIGARETTES, VEGAN NOODLES, HALF AN EASTER EGG, ANOTHER NIGHTMARE

[FRI] MUSICIANS THAT ARE TOO YOUNG FOR ME:
1. YUNG LEAN
2. POST MALONE
3. LIL PEEP RIP
4. LIL XAN

[SAT] SCRAMBLED EGGS, ALMOND MILK FLAT WHITE EXTRA BREAD ROLLS

[SUN] WOKE UP AT 4AM COULDN'T MOVE, PANIC ATTACK OR AGGRESSIVE INDIGESTION DEPENDS WHO U ASK

3.9

A woman fainted on my train today and I just stood and watched. Everyone mobbed her and I was worried about being late.

64.平衡冗杂的材料

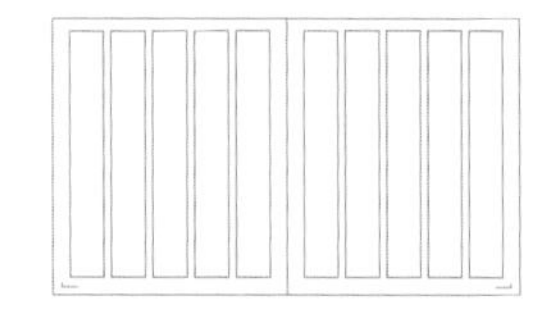

某些信息的表达需要用设计来平衡。在新闻报纸中，篇幅往往是最重要的，尤其是对非盈利性的报纸而言。将所有内容都放入一定数量的页面中（通常是4个或8个页面）会受到很多限制，但这些限制反过来又有助于确定结构。

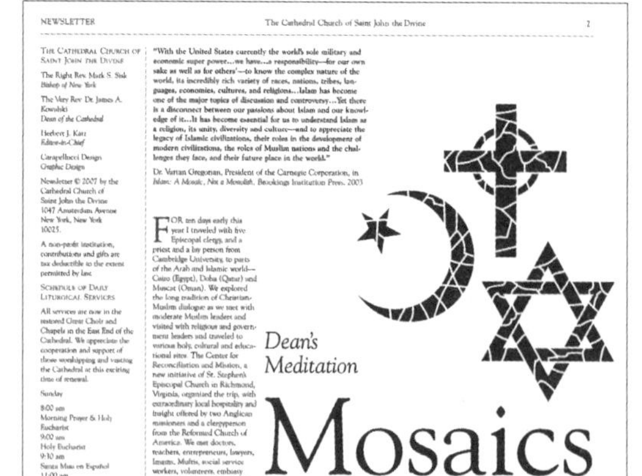

NEWSLETTER　The Cathedral Church of Saint John the Divine　2

Dean's Meditation

Mosaics

在上图这个页面上，侧边栏是一个实用区域，放置了人员名单、联系信息和使用说明。一条纵向的线把侧边栏和其他栏目分开，其余的分栏包含了一篇文章、几个艺术图案和引言，让读者在阅读这篇文章时可以有一点儿思考的空间

The Cathedral Church of Saint John the Divine

Newsletter

Summer 2007 Volume 12 Number 37

ESTABLISHED in 1991, the Cathedral's Public Education & Visitor Services department welcomes all visitors to the Cathedral. From its Visitor Center in the West End of the Cathedral, the department provides hospitality by offering information on the Cathedral's history, liturgical services, performances, and events.

Cathedral Joins New York Transit Museum to Honor Architects Heins and LaFarge

Newsletter Spring 2007

网格结构延续到了背面，这个页面对折后就是一封广告邮件

项目
《时事通讯报》
（*Newsletter*）

客户
圣约翰大教堂（Cathedral Church of St.John the Divine）

设计指导
Pentagram公司

设计公司
卡拉佩鲁齐设计公司

这份由非盈利性组织创办的《时事通讯报》完美地展示了五栏网格的多功能性

Events Calendar

The Cathedral Church of Saint John the Divine
1047 Amsterdam Avenue
New York, NY 10025
212 316-7540
www.stjohndivine.org

Services
HELD IN GREAT CHOIR AND EAST END CHAPELS DURING CURRENT PHASE OF CATHEDRAL CLEANING AND RESTORATION

✚ Monday–Saturday
8:00 a.m. Morning Prayer
8:30 a.m. Morning Prayer and Holy Eucharist
12:15 p.m. Holy Eucharist
5:30 p.m. Evening Prayer

✚ Sunday
8:00 a.m. Morning Prayer and Holy Eucharist
9:00 a.m. Holy Eucharist with Hymns and Sermon
11:00 a.m. Sung Eucharist with Hymns and Sermon
1:00 p.m. *La Santa Eucaristía en Español*
6:00 p.m. Evensong

December

12th Annual Crafts Fair
To Benefit the Congregation of St. Saviour
Synod House
Friday, November 30, 5–9 p.m.
Saturday, December 1, 11 a.m.–6 p.m.
Sunday, December 2, 11 a.m.–5 p.m.
Admission $6 per day, $7 weekend pass.
An extraordinary show of talented crafts people selling their wares—a 'divine' holiday shopping opportunity! Volunteers welcome to help with the show. Contact Rich Whaley at richawhaley@aol.com. See a preview of the crafts online at www.craftsatthecathedral.com.

Installations
Saturday, December 1, 10:30 a.m. with reception to follow
Installation of The Rt. Rev. Frank Griswald as Canon for International and Interfaith Mission and The Rev. Victoria Sirota as Canon Pastor and Vicar of the Congregation of St. Saviour

SUNDAY RECITAL SERIES PERFORMANCE
TetraWind Quartet
Sunday December 2, 2 p.m.
In the Chapel of St. Saviour
A woodwind quartet comprised of flute, oboe, clarinet and bassoon, TetraWind is a classical music ensemble whose thought-provoking, theme-based programs combine with interactive presentations to create a revolutionary concert experience.

Paper Making Workshop
Saturday, December 8, 12:30 p.m.–4:30 p.m.
Join artist and educator John Currie to learn the basic history of paper by examining both Western and Eastern paper making techniques. Cotton and other fibers will then be used to create sheets for bookmaking, collage, and paper casts. Emphasis will be placed on how to manipulate fiber to create simple and elegant visual art pieces. Participants are encouraged to bring collage or plant materials of their own to potentially incorporate into their work. Space is limited to 10 people, and reservations are required. $75 per person, materials included. For information and reservations, call 212 932-7347.

The Cathedral Christmas Concert
Saturday, December 8, 7:30 p.m.
Join the Cathedral Choristers and choir, Stewart Brass Works, and special guest artists Three Mo' Tenors as they usher in the Christmas season with festive carols and seasonal music. Johnson Flucker, conductor, and Timothy Brumfield, organist.

Paul Winter's 28th Annual Solstice Concerts
Thursday, December 13, 8 p.m.
Friday, December 14, 8 p.m.
Saturday December 15, 2 p.m. & 8 p.m.
The Paul Winter Consort's 2007 Winter Solstice Celebration will showcase music from two new albums, and present special guests Native American singer/drummer/flutist John-Carlos Perea, and Brazilian singer/guitarist Renato Braz. The program will premiere music from the Consort's recently released new album, CRESTONE, as well as songs from Renato Braz' forthcoming North America debut album, RENATO. Featured in the Solstice concerts will be long-time Consort players Paul McCandless, on oboe and bass clarinet, and cellist Eugene Friesen. Other Consort players will include pianist Paul Sullivan and Cathedral organist Timothy Brumfield.

"Fool's Mass"
Sunday December 16, 7 p.m. in St. James Chapel
"Fool's Mass" is staged by Dzieci (pronounced "djyeh-chee"), an international experimental theater ensemble based in New York. Midnight Mass is about to begin. The congregants are arriving. Nervous choir members escort people to their seats and wander about anxiously dusting the seats and muttering to themselves. Then one discovers the priest has just died. So begins "Fool's Mass."

A Dutch Christmas
Kerstmis in Holland–1550–1650
Early Music New York
Sunday, December 16, at 3 p.m.
Sunday, December 23, at 3 p.m. & 8 p.m.
Tuesday, Christmas Day, at 3 p.m. & 8 p.m.
Early Music New York's soon to be released CD album, produced in association with The Metropolitan Museum of Art in tandem with the exhibition "The Age of Rembrandt," will be debuted in live performance—seasonal music Dutch settlers of Nieuw-Amsterdam would have known. *Cantiones natalitiae* (Latin songs) and *kerstliedjes* (carols) for voices, variations for recorders by Jacob van Eyck and Sweelinck, jaunty dances by Susato and Phalese, as well as rowdy *drinken liedeken* ring in the New Year. Join the artists in EM/NY's traditional post-concert New Year toast in the festooned Cathedral—*gelukkig Nieuw Jaar!* For tickets, go to www.earlymusicny.org.

Service of Lessons and Carols
Monday, December 24, 4 p.m.
The Christmas story for the whole family told through lessons, carols, and puppetry, with music from the Cathedral Choristers.

Christmas Eve Eucharist
Monday, December 24, 10:30 p.m.
Celebrant and preacher: The Right Reverend Mark S. Sisk, Bishop of New York. Music for choirs, brass and organ. Anthony Piccolo: *Canterbury Mass*; James Bassi: *Quem Pastores*; and other carols. With special guest Paul Winter.

Christmas Day Eucharist
Tuesday, December 25, 10:30 a.m.
Celebrant and preacher: The Very Reverend James A. Kowalski, Dean of the Cathedral Church. Please note that this is the only service, and that there will be no tours on Christmas Day.

Annual New Year's Eve Concert for Peace
Monday, December 31, 7:30 p.m.
Join host Harry Smith, conductor Glenn Cortese, opera star Lauren Flanigan, and guest artists for an evening of music, readings, and interfaith prayer as we look forward to a time of peace on Earth. The featured work will be Aaron Copeland's *Lincoln Portrait*.

New Year's Eve Eucharist
Monday, December 31, 11:00 p.m.

Spotlight Tours

On Common Ground: Spotlight on Faith Traditions
Sunday, December 9, 2 p.m.
Experience how the Cathedral stands as a "house of prayer for all people." View and discuss the parallels between holidays in Christianity, Judaism, Islam, Buddhism, and the Vedic traditions as they are reflected in the Cathedral's architecture and artwork. Led by Cathedral Educator John Brooks. $10 per person, $8 for students and seniors. For more information and reservations please call 212 932-7347.

Brilliant Walls of Light: Spotlight on Cathedral Windows
Sunday, December 16, 2 p.m.
Each chapel contains a unique style of stained glass. Explore beautiful narrative and geometric windows by English and American firms and view the memorial to a stained glass artist. Led by Senior Cathedral Guide John Simko. $10 per person, $8 for students and seniors. For more information and reservations please call 212 932-7347.

Children's Programs

Medieval Arts Children's Workshop
Saturday, December 1, 10 a.m.–noon
In this signature workshop, children carve a block of limestone; create medieval illuminated letters; design gargoyles, weave and more! Recommended for ages 4 and up. $5 per child, with accompanying adult. For more information and reservations please call 212 932-7347.

A Season of Lights: A Winter Solstice Celebration
Saturday, December 15, 10 a.m.–noon
In this special workshop, families brighten up their winter with a reading of Nancy Luenn's Celebrations of Light. They are then off to the workshop to construct paper lanterns, drums, beeswax candles and more! Recommended for ages 4 and up. $5 per child, with accompanying adult. For more information and reservations please call 212 932-7347.

January

New Year's Day Eucharist
Tuesday, January 1, 10:30 a.m.

Mosaic Workshop
Saturday, January 12, 12:30 p.m.–4:30 p.m.
Join artist and educator Danielle Oteri for this special workshop to learn about the craft of mosaic. Begin with a history and overview of mosaics and then learn how to work in this ancient medium. Using vitreous colored glass tile, participants will design, cut, place, and grout a decorative mosaic. Space is limited to 10 people, and reservations are required. $70 per person, includes all necessary materials. For more information and to make reservations, please call the Public Education Department at 212 932-7347.

Let My People Go: A Service of Liberation
A Celebration of Lessons and Carols Commemorating the Abolition of Transatlantic Slavery
Sunday, January 13, 3 p.m. followed by a Multi-Media Display
6 p.m. Evensong and Refreshments
Bishop Mark Sisk will be joined by actors, dancers, musicians, readers and a mass choir in an interdenominational, cross-diocesan celebration of the bicentennial abolition of the transatlantic slave trade. On display will be scenes from slavery in New York from the New York and Brooklyn Historical Societies and clips from the New York Reparations Committee's resource guide to Katrina Browne's documentary, *Traces of the Trade*. Sponsoring organizations include The Diocese of New York, The Diocese of Long Island, The Diocese of Newark, The Diocese of New Jersey, Social Justice Ministries, The Episcopal Church's Washington office, The Riverside Church, and the Cathedral of St. John the Divine.

Spotlight Tour

Signs and Symbols: Spotlight on Symbolism
Sunday, January 13, 2 p.m.
Explore the signs and symbols in the Cathedral and discover the paintings, glass and stone. Learn how the legends have inspired artists through the centuries. Led by Cathedral Guide Becca Earley. For more information and reservations please call 212 932-7347. All tours meet for registration at the Visitor Center inside the Cathedral entrance, at 112th Street & Amsterdam Avenue. Spotlight Tours are $10 per person, $8 for students and seniors.

Children's Programs

Camels & Kings: Celebrating the Epiphany
Saturday, January 5, 10 a.m.–noon
Families gather to explore the story surrounding the famous journey of the three wise men, celebrated around the world. The workshop begins with a story and then families make gift boxes, costumes and sparkling crowns. Recommended for ages 4 and up. $5 per child, with accompanying adult. For more information and reservations please call 212 932-7347. All programs meet for registration at the Visitor Center inside the Cathedral entrance, at 112 Street & Amsterdam Avenue.

Medieval Scribes: A Children's Illumination and Bookmaking Workshop
Saturday, January 19, 10 a.m.–noon
Utilize the tools and techniques of the medieval scribe to learn about the evolution of book making. Bind a book, create an illuminated letter, practice calligraphy and more! Recommended for ages 4 and up. $5 per child with accompanying adult. For more information and reservations please call 212 932-7347. All programs meet for registration at the Visitor Center inside the Cathedral entrance, at 112 Street & Amsterdam Avenue.

General Information

The Cathedral is in the midst of an ambitious and comprehensive cleaning and restoration after the 2001 fire. The Cathedral will remain open, but public access will be limited in some areas. We appreciate the cooperation and support of those worshipping and visiting the Cathedral during this exciting time of renewal. Please check the Cathedral's website, www.stjohndivine.org, or call the Visitor Center, 212 316-7540, for additional information.

TICKETS AND RESERVATIONS
Unless otherwise noted all events are free and do not require tickets or reservations. Performances: Tickets for all performances other than free or "suggested contribution" events may be purchased from Ticket Web, www.ticketweb.com or 866 468-7619.

Ongoing

Highlight Tours
Tuesdays–Saturdays 11 a.m.–noon & 1–2 p.m., Sundays 1–2 p.m.
Explore the Cathedral's newly cleaned and magnificently restored East End, Crossing, Great Choir, and seven Chapels of Tongues. Learn about the art, architecture and history of this great sacred space from 1892 to the present. No reservations necessary. $5 per person, $4 per student/senior. Meet at the Visitor Center inside the Cathedral entrance, at 112th Street & Amsterdam Avenue.

Vertical Tours
Saturdays, Noon–1 p.m. & 2 p.m.–3 p.m.
Climb 124 feet through spiral stone staircases to the top of the Cathedral. Get a look at the huge scale of the nave restoration and study the grand architecture while standing on a buttress. The tour culminates on the roof with a wonderful view of the Morningside Heights area of Manhattan. Space is limited to 10 people, and reservations are recommended: 212 932-7347. $15 per person, $10 per student/senior. Meet at the Visitor Center inside the Cathedral entrance, at 112th Street & Amsterdam Avenue. Bring a flashlight.

Spotlight Tours
Select Sundays, 2 p.m.–3 p.m.
Spotlight tours are especially created by senior Cathedral guides to give visitors a closer look at some of the many wonderful, unique aspects of the Cathedral's extraordinary architecture. Please see individual descriptions listed under each month for more information. Reservations recommended. $10 per person, $8 per student/senior. Meet at the Visitor Center inside the Cathedral entrance, at 112th Street & Amsterdam Avenue.

Medieval Birthday Parties
Saturdays & Sundays, by availability
Celebrate your child's birthday with a two-hour party in the Medieval Arts Workshop, where children sculpt gargoyles, weave, make brass rubbings, carve a block of limestone and much more! For children ages 5 & up. Call the Public Education & Visitor Services Department at 212 932-7347 for more information and reservations.

February

Ash Wednesday
Wednesday, February 6
Services with Imposition of Ashes at 8:30 a.m., 12:15 p.m., and 7 p.m.

Someone to Watch Over Me: Spotlight on Saints
Sunday, February 10, 2 p.m.
Study the Cathedral's extraordinary sculptures, and discover the unique attributes that characterize saints, martyrs, and angels. Led by Cathedral Guide Becca Earley. For more information and reservations please call 212 932-7347.

Spotlight Tours

Diversity United: Spotlight on the Chapels of Tongues
Sunday, February 3, 2 p.m.
[illegible]

Children's Programs

Medieval Arts Children's Workshop
Saturday, February 2, 10 a.m.–noon
[illegible]

Kids' Cathedral
Tuesdays, February 5, 12, 19 & 26; 10:30–11:30 a.m.
[illegible] adult. For more information and reservations please call 212 932-7347. All programs meet for registration at the Visitor Center inside the Cathedral entrance, at 112 Street & Amsterdam Avenue.

Arches Everywhere: A Children's Architecture Workshop
Saturday, February 23, 10 a.m.–noon
Learn how the huge Cathedral stands by [illegible] programs meet for registration at the Visitor Center inside the Cathedral entrance, at 112 St. & Amsterdam Ave.

A.C.T.
INFO.: 212 316-7530, www.actprograms.org

Summer Camp for All? Help us continue to assemble children from many countries, religions and economic level under the shadow of our beloved Cathedral. While any amount will help, $1,900 enables a child from a low-income [illegible]

School Holiday Camp A respite for children and parents. A.C.T. is open when schools are closed. Select a single day or all days. Choice of trips or theme activities. Program hours vary. Call for catalog or visit www.actprograms.org.

A.C.T. Pre-school Registration and Open House Now registering for [illegible]

July Day Camp and August Specialty Camps.

The Congregation of St. Saviour

Congregation Christmas Party and 15-Year Anniversary Party *Sun., Dec. 16, following the 11 a.m. service.* Join us for Christmas fun and to celebrate 15 years as a congregation at this wonderful Cathedral. For more information please send email [illegible]

Winter 2007/2008

大事记日历充分利用了网格的优势，根据信息的不同，将一周中的每一天划分为不同宽度的栏目。将分隔线作为分界，粗分隔线作为容纳文字的容器，空白处作为侧边栏，用大标题为页面带来变化和质感

Griswold Appointed Canon

Blessing of the Bicycles Rolls Into Cathedral on May 12

Junior League Honors Brock

Henry L. King to Receive Diocesan Lawyers' Award

Key Roof Work Finished

Temporary Play Area Opens on the Pulpit Green

文章及其标题可以填满单栏、双栏或三栏网格。图片也可以填满这三种分栏，小插图给这个条理清晰的网格结构带来了一点儿调剂

65.密度与活力相结合

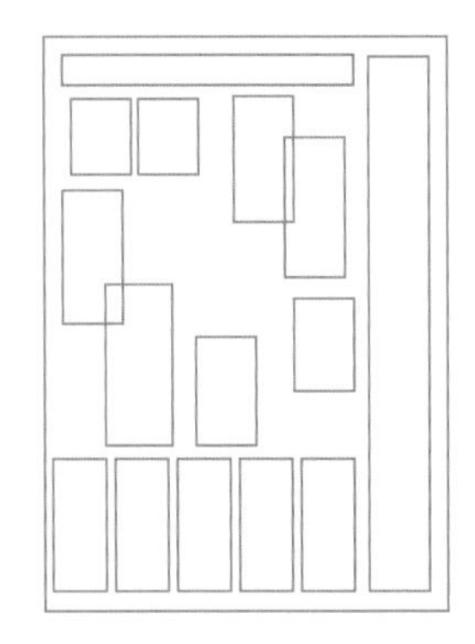

有时，一个项目的主要目标是让所有内容都具有可读性。对目录、术语表或索引来说，最佳的处理方式是想清楚它们与正文的搭配方式。

GOOD TROUBLE

"DO-IT-YOURSELF CULTURE MEANS TO LOOK AFTER YOURSELF AND THOSE ALL AROUND YOU... CREATE YOUR OWN FUTURE, WORK TO MANIFEST THE EQUALITY, LIBERTY AND KIND OF SOCIETY THAT YOU WOULD WISH TO BE A PART OF."

"IT'S A TIME FOR PEOPLE TO MAKE THEIR VOICES HEARD AND TO SAY NO."

JUST SAY NON! P12

@-GOODTROUBLEMAG

INSIDE

LARA SCHNITGER / ADAM BROOMBERG P2
PETER KENNARD: LOUDER THAN BOMBS P3
EMEL: VOICE OF THE ARAB SPRING P4
DOSSIER: ART OF RESISTANCE P5
JARRETT GREGORY / JIMMY CAUTY P9
RAVE + RESIST: 90s UK YOUTH PROTEST P10
MALWARE: GRIZZLY STEPPE P11
CHINO AMOBI / TAUBA AUERBACH P12

"BEING A HUMAN IS NOT ABOUT VIOLENCE AND HATRED... ART IS HUMAN, A BEAUTIFUL THING THAT ONLY A HUMAN CAN DO."

CUT-OUT-AND-KEEP! RUSSIAN CYBERWEAPON

GOODTROUBLEMAG.COM

GOODTROUBLEMAG@GMAIL.COM

Issue Edición Édition Edizione 23

1000

START AS YOU MEAN TO GO ON

PROTEST p2,3,8,9
ART p2,3,4,5,6,7
MUSIC p8,10,12
MORE ART p8,9,11,12
HACKING p11

CHANGE IS IN THE AIR
"WORK AS IF YOU LIVE IN THE EARLY DAYS OF A BETTER NATION"
NO ONE HAS THE RIGHT TO OBEY

"IT IS POSSIBLE FOR CULTURE TO BE AN INTEGRAL PART OF THE WORLDWIDE MOVEMENT FOR CHANGE."
(PETER KENNARD, P3)

本页图：在这本杂志的第一页上，水平排列的标示牌让很多视觉上的复杂元素变得清晰

下页图：杂志的各栏没有严格限定，在独立的、位置不固定的方框里穿插着问答内容

项目
《麻烦不烦》（*Good Trouble*）杂志

客户
罗恩·斯坦利（Ron Stanley）

设计师
理查德·图利

这个页面改变了规整的版式布局，让栏目从网格中跳脱了出来

12

NON WORLDWIDE is a group of African or diaspora musicians and artists. Their tagline is EXORCISE THE LANGUAGE OF DOMINATION. And they're here to shatter every illusion.

JUST SAY NON!

Even by the platform-agnostic standards of today, NON's activities are dizzying. They're a label, releasing stunning, radical music by musicians from Cape Town to Egypt to Virginia to Brixton. They're also magazine publishers who organise talks that run into all-night raves, who have opened a NON-branded range of travel merchandise in a duty-free store in downtown New York. Yet their ambitions go beyond these: NON is a borderless country open to all, a dissident political faction and a tight group of creative idealists…

Though the collective is sprawling and dozens of artists have released through their compilations and EPs, the core group is three DJ-producers-artists – the South African Angel-Ho, the Belgian-Congolese Nkisi, and Nigerian-American CHINO AMOBI. After a series of incredible mixes and mixtapes, Amobi has just released his debut album proper – the epic, collisionist double album Paradiso. It's an epic, complex, urgent, thrilling album, themed around an apocalyptic Edgar Allan Poe poem, and a radio station that flickers through moments of hellishness and total beauty. Kind of what life in America feels like right now.

"I like the chaos, throwing different variables in there, letting the chips fall where they may, shattering and breaking the canon in a way," Chino told us over Korean food in Berlin. "The depth and scale of the narrative is wider and deeper than one thinkpiece. The idea of NON is a constant rejection of definition. We're going to tell it ourselves."

RIGHT NOW

Your album is called Paradiso. Are you optimistic?
This time we're in has been growing. Trump is a benchmark, but a certain politicised feeling has been festering for time, with people like Black Lives Matter, the LGBTQ rights community, immigration, terrorism, home-grown terrorism and the way information is disseminated online,: it's all come to a head. it's like a boiling point.

Sometimes I think like the whole thing has to burn down in order for new life to be birthed. I say that optimistically. I'm not talking about masses dying – I don't want that – but destruction causes creation. It's always darkest before the dawn: in my life, the good things have happened directly after the bad.

I feel good about the future, about the youth, the spirituality in youth, the love. I think that the good will triumph. You can strengthen and pressure each other through productive measures. I'm all about shattering illusions, and the more you shatter, the better.

CHRIST

Your album is soaked in Christian allusion. Why is this so important today?
In times of strife which feel very dark, people go to faith to reconcile with what's going on, and [communicate] with something that's larger than themselves. Sometimes Christianity is represented like this fluffy thing, but the bible is super dark. It's gothic as hell. There's a mystery in those words. I'm just more malleable with data than some people are.

I identify as a follower of Christ, but I also identify as a queer body. That's often seen as a contradiction in the world, but the way I think about it, is it's about queering time and space. I really feel like The Bible did that.

Jesus entered time and space in a body that was queer, because only a queer body can transcend time and space, and change it in physical space. I believe that the body of Christ is here with us now and is changing who we are and our hearts. There's a sacred Blood that unites us together in that way, but that bond becomes more than just me, which connects me to other people, which is The Body. I know – it's a lot.

There's a certain magic element of faith that's important. A leap into that magic, I think, can change hearts. I put that into my music, and it's something that brings me closer to NON artists. Two become one. Transindividuation was something I was thinking about heavily on this album.

THE NATION

You've issued passports to concert attendees in the past. Is NON a nation?
Yes it is. It's a nation, it's a platform, it's an identification. We use the word NON because NON is everything and nothing. It's not limited to one thing. We can do anything. We can work with scientists, non-profit organisations, dancers, mathematicians, publications, designers. We can reflect our interests without things getting watered down. We have citizens all over the world, and I believe in multi-citizenship – so a NON citizen can be also be a citizen of the UK, or Nigeria, or the US, but NON citizenship augments the citizenship of the location, where they are able to utilise that citizenship for creative intervention in their community, online, and the world.

I love that quote, "Work as if you are living in the early days of a better nation."
I love that too. NON is very nascent. I'm very concerned about giving ourselves space to develop. It's like a garden. Say there's water in the garden. If people hoard the water, the garden suffers. The power of the garden is its diversity. It's important to have that multiplicity of voices. The water is data. And the garden is the systems and infrastructures we work in.

AIRPORTS

You work a lot with air travel – you released an album called Airport Music For Black Folk, and opened a duty free-style shop of NON-branded travel accessories in New York. What's behind this?
I always come back to airports because of what they represent to me. It's a liminal space between cities and countries, and it's a trans space, where we literally are preparing to change our bodies, inside and out, by getting on a plane. It's a very democratic space, but there's so much class things. There's so many codes of society and ideology that's brought to surface this really transparent way. It's almost like a no space. It's like someone took white infinity and made a building out of it. I'm very drawn to that in a very tactile way.

My parents are from Nigeria, and oftentimes Nigeria is on the list of countries for Americans not to visit. So, sometimes I've been questioned and searched heavily, as have other NON artists. I've also had really good experience at airports. I love to people watch. There's multiple things going on: migrants, workers, amazing-looking dogs, the richest people in the country. There's a lot of spontaneity. But spontaneity in this formal way. When statements are isolated in a way, sometimes very mundane actions are way more powerful, because there's some much space around them. Airports are a very "NON" space.

It's between countries, but it's also the only place you can literally point at what a country is. It's a man with a gun saying "You can enter, and you can't." Everything else is scenery. You can really tell a lot from a country by its airport.

DIASPORA

For many of the global south, long-distance air journeys are an integral part of life – not a luxury, as in the global north. This may be an obvious point, but it blows my mind.
The diaspora has given people of the global south this fluidity. This, I think, changes how we create. The ability for our creativity to cross cultures, and also have enough being to assimilate to where we are. People of the diaspora learn to speak in many languages and touch on countries they're in and where they're from. It forces you to think in a way that's multi-levelled, very abstract, and highly conceptual. It's a trans idea.

You've said before that you make music to reject passivity. If you're a migrant, you took the most incredibly active step a person can. Take a lot of guts.
Heavy guts. And urgency. And you can see that urgency, in the work and the conversations. Like, they have so much life. Because you have to have that life – and light, because it can get super dark. And you have to do it together, because your take your family and culture and identity to survive, if you fail a little bit, you have at least that. There's this double consciousness.

MONEY

You're a corporation, rather than non-profit. You had the Duty Free shop, work with Red Bull, and set up Buy-Black Friday. How does money fit in?
I always go back to Robin Hood, man. Steal from the rich to give to the poor. Divide it as equally as we can. We believe in walking in the building and saying "We here. We don't believe in everything you believe, but: We. Are. Here. You need us, we don't need you." We're not playing around, we're smart, you know. Infiltrate and subvert culture in whatever ways we see fit.

It's more honest to operate in these spheres and to politick in them, than go back into the echo chamber and only be around voices that agree with me. Nah. We need a multiplicity of voices, and we deserve to be heard ●

CHARLIE ROBIN JONES

Photography by Johnny Utterback, Live photography by Brian Whar

"INSTRUCTIONS FOR NON-CITIZENS"

1 — Volunteer at an organization which benefits the quality of life of marginalized people.
2 — Feed your friends. Share your resources with one another.
3 — Spread the message of The Non State.

'VIOLENCE= NO CHANGE'

CALLIGRAPHY

Over the last few months, artist TAUBA AUERBACH has written out the word 'Persevere' thousands and thousands of times.

A series of posters and public installations are now aiming to raise money and awareness for organizations including the Committee to Protect Journalists and GEMS (Girls Education and Mentoring Services).

"My favorite exercise in Daniel T Ames' Compendium of Practical and Ornamental Penmanship shows the word persevere written in lowercase script. Each letter is surrounded by a loop, similar to the a in the @ symbol. The loops are all the same but the letters are different, so the exercise teaches you to maintain a rhythm amidst otherwise varying circumstances."

"Calligraphy has become the activity during which I reflect on what's happening in the world, what's at stake, and what I'm willing to do about it. Maybe I've just needed something to do with my hands while I think. Until now, my politics have manifest mostly in quotidian, domestic choices like being vegan, composting and riding a bike. Feel free to roll your eyes. I support a few organizations. Big deal. I've always spoken my mind, but probably too politely. Besides, all of these choices are luxuries, and none of them registers as a sacrifice because they actually make my life more enjoyable. They are also, clearly, not sufficient."

"While doing calligraphy I've listened to a lot of speeches made by activists and philosophers. I've asked myself frequently if revolutionary change can take place without violence, and I've heard many sound arguments for why it cannot. Nonetheless, I remain certain that violence = no change, and that it is a doomed methodology for achieving it. In my view, violent means not only don't justify but also don't result in peaceful ends because the notion of an "end" is flawed. Now is the end. Every moment is the end. Civilization will always be in a state of becoming, so how we become what we want to be is what we are."

"Over the last few months, I've probably written the word persevere thousands times and in of hundreds ways. I've needed the time to think about what I can truly offer, about what a real contribution might be. I have some ideas, but I don't yet know if any of them are any good. In the meantime, I'm offering these drawings to support and thank some of the people I've held in my mind as I've written the word." ●

TAUBA AUERBACH

Persevere posters are available from diagonalpress.com for $25. 100% of profits benefit the Committee to Protect Journalists, GEMS (Girls Education and Mentoring Services), Chinese American Planning Council, and PLSE (Philadelphia Lawyers for Social Equity)

CLOSING SHOT

MARTIN SKAUEN

GANG LIFE

Q&A: RICHARD CABRAL

Born in the mid-80s into a family of East LA gang members, RICHARD CABRAL did his first time aged 13, going back to jail every year until he was 25. His longest stint, for attempted murder, was his last. On getting out, he left the gang he had grown up in. With the help of Christian organisation Homeboys Industries, he began mentoring those still caught up in gang life and prison, and embarked on a new career as an actor. He secured an Emmy nomination for his portrayal of Hector Tontz, a former gang member struggling to go straight, in the excellent ABC series American Crime, now in its third season.

"People see me how they see me, and that's all they see," Cabral's character says at one point. And Cabral's own story is one of identity and acceptance – of how the marks of a tough, violent past impact the present. But his story is also one of of how hard history can be held close, and how loyalty – to himself, as well as his fellow former gang members – can allow radical honesty to help others. "I witnessed guns, and violence, and everything people growing up there witness," he says, as we speak for an hour about prison reform, power and acting. "I finally came home at 25. And then it turned to what it is now."

GOOD TROUBLE: Tell us about life in LA.
Richard Cabral: I'm a second-generation Mexican-American, raised by my mom in East Los Angeles. I grew up in a metropolis of just Mexicans. The inner cities of Los Angeles have been riddled with guns and drugs since the beginning – it was poor, and law enforcement just didn't care. I was born in 1983, when the crack epidemic hit. So, I guess you could say I was a product of that energy, that time, and that sickness. Gangs, murder, mass incarceration.

LA historian Mike Davies said this explosion of gang violence from the 80s onwards is the result of deindustrialization. You have places where jobs were disappearing, so people were hanging around instead of working. And this coincides with the arrival of crack…
It was like these two forces that coincided at the same time. Boom. In the south side and in East LA, you have these cities which are all industrial. Right along the LA River, it's all factories and warehouses. So, you those kids with the mind to work, but all you have are drugs. The knowledge now is methamphetamine, and has been for the last 15 years. And while it's not as visible as the crack epidemic, it's taken its toll on the communities. The craziness of the stories, mothers killing babies and shit, all that has to do with drugs. The drugs really fucked things up.

One thing I heard about solving gang violence was that only warriors can end the war.
Yeah, that's a good one. For sure, for sure. To talk about the war, you have to know the war. To talk about death, you have to know death. There's a normality to it. It's the philosophy of a warrior, or a man in the army. It's not abnormal to know you might die, because there's a gang of other motherfuckers that might die with you. They all get it: we talk about death, and we talk about jail. The first time I went to jail as a kid, I looked around and thought 'Oh! There's hundreds of others like me.' I remember being young and seeing my uncle go to prison. My uncle has been a gang member since before I was born. You look outside and see gang members. You know the violence and craziness it carries, but you know they're not bad people… They're people.

What was the thing that turned it round?
The truth was I didn't want to spend my life in prison. I spent a year in jail. I had a whole year to think. And through my prayers or whatever, I got five years. But for that whole fucking year, you're thinking you might never come home.

What is the effect of all these years getting handed down by the state on the various communities affected?
You fuck up the community by having kids grow up without their fathers and mothers. You destroy the community. My best friend was 15 when he got life. Fifteen! California gives you life.

Why do you not hide from the past you had?
If I don't stand behind it, and say this is what made me, I cannot be inspiration. I cannot go into prisons and talk to people. Embracing it has been the most powerful thing.

Was getting the Emmy nomination for your acting a validation?
Yeah, but a validation I wasn't seeking. I'm happy now. I was in a cell eight years ago. Now I'm out and working and seeing my kids. But it was a surprise, because I just concentrate on the work, and this just meant people recognised the work.

What are your feelings about Trump?
Well, during Obama's reign, he deported more than any other president in history, so we've always been in the shit in a way! But when the threat becomes real evident, it makes people united. If I let someone piss me off, I've given them power. This. Too. Will. Pass. As a prison reformer, we're in a good place. In California, laws are getting passed, and we just need to push on ●

CHARLIE ROBIN JONES

Photography by James Mooney

66.把握节奏①

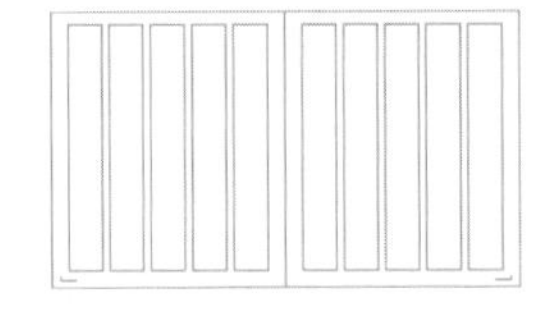

排版是一种叙述方式，尤其是在一个包含许多插图的多页项目中。许多书籍的章节和杂志中的专题文章都涉及多个页面或屏幕的设计布局。

开阔的对页可以使用全出血版式。这种对页为接下来的页面设定了背景，就像电影的标题为电影设定了基调一样

一个页面或一个对页与另一个页面或对页使用不同的字号和字体、分栏、图片和色彩，引导着故事的发展，从而产生了戏剧性的效果

项目
《伊甸园图》（*Portrait of an Eden*）

客户
费拉本德（Feirabend）

设计师
丽贝卡·罗斯（Rebecca Rose）

这是一本详细描述一个地区的历史和发展的书，采用了不同的对页布局来引导读者"穿越时空"

① 由于原英文版书中的第66个案例不适合本书，经版权方同意进行了删除，故此处案例应为原书中的第67个案例。本书此后案例序号已重新排列。

Opposite:
Gertrude leaning against a coconut palm in Lummus Park wearing a playsuit, 1938. A hedge of *Malvaviscus arboreus*, a relative of the hibiscus, is in the background. Stretching the length of Ocean Drive from Sixth Street to 14th Place, Lummus Park was donated to the City in 1912 by the Lummus Brothers' Ocean Beach Realty Company.

A Bermuda grass lawn was immediately planted with the hope that its aggressive root system would supply strong underground runners to hold the sandy soil in place. Coconut palms were planted as well, to provide inviting shade and a sense of site. Finally, a ten foot-wide sidewalk was installed. From 1912 to 1917, the Lummus Brothers spent $40,000 to create and maintain Lummus Park for the people of Miami Beach.

Left:
Barbara June Oka poses by the Shower of Gold (*Cassia fistula*), late 1940s. Her right arm mimics the smooth barked limb. Joints of movement and growth, the elbow and node are parallel structures.

Healing Plant
The Shower of Gold tree was critical to early people dependent on the properties of nature to heal. *Cassia fistula* was valued by ancient Egyptians for its long cylindrical pods or fruits. Ripening from fresh green to burly [illegible] these pods grow up to two feet in length. When mature, they contain a sticky brown pulp and several seeds used to cure a multitude of afflictions. Cuttings were carried with the Spanish conquistadors to the New World and firmly planted on the island of Cuba. English naturalist Sir Hans Sloane, practicing medical botany in Jamaica almost three hundred years ago, discussed the curative powers of Cassia and gave a potion for a purgative cure: "The pulp purges [illegible] much if powdered Sassas be added." Today the medicinal properties of *Cassia fistula* are being rediscovered.

Miami Beach of the Orient
Mayor Kenneth Oka fostered Miami Beach's participation in President Eisenhower's newly-minted People to People Program and instituted the city affiliation between Miami Beach and Fujisawa, Japan. The bond gained national and international publicity for Miami Beach.

In recognition for his outstanding work in foreign relations, Oka received the annual People to People Award in New York City from United Nations Ambassador James Wadsworth. Gertrude took up the brush while spending several months visiting all of her new Japanese friends in the "Miami Beach of the Orient."

Ink drawing by Gertrude Oka, c. 1960

● 67.创造一片“绿洲”

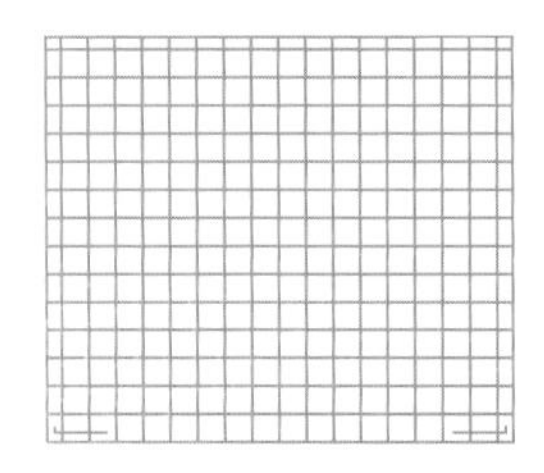

如果要表现出权威感和聚焦感，记住，少即是多。留白能让读者更专注。

另见128～129页

The driving principle of Cuadro Interiors is to ensure the building process works professionally and efficiently. Projects are executed on an individual basis, utilizing a skilled team of long-term employees and tradespeople. Our commitment is to produce the highest quality project in a reasonable and honest manner.

Founded by Raphael Ben-Yehuda and Mark Snyder, Cuadro's approach reflects its partners' backgrounds in fine arts. Their combined forty plus years of building experience includes projects ranging from wood boat building to faux finishing.

Today we are a company with extensive experience in a broad range of project types, from historically accurate prewar homes to modern offices to contemporary residences in a range of materials.

一个模块化的图形引出了这个作品

项目
《卡德罗室内设计宣传册》

客户
卡德罗室内设计（Cuadro Interiors）

设计公司
杰奎琳•陶设计公司（Jacqueline Thaw）

设计师
杰奎琳•陶

首席摄影师
伊丽莎白•费利切拉（Elizabeth Felicella），安德鲁•祖克曼（Andrew Zuckerman）

这是一家室内设计公司的宣传册，其版式设计基于模块化的网格，但内容化繁为简，只专注于介绍有特色的家居和办公室设计案例

充分留白的页面犹如一片让读者流连忘返的绿洲，读者可以专注于图片和信息的方方面面

OXO
Industrial design firm

● 68.让图片“呼吸”

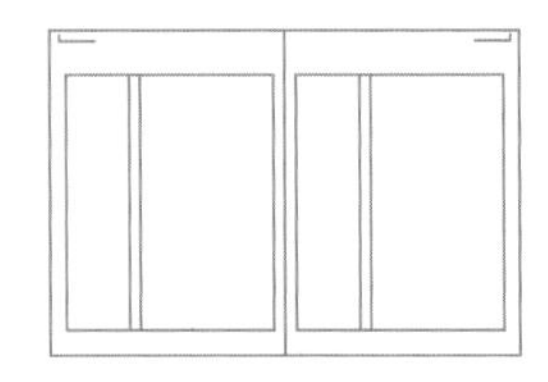

留白较多的页面可以迅速将读者的目光吸引到重要的照片或插图上来，让读者可以集中注意力。

留出空间

通常，素材的内容是设计师分配文本和图片空间的依据。如果文本涉及特定的照片、画作或图表，那么，将图片放在相关文字的附近对读者来说是最清楚的。如果读者要向前或向后翻页才能将图片和文字相匹配，会导致阅读效率降低。

图片的大小也很重要。放大一个图片的细节可以为页面增添活力。周边留白的图片往往比与许多元素组合在一起的图片更能吸引读者。

另见126～127页

项目

《马扎尔集市：巴基斯坦的设计与视觉文化》（*Mazaar Bazaar: Design and Visual Culture in Pakistan*）

客户

牛津大学出版社巴基斯坦分公司（卡拉奇）（Oxford University Press, Karachi），海牙克劳斯王子图书馆（Prince Claus Funds Library, The Hague）

设计师

赛马•扎伊迪（Saima Zaidi）

这是一段关于巴基斯坦设计的历史，设计上使用了严谨的网格排列巴基斯坦的艺术品，页面中有充足的留白

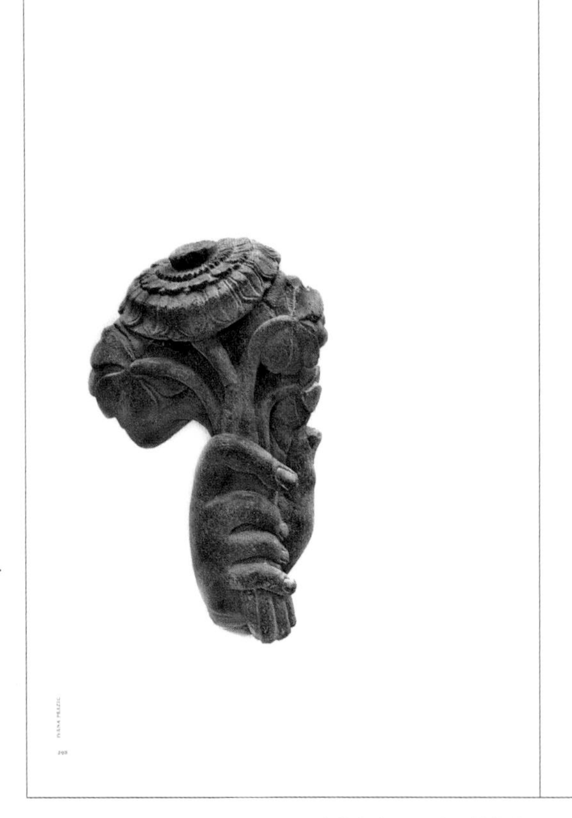

STORYBOARDS IN STONE

THE BUDDHIST ART OF GANDHARA

IVANA FRAZIC

Gandhara, an ancient region located in the North of Pakistan is the alleged art-historical birthplace of the first known sculpted anthropomorphic image of the Buddha.[2] Legend has it that Buddhism reached Gandhara soon after the Buddha's death.[i 3] Even though there is no archaeological evidence to support this possibility, unearthed materials have proved that Buddhism was present in the region, at the latest, from the time of the reign of the Mauryas (300-150 BC);[4] according to travel accounts of two renowned Buddhist pilgrims from China, the religion thrived there for several centuries.[ii 5]

The land of Gandhara was practically studded with *caitya* halls, *viharas* and *stupas*[iii 6] when Fah-Hian, one of the Chinese pilgrims, visited it in the 5th century BC, Taxila[v 7], a highly important Gandharan town was perceived as a site of particular religious significance:

Buddha, when he was bodhisattva,[iv] in this place [Taksasila] gave his head in charity. [...] Still going to the eastwards two day's journey, you come to the place where he gave his body to feed a starving tiger. In these two places they have raised great towers, adorned with all the precious substances. The kings, ministers, and people of all the surrounding countries vie with each other in making religious offerings at these places, in scattering flowers, and burning incense continually.[8]

Turbulent historical events caused slow destruction of many of these monuments, just as they brought about gradual fading of Buddhism in Gandhara. By the time when yet another Chinese Buddhist, Hsüan-tsang, went on a pilgrimage in the 7th century, Gandhara had been left almost desolate, its sacred places and monasteries in ruins.[9] The huge output of art in the Buddhist Gandharan style (first to sixth centuries AD) is believed to have ended with the devastations by the Central

i. Precise dates not known, various sources mention years 544 BC, 487/486 BC, 368 B.C.
ii. Fah-Hian in the 5th and Sung-Yun in the 6th century
iii. From Sanskrit bodhi, or enlightenment/supreme awakening, and sattva, meaning pure and fine. He is a future Buddha, 'a being who is in process of obtaining but has not yet obtained Buddhahood'.
iv. Caitya, a place of worship; vihara, buddhist monastry; stupa, monument built over a sacred place or relics of the Buddha
v. Ancient Takshasila, from Sanskrit sila, stone or rock, and Taksa, which according to the *Ramayana* was the name of the son of the founder of the town. Thus, its meaning is '*the rock of Taksa*'.

Figure 1. HAND HOLDING LOTUS (*Left page*)

2nd - 3rd century. Sahri Bahlol, Mardan, North West Frontier Province (NWFP). Schist, H 19 cm

National Museum of Pakistan, Karachi

N.M. 1974

这篇名为《石头里的故事》(*Storyboards in Stone*)的文章中介绍了一只拿着一朵莲花的手。这篇文章有充足的留白，并用标题、正文和对页的脚注来保持版面平衡

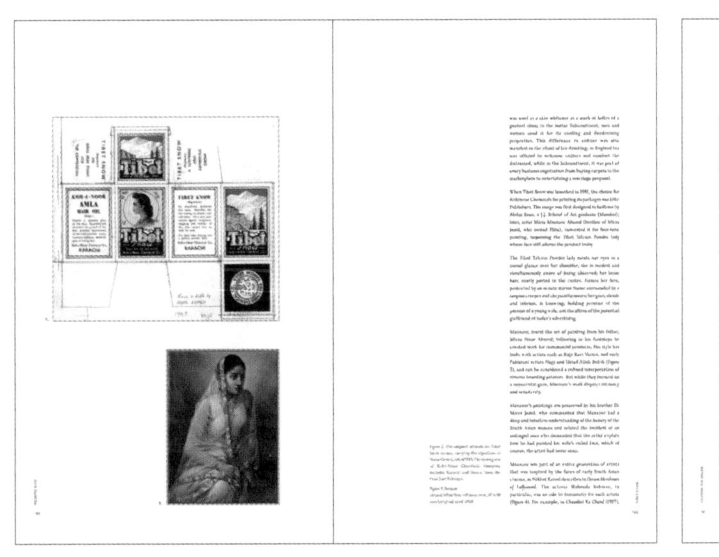

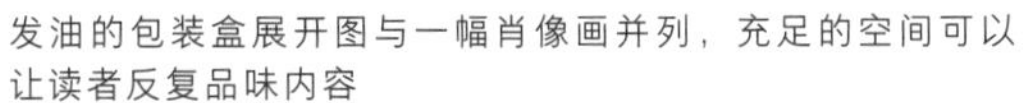

发油的包装盒展开图与一幅肖像画并列，充足的空间可以让读者反复品味内容

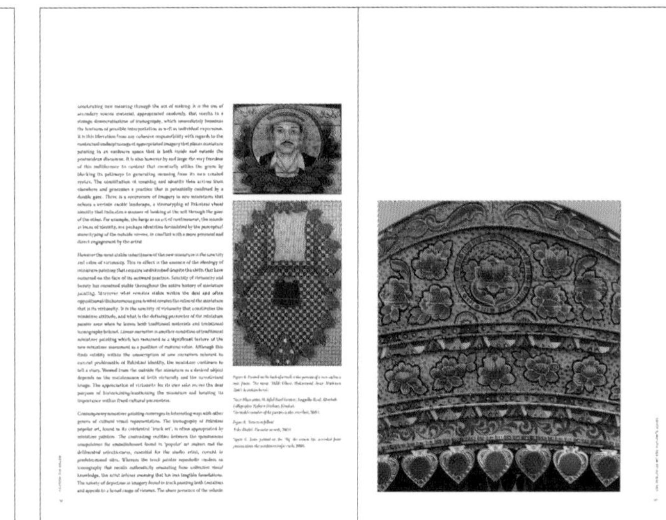

取自卡车尾部的图画与图案使版面色彩艳丽，有质感

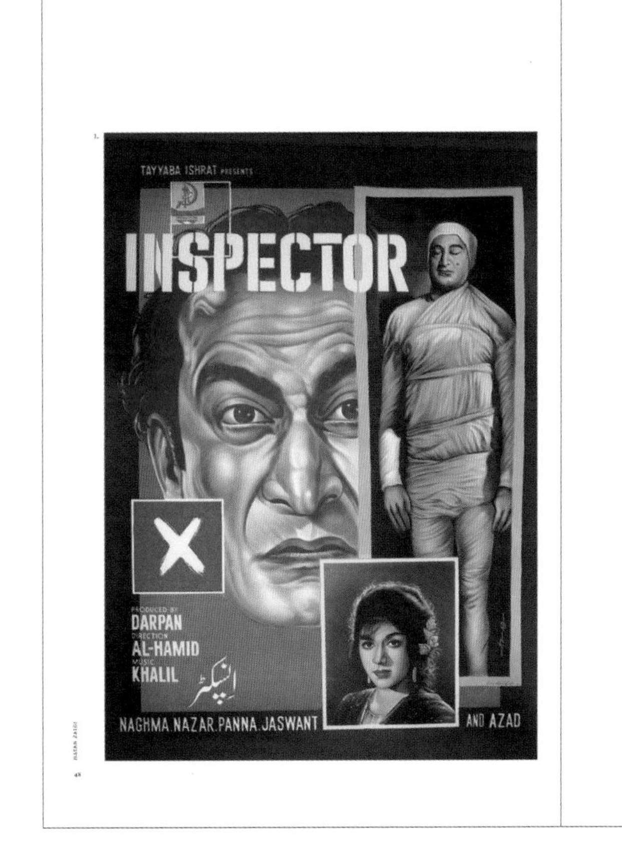

Ama: PUBLICITY DESIGN AS A REFLECTION OF LOLLYWOOD FILM CONTENT

HASAN ZAIDI

Buxom belles in scarlet and gold, *gandasa*[i]-clutching heroes in beige and black, blood—or sweat-drenched villains in diabolical green and purple: nothing captures the essence of commercial film in the subcontinent better than its medium of publicity. Literally larger than life, cinema hall hoardings depict as much action and emotion on canvas as film-makers squeeze on to three hours of celluloid. As a result, they inevitably evoke the visual chaos we have come to associate with Lollywood,[ii] the Pakistani film industry.

Hoardings and posters are the first point of contact between the film and its target audience. The 'reel 'em in' splash of colours depicting dramatic scenes of sex and violence is calculated to have the masses lining up outside ticketing booths. Before television promotions and mass marketing made film publicity itself an art, these melodramatic hoardings were solely responsible for enticing audiences into darkened theatres screening black and white films. Even now, no cinema district would be worth its name if it did not offer an array of exaggerated, multihued interpretations of silver screen fantasy.

Much like film posters on the back of horse-drawn *tongas*[iii] and motorised rickshaws—which long preceded mobile advertising in the West—cinema hoardings came to be identified as a uniquely Subcontinental medium of film promotion, even though the films being promoted were often Hollywood musicals adapted to suit the indigenous taste for *nautanki*.[iv] Echoing the diversity of the region, the singular film poster—a staple of film publicity worldwide—was translated into two- or three- storey-high works of art for the street. As such, cinema hoardings or street art in the real sense of the phrase became popular. Inviting, arousing, brash, and disposable, the hoardings survived only as long as the films they represented screened

Figure 1. (Left page) Poster

FILM Inspector DESIGN Galib MEDIUM Lithography PRINTERS Nawa-i-Waqt, Lahore.

Aina meaning mirror, the title of an epic film from the 1970's.

i. A wooden baton

ii. The Pakistani cinema industry primarily based in Lahore is popularly termed Lollywood, referring to Bombay's Bollywood derived from the Hollywood Studios.

iii. Horse drawn carriage

iv. Folk theatre

一张极具震撼力的图片引出了一篇文章

● 69.绘制草稿

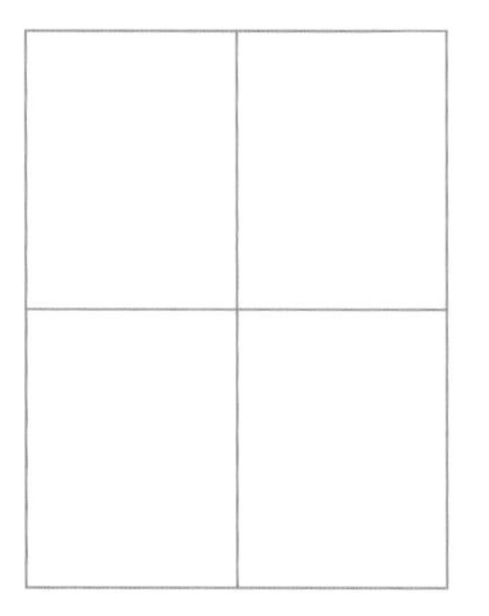

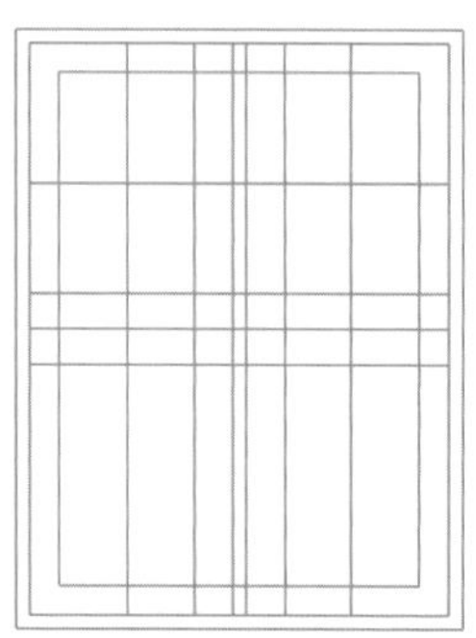

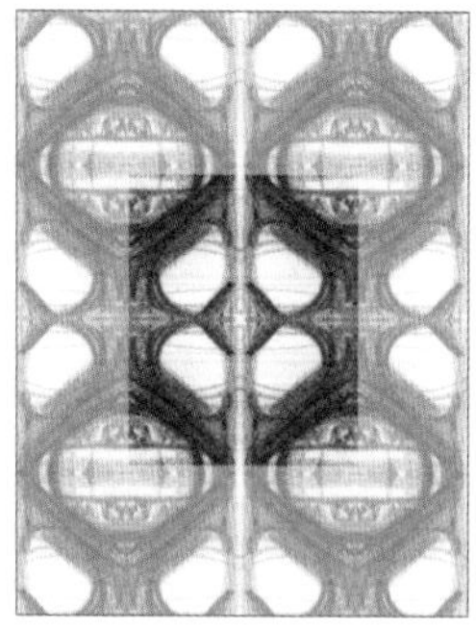

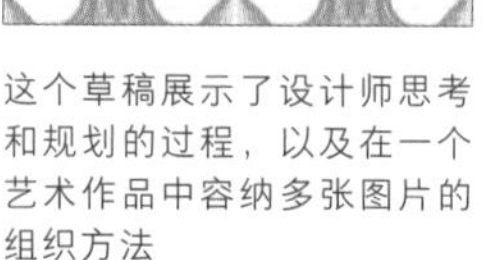

这个草稿展示了设计师思考和规划的过程，以及在一个艺术作品中容纳多张图片的组织方法

有了全局的考虑和规划，每个独立的部分都可以被设计出来

项目
《麦克斯威尼》杂志（第23期）（*McSweeney's Issue 23*）

客户
麦克斯威尼出版公司

执行编辑
埃利•霍洛维茨（Eli Horowitz）

设计师
安德里亚•德佐（Andrea Dezsö）

这是《麦克斯威尼》杂志（第23期）的护封，艺术家安德里亚•德佐设计的手绘的镜像和重复的图案将各载体上的作品统一了起来。铅笔画、手工刺绣、手工制作的三维皮影戏照片、蛋彩画（一种古老的绘画技法，是用蛋黄或蛋清调和成的颜料绘成的画，多画在表面敷有石膏的画板上）在一个强大的框架内共存。在这个项目中，德佐仅在扫描及合成的时候使用了电脑

画草稿有助于构思，并帮助设计师规划出版物或页面的布局。最初的草图可能看起来更像涂鸦，而不是易于识别的元素，但草图可以形成一个整体计划或概念。当在一个较大的概念中包含一个或多个图片时，最好的方法是组织一个模板和网格，让一个作品中的各种元素布局恰当，互相搭配。

草拟出一个想法和模板可以节省很多工作，很少有人有时间重复每一个步骤。事先的规划非常重要，无论这个规划是只包含文字或图片，还是两者的组合。

该项目讲的是图案和规划，探讨如何将许多封面图案融合在一个大的护封上

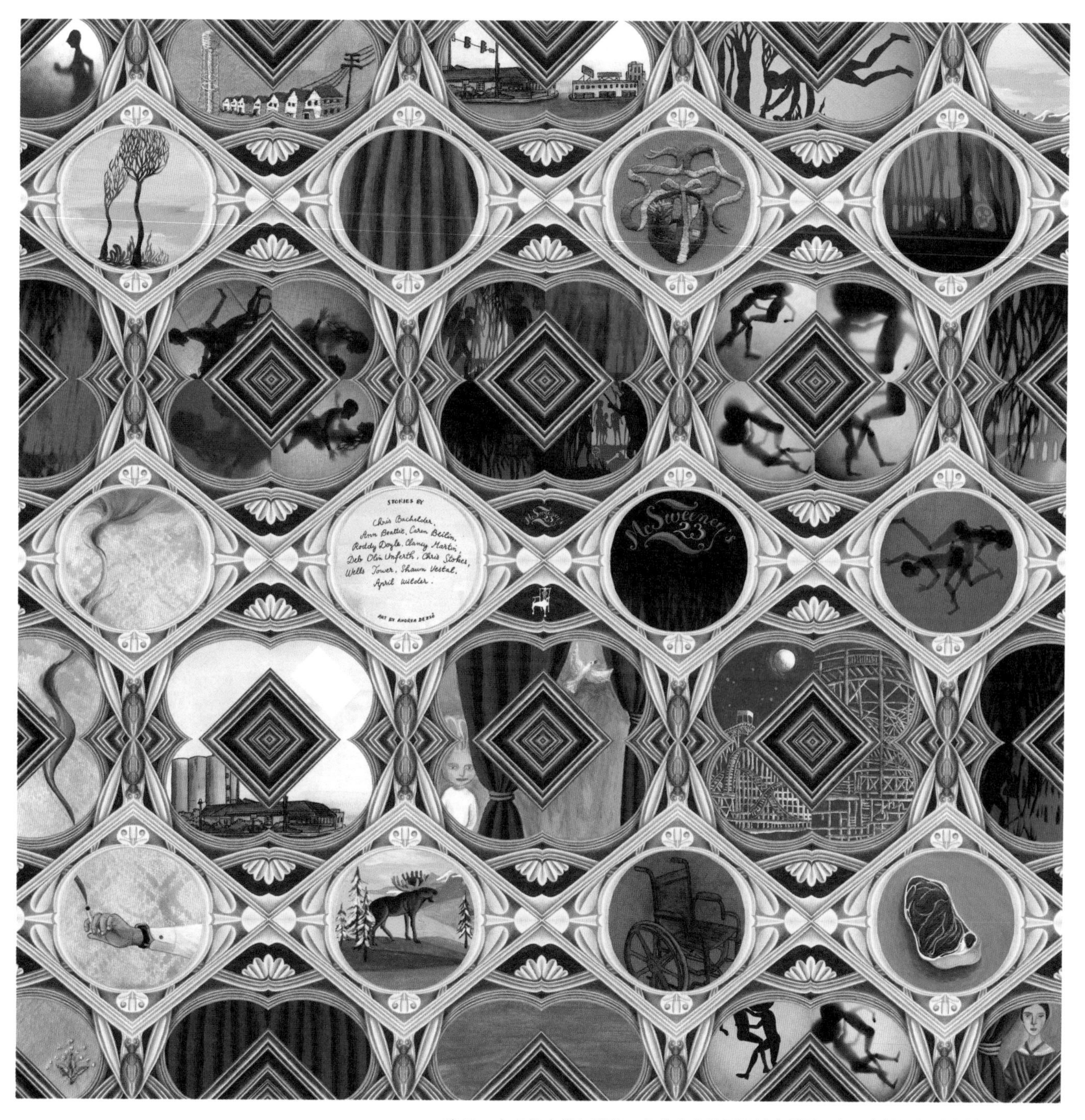

这是一个“框中框”设计，包含十个封面和封底插图，每一个插图都对应着《麦克斯威尼》杂志（第23期）中的一个故事。十个封面可以组合成一个大护封，护封展开后可以形成一个全尺寸海报。手工绘制的视觉框架成功地统一了各元素，使独立的艺术作品组合成一个更大的整体

● 70.利用层次梳理材料

一个前后一致的简单条形框设计可以将数据分成不同的层级

过多的数据会使页面变得复杂、混乱。一个简单的横条与清晰的解释和说明可以将杂乱的内容理顺。

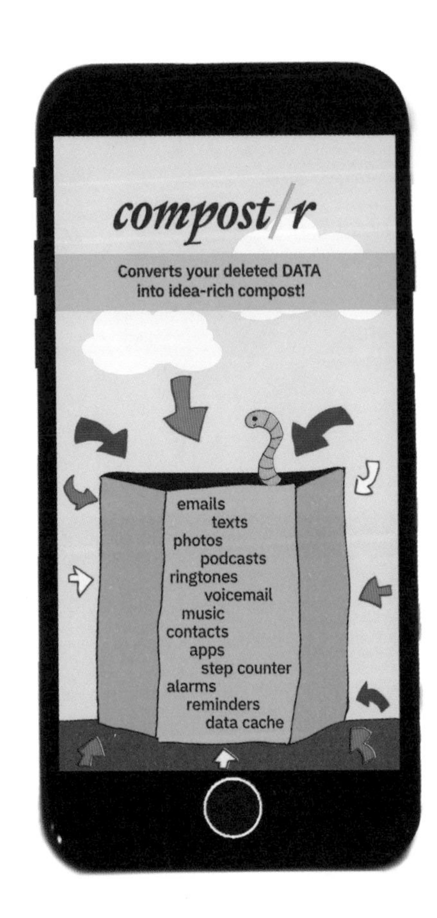

项目
compost/r应用程序

客户
多坡多马尼（Dopodo-mani）

设计师
苏珊娜•戴尔•奥尔托

插画师
尼娜•劳森（Nina Lawson）

compost/r是一款应用程序，它借用了现实世界中堆肥的概念，将手机中删除的数据转化为诗歌、图案、铃声和音乐

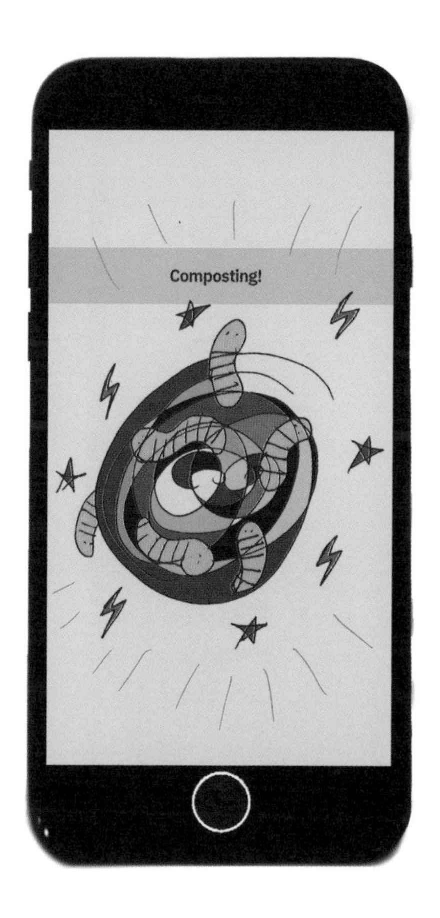
Composting!

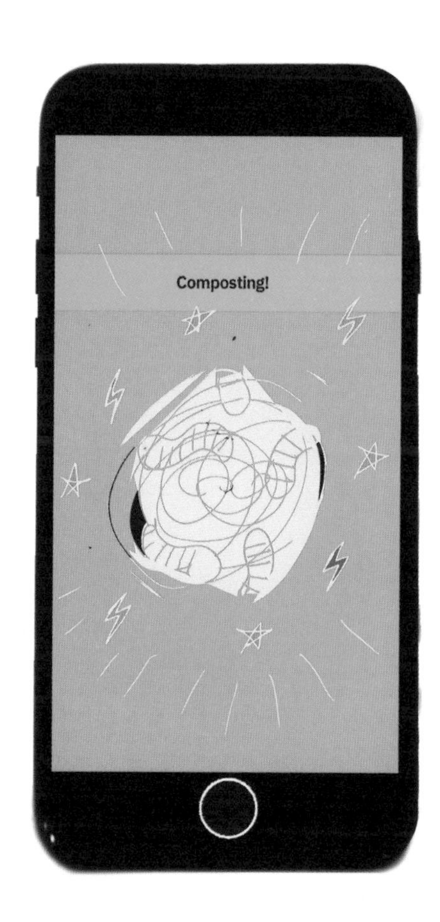
Composting!

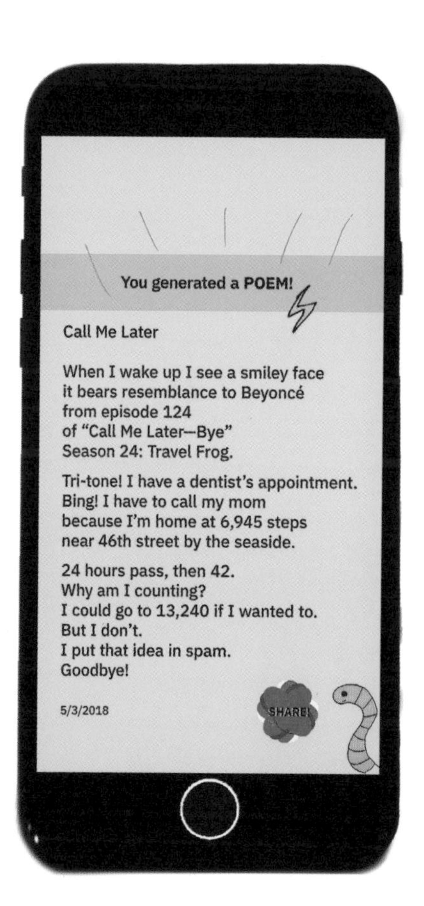
You generated a POEM!
Call Me Later
When I wake up I see a smiley face
it bears resemblance to Beyoncé
from episode 124
of "Call Me Later—Bye"
Season 24: Travel Frog.
Tri-tone! I have a dentist's appointment.
Bing! I have to call my mom
because I'm home at 6,945 steps
near 46th street by the seaside.
24 hours pass, then 42.
Why am I counting?
I could go to 13,240 if I wanted to.
But I don't.
I put that idea in spam.
Goodbye!
5/3/2018
SHARE!

● 71.使用网格的原则

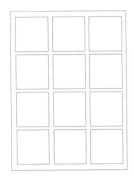

即使设计师以为自己并没有使用网格，他们也会下意识地运用网格的基本原则。有些设计可以明显地看出是在网格上规划的，有些则是依靠“视觉网格”规划的，还有一些设计只有一点儿网格的痕迹。

另见19页

项目
《一些有趣的书》（*Some Fun*），《我是特别的》（*I'm Special*），《美国呆子》（*American Nerd*）

客户
西蒙和舒斯特公司（Simon & Schuster,Inc.），斯克里出版社（Scribner），隶属西蒙和舒斯特公司

《一些有趣的书》
艺术总监
约翰•富尔布鲁克（John Fulbrook）
设计师
杰森•海尔（Jason Heuer）

《我是特别的》
艺术总监
杰基•苏（Jackie Seow）
设计师
杰森•海尔

《美国呆子》
艺术总监
约翰•富尔布鲁克
设计师
杰森•海尔
插画师
沙斯蒂•奥莱利•索萨特（Shasti O'Leary Sounclat）

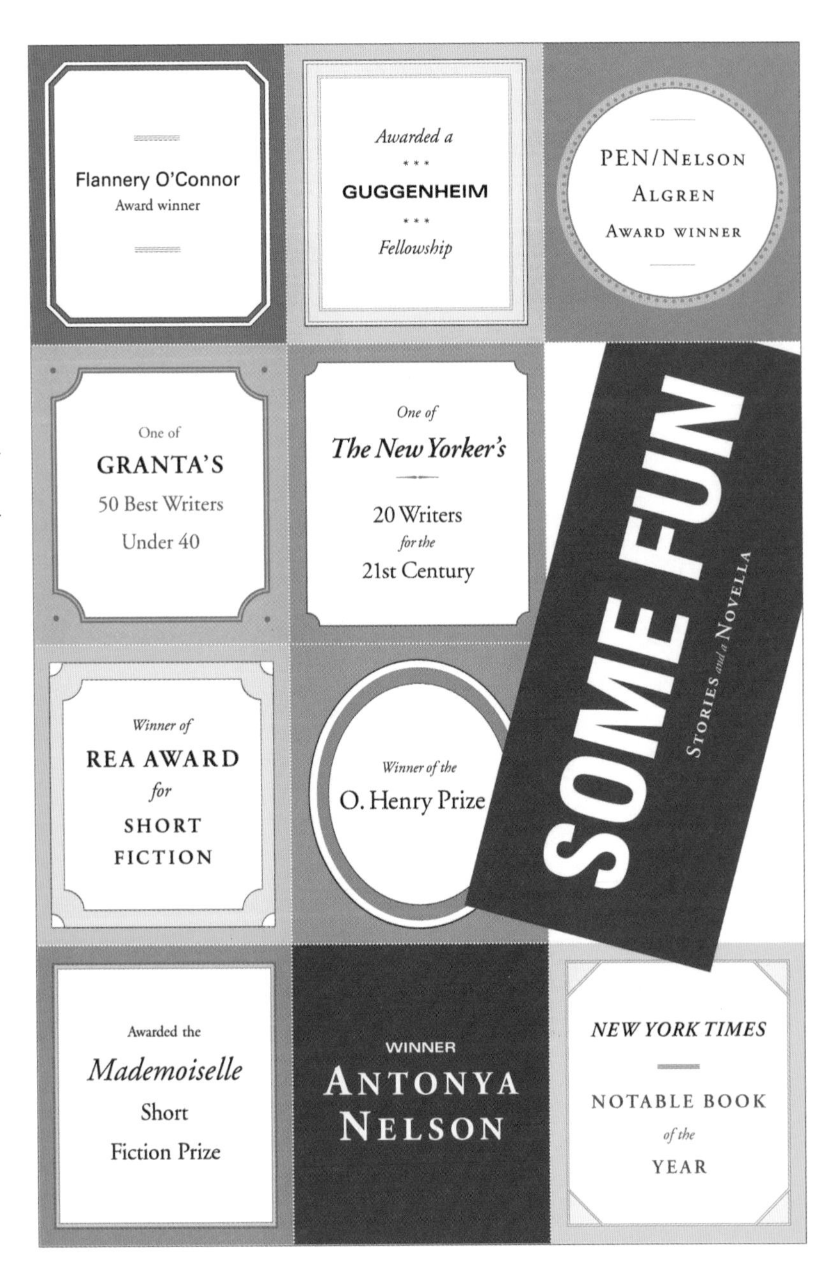

三个护封展示了在网格使用方面不同的严谨程度

《一些有趣的书》使用了一套严谨的网格，但又用标题打破了这个网格，十分有趣。

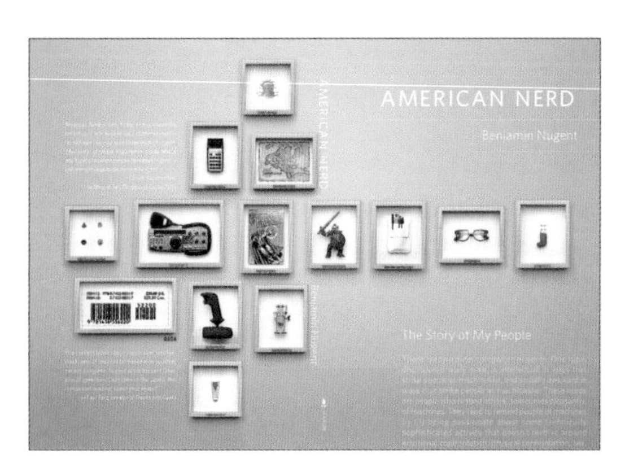

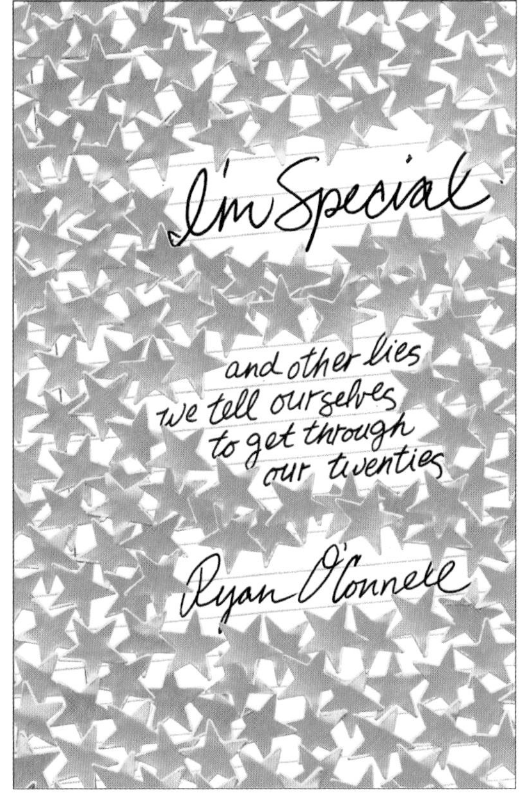

《我是特别的》的设计非常独特，为了保持有机的感觉，它抛弃了网格，以带有蓝色线条的纸张强调其摒弃网格后的活力感

如封面和封底的内容和图片所示，《美国呆子》使用的是视觉网格，而不是数学上的网格

视觉网格：依靠直觉而不是数学计算

这些案例的设计师认为，艺术家除了要有目的地使用网格和黄金比例之外，还要会自然地使用网格。与许多设计师一样，杰森•海尔在设计时使用了视觉网格，但在最后使用了经过数学计算的网格来排列设计元素。

● 72.保持流动性

一个结构合理的设计即使是在框架不是很明显的情况下，也有坚实的基础。

项目
杂志插画

客户
《印刷》（*Print*）杂志

设计师
玛丽安•班杰斯

这个为杂志设计的页面使用了精细的印刷工艺

“我用视觉编排的方式设计。我要确保在自己的作品中有结构可循。在标题中，我会利用图片和垂直线编排素材，然后反复修改细节，确保排版效果。我还很在意逻辑结构，注重信息的层次感和一致性。我相信优秀的排版和设计就像一件裁剪得体的西装：普通人可能不会特别留意手工缝钉的纽扣（字距调整）、剪裁合身的衣褶（完美的字距），或者高级的布料（完美的字号）……他们唯一知道的是，这件衣服看上去价格不菲。”

本页图和上页图：玛丽安•班杰斯尤其关注排版的细节，她认为，合理的段落应包含连贯的字母和恰当的字间距，以及来自特定时期的字体。她拥有敏锐的眼光，经她设计之后，这些字体看起来会更鲜活。不过，这两页真正的点睛之笔是她设计的书法式的插图

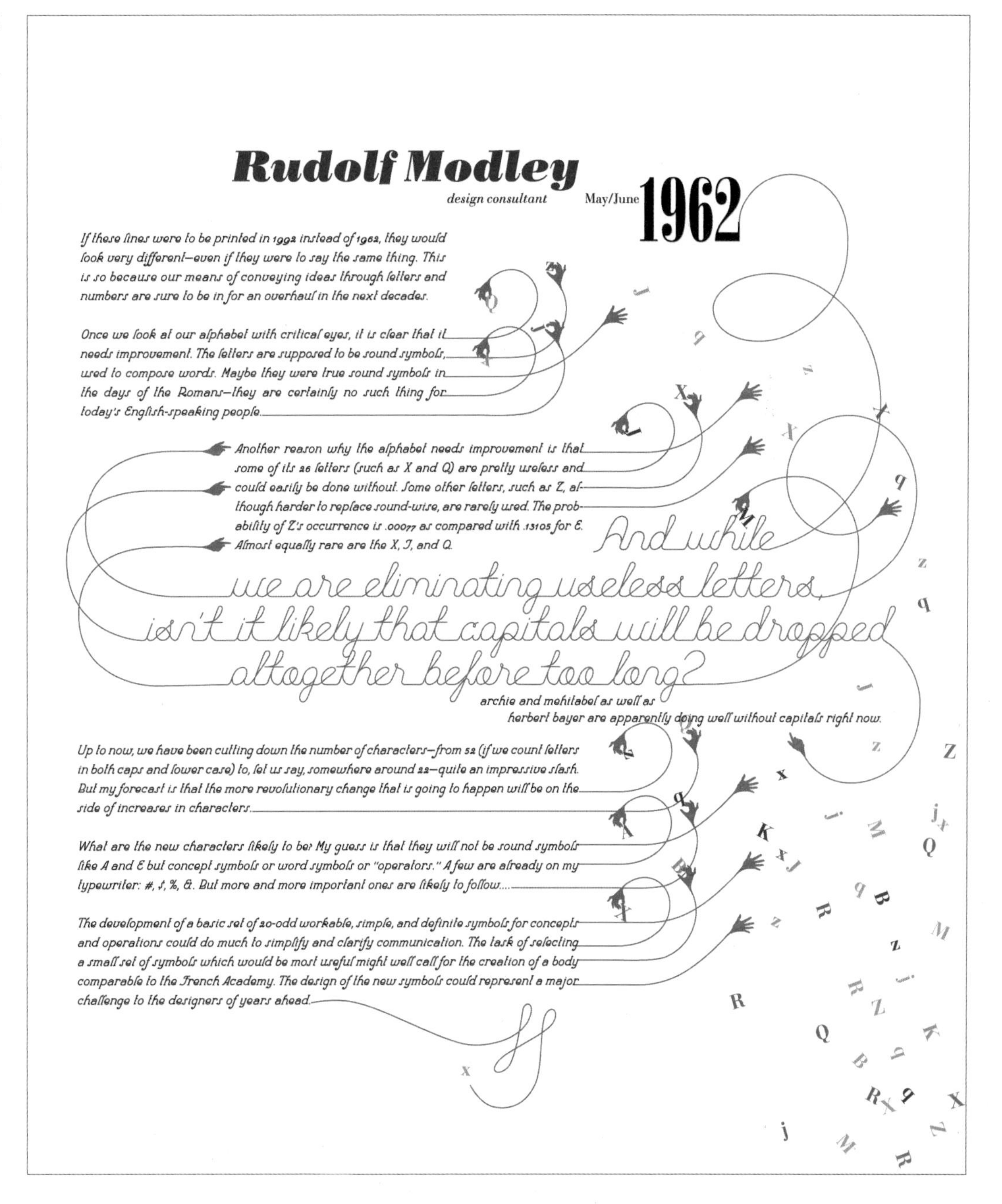

Rudolf Modley

design consultant

May/June **1962**

If these lines were to be printed in 1992 instead of 1962, they would look very different—even if they were to say the same thing. This is so because our means of conveying ideas through letters and numbers are sure to be in for an overhaul in the next decades.

Once we look at our alphabet with critical eyes, it is clear that it needs improvement. The letters are supposed to be sound symbols, used to compose words. Maybe they were true sound symbols in the days of the Romans—they are certainly no such thing for today's English-speaking people.

Another reason why the alphabet needs improvement is that some of its 26 letters (such as X and Q) are pretty useless and could easily be done without. Some other letters, such as Z, although harder to replace sound-wise, are rarely used. The probability of Z's occurrence is .00077 as compared with .13105 for E. Almost equally rare are the X, J, and Q.

And while we are eliminating useless letters, isn't it likely that capitals will be dropped altogether before too long?

archie and mehitabel as well as herbert bayer are apparently doing well without capitals right now.

Up to now, we have been cutting down the number of characters—from 52 (if we count letters in both caps and lower case) to, let us say, somewhere around 22—quite an impressive slash. But my forecast is that the more revolutionary change that is going to happen will be on the side of increases in characters.

What are the new characters likely to be? My guess is that they will not be sound symbols like A and E but concept symbols or word symbols or "operators." A few are already on my typewriter: #, $, %, &. But more and more important ones are likely to follow....

The development of a basic set of 20-odd workable, simple, and definite symbols for concepts and operations could do much to simplify and clarify communication. The task of selecting a small set of symbols which would be most useful might well call for the creation of a body comparable to the French Academy. The design of the new symbols could represent a major challenge to the designers of years ahead.

● 73.为插入做好规划

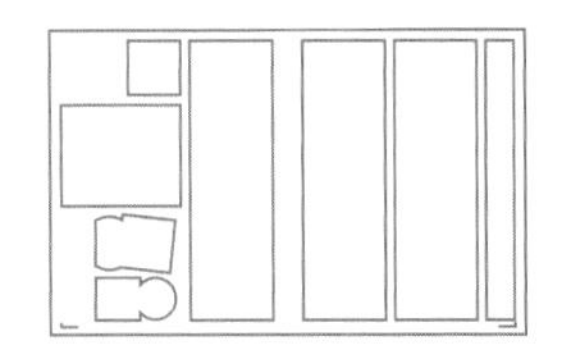

规划是设计最重要的原则之一。格式是规划，网格也是规划。插入的元素可以是规划的主要部分之一，合理的排版可以使插入元素成为规划中清晰的一部分。设计师通过确定哪些名称或有特色的部分需要设置得较大或较粗，哪些部分需要颜色，首字下沉是否必要来决定哪些元素可以插入版面。

不同大小的图片也可以插入页面，给一个作品或一组对页带来活力。

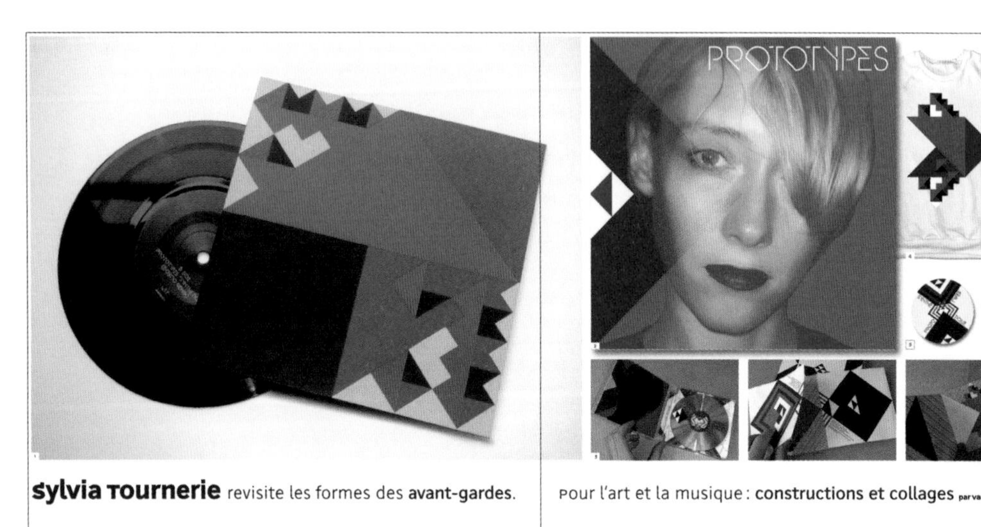

sylvia Tournerie revisite les formes des **avant-gardes.**

Douze ans de création graphique française : la production de Sylvia Tournerie est un échantillon symptomatique d'une génération, tout en étant singulière. En 2008, la graphiste peut s'enorgueillir d'une livraison clairsemée mais régulière. Douze ans d'indépendance, et cette liberté s'affirme aujourd'hui avec une même fragilité davantage décomplexée. Son dernier travail est un "plaisir" de simplicité, elle a composé un univers cohérent, tout en variation, pour le groupe Prototypes. Pour leur dernier album, *Synthétique*, Sylvia Tournerie manipule un répertoire formel basique, un triangle, deux couleurs, le noir et blanc. Peu d'éléments, mais l'infini s'esquisse. Son plaisir d'orchestrer une communication avec un minimalisme rythmé ravive ses envies sur ce métier, et confirme ses audaces typographiques.

Retour sur un parcours

À sa sortie de l'école Penninghen, ses commandes s'enregistrent toutes sur le répertoire de la musique électronique. Non pas que le culturel ou l'espace public ne l'attirent pas – ils furent son sujet de diplôme[1] –, mais la musique lui offre l'occasion d'amorcer ses premières manipulations. Le groupe Bosco lui demande de créer son image et lui assure ainsi une prodigieuse liberté. Suivront d'autres invitations de labels indépendants qui lui ouvriront un espace d'expérimentations à même de construire son répertoire, de s'affranchir d'une certaine pudeur, puisqu'elle se rend compte que *tout est possible*. Les pochettes se suivent et donnent le jour à une palette variée, allant d'un registre construit et typographique (Bosco, *Novo Screen*) à des collages provocants et plus trash (Bosco, *New Pux*). Milieu des

42 : 3.2008

Pour l'art et la musique : **constructions et collages** par Vanina Pinter ■ ■ ■

années 1990, la musique électronique est un territoire expérimental et propice aux compositions graphiques. À l'image des samples, les mariages cocasses de sons, le tempo électronique trouve une résonance en connivence avec les motifs géométriques, répétitifs et anonymes, et/ou les collages dissonants.

Transferts de savoir-faire

En 2004, la prise en charge de la revue *02*[2] marque un tournant dans le parcours de Sylvia Tournerie. Sa mise en forme de la revue gratuite et pointue d'art contemporain la fait remarquer par d'autres commanditaires, une brèche s'ouvre dans le monde de l'art contemporain. Cette brèche rappelle la difficulté pour les graphistes à aborder d'autres genres s'ils n'ont pas été invités et repérés. Pour le deuxième numéro, elle décide de travailler avec son ancien camarade d'école Gilles Poplin, avec lequel elle avait déjà créé des clips. *02* est ainsi à l'origine d'une fructueuse collaboration, un travail en duo, une joute en complicité, donnant naissance à une dizaine d'alphabets. En effet, pour chaque numéro, ils dessinent une typo, cette dernière fournira une allure et une unité à la revue, qui, pour le reste, affiche une maquette sobre, une intégration classique des images. *Nos typos de titrage sont un outil pour construire un mot et pour que visuellement, elles aient la force d'une forme.* Leur travail du signe est leur réponse pour ne pas ajouter du visuel à la création artistique contemporaine. Réponse qui, pour des catalogues d'expositions, lui fera adopter des postures de retrait : la graphiste manie les lettres et cherche la lisibilité des images des artistes.

1. disque promotionnel. 2. verso. 3. intérieur de la pochette CD. 3. stickers pour Synthétique. Sylvia compose autour d'un triangle des motifs aux évocations "péruviennes ou constructivistes". Elle dessine le "typotype", sur la variation de sa gamme. 4. Sylvia habille une série de vêtements, qu'elle souhaite voir prendre son indépendance et réfléchit à les autoproduire. ces motifs prendraient ainsi leurs aises dans l'espace quotidien.

3.2008 : 43

大幅彩色的图片与干净的网格形成对比

项目
《阶段》杂志

客户
金字塔/《阶段》杂志

设计师
安娜•图尼克

法国《阶段》杂志的内页展示了如何利用一张大图片、一个剪影图和大量的留白让页面显得不呆板

6. Pochette du maxi-vinyle "Novo Screen" pour le groupe Bosco, 2002.
7. Pochette CD pour Panti Will, album "H.E.L.L", 2003.
8. Pochette de "Fastback" maxi-vinyle pour Sodex, (pour le label client 2000st), utilisation d'une typo originale, la Copland.
9. Pochette CD pour Experience, album "Hémisphère gauche", 2004.

ses "gimmicks"

À l'incontournable – et douloureuse – question sur l'auto-définition de son style, Sylvia Tournerie évoque deux éléments signifiants. L'école s'étant équipée d'ordinateurs à la fin de ses études et le recours à la photocopieuse étant également plus facile, cela a entraîné un style repérable, économique, un jeu de découpes. *Mon travail est marqué par des grosses masses noires avec des couleurs primaires.* Difficile de ne pas faire allusion à l'empreinte de Cieslewicz. Sylvia Tournerie a étudié à l'ESAG-Penninghen au temps où Roman Cieslewicz y enseignait[3]. Il fut son maître de thèse. De lui, elle se souvient d'un rire qu'il eut, durant un stage, alors qu'il manipulait des formes et concevait un hors-série pour *Le Monde*. Cette excitation, cette légèreté, qui ne s'essouffle pas malgré les années, cette ouverture d'esprit face aux étudiants, n'excluant pas la sévérité, sont les "outils" qu'il lui légua. L'attitude de Cieslewicz, entre détachement et jouissance personnelle d'une affirmation, semble être une aspiration, comme un moteur pour la graphiste. Son style se forgea aussi en raison des contraintes financières qu'elle subit. Les labels n'ayant pas de budgets pour une production photo, jugeant que ses propres photos ne peuvent se suffire à elles-mêmes, elle transforme celles qu'elle reçoit ou qu'elle prend en paysages. Ainsi, ses photos sont-elles plus à l'aise avec l'esprit décalé provoqué par les collages. Dans ces conditions naît la mémorable et si furtive identité de *Point éphémère*, où elle transforme en une toile de Jouy, *les acteurs de la musique*.

émergence

Sylvia Tournerie ne compose que sur ordinateur, et parle de la légèreté de l'outil, puisque, au propre comme au figuré, les données ne pèsent rien. Sur son Mac, un dossier vrac regroupe ses premières sessions de travail peu organisées, *une étape de vidage, suite à ma rencontre avec le commanditaire*. Dans un état presque hypnotique, où *l'important est de se laisser aller*, elle façonne une matière formelle abstraite. Elle la pétrit jusqu'au moment où se manifeste la première émotion, cette émotion, qu'elle peut perdre en cours de route, mais qu'elle *n'a de cesse de faire vivre, de conserver jusqu'au bout du projet*. Tout est dans le doigté et dans ces ressentis impalpables. Sylvia Tournerie parle avec sensibilité, avec intelligence de cette étape de travail, capitale, qui l'interroge douloureusement aussi. Elle évoque son incapacité à décrypter ses convictions. Cette étape est de l'ordre de l'émotion. *J'ai rarement une idée avant de faire les choses*. Ainsi, l'objet graphique émerge-t-il de son façonnage. *Je justifie les formes une fois qu'elles sont là*. Pendant longtemps, il lui fut difficile d'assumer cette prétendue gratuité, aujourd'hui, Sylvia Tournerie se dit plus sereine face à sa façon de composer[4]. Ses formes ne sont pas le fruit du hasard, avec l'expérience, toutes relèvent d'un choix. Sylvia Tournerie agit dans la traduction – le graphisme avec ses composants parle à l'âme directement de la même manière que la musique parle avec ses notes et ses gammes –, elle n'est pas sur le territoire des intentions. Ses identités visuelles ne sont pas des chartes, mais des pulsations, des vibrations, concentrées ou fragmentées.

Peu d'affiches, pas de théâtre, ni d'identité institutionnelle (excepté sa participation avec Gilles Poplin à l'identité du CNAP *é : 126*), pas de gros chantiers, ni de régularité (cette situation qu'on retrouve chez d'autres de ses contemporains devrait inciter les commanditaires à défier ces graphistes sur ces terrains balisés). Pourtant, les gammes de Tournerie marquent leur empreinte dans le

Poster pour la styliste Andrea Crew (avec la participation de Leslie David, 2006). Au recto, les mannequins présentaient la collection de la saison et des motifs géométriques auréolaient chaque modèle et accentuaient leurs postures irrévérencieuses. Au recto, le processus de travail d'Andrea Crews se révèle dans un vaste désordre recollé : la styliste élabore ses pièces uniques à partir d'habits récupérés et recyclés.

44 : 3.2008

3.2008 : 45

剪影的形状和精心选择的艺术品图片为井然有序的页面带来了活力

● 74.追求戏剧性

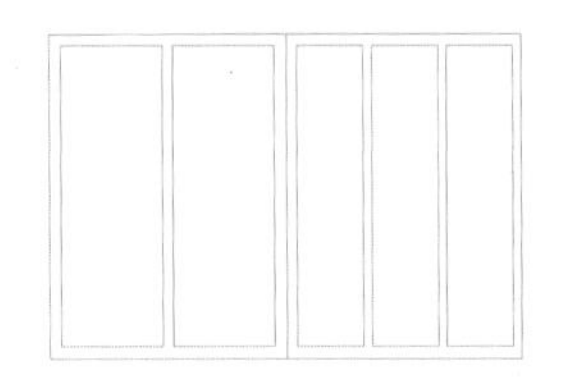

裁剪图片可以创造戏剧性的效果。使用未经编辑的照片能够还原故事，但是将照片裁剪一下可以表达出特定的观点和看法，还可以激发读者的恐惧、兴奋等情绪。另外，通过裁剪还可以去除多余信息，引导读者关注照片中某个特定的方面。

注意限制条件

在裁剪某些图片时，注意限制条件。许多摄影师、博物馆对艺术品的再加工都有非常严格的规定。有些图片，特别是著名的绘画或雕塑图片是不能随意剪裁的。

本页图： 圆形文本框内的标注为网格内的细节照片增加了动感

下页图： 这本书的封面将重要信息的文本框设置为圆形（象征聚光灯）。拼贴照片会减小照片的冲击力，但照片的剪裁让这个作品非常精彩

项目
《百老汇：从出租到变革》

客户
作者：德鲁•霍奇斯，出版商：里佐利

创意总监
德鲁•霍奇斯

设计师
内奥米•米祖萨基

这本书籍的设计有明确的网格基础，图片的形状和裁剪，以及精彩的排版还为这本书增加了戏剧性，使它带有强烈的爵士风格

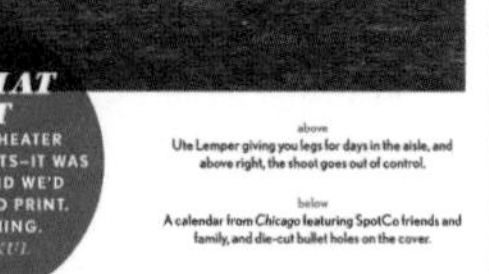

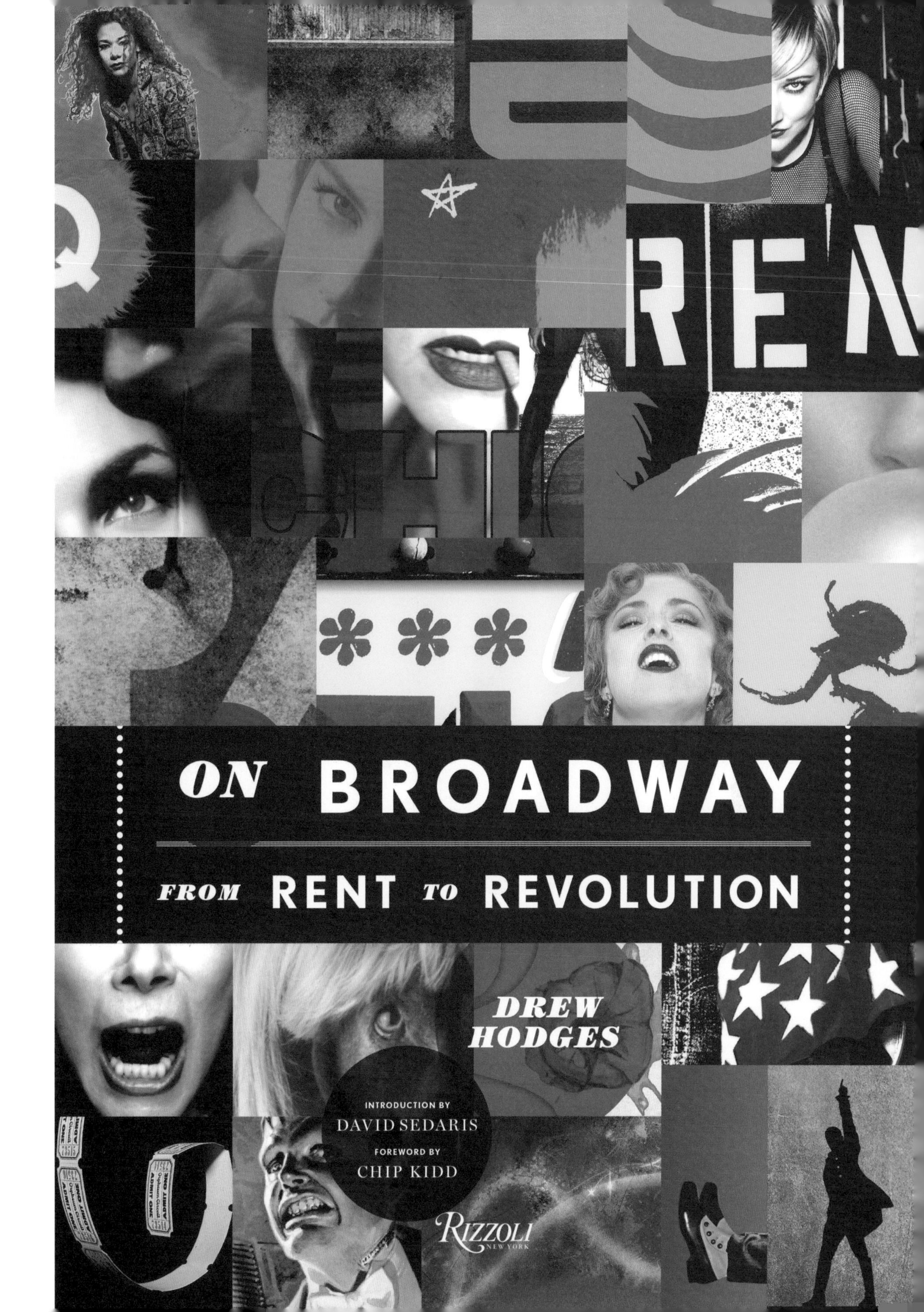
ON BROADWAY
FROM RENT TO REVOLUTION
DREW HODGES
INTRODUCTION BY
DAVID SEDARIS
FOREWORD BY
CHIP KIDD
RIZZOLI
NEW YORK

● 75.裁剪图片

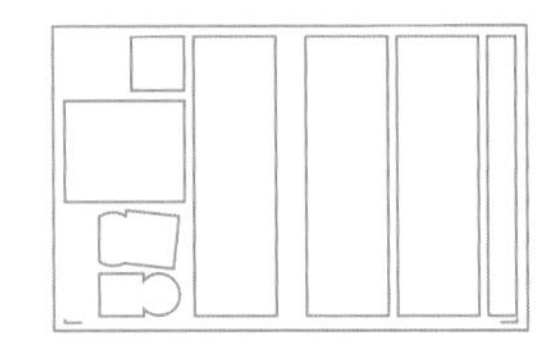

轮廓图可以避免页面过于死板。在版面设计中，轮廓图是指一张删除了背景的图片。轮廓图可以是一个有机的形状，如叶子的形状，也可以是像圆形这样常见的形状。轮廓图流畅的线条可以为页面增加动感。

旬の、相性のいいもの同士が醸しだす味わいの深いことといったら。

新じゃがの花まぶし

ふきの炊いたんと鯛の子

若ごぼうの炊いたん

糸こんちりめん

若竹汁

わけぎとイカの辛子酢味噌和え

垂直和水平分隔线清楚地划分出了包含标题、导言和信息的区域。这些页面的说明指导内容表达得很清楚，而轮廓图的有机形状又赋予了页面新的活力

项目
《可颂》杂志

艺术总监
马场诚子

设计师
高梨裕子（Yuko Takanashi）

这张对页取自一本日本的杂志，展示了在一个条理清晰的文本中，如何通过轮廓图增加故事性

おいしいものは、端っこまでおいしい。手を抜かず、手間かけて、余さず食べる知恵と工夫。

「私の母はつくづくすごい人やったな、と思います。魚を煮たおつゆや、野菜を煮たおつゆをためておいて、合わせておからを炊いてました」

何種類かの煮汁が混じり合って複雑さを増したうまみ。それをたっぷり吸った卯の花炊きの味はさぞや……。

「これが〝始末〟のお手本です」

始末とは、食べられるものは余すところなく使いきる知恵のこと。だしをとった後の昆布や削り節を細かく刻んだもので作るふりかけ、捨ててしまうような古漬けのたくあんの端っこを使って作る煮物。

その首藤さんにもふきの葉伝説がある。捨てるものとされていたふきの葉が、京都ではあるときからゴミとして出なくなったというのだ。

「ふきの葉をゆでて刻んで、じゃこと合わせて甘辛く煎りつけてふりかけにしてみたんです」

それが広がって、八百屋さんの店先で葉を捨てずにふきを買っていく人が増えたんだそうだ。

「その頃だったかな、娘が知人からこんなこと言われたんだそうです。『あんたんとこのおかあちゃん、余りものを使った料理教室、開いてるんやて？』って」

〝ちゃんとした料理も教えてる〟って言い返したそうだが、ふきの葉のふりかけ、ほろ苦さがうまく残っていて、春の息吹を感じさせるちゃんとした料理だ。残りものを始末するからといって手は抜かず、おいしく食べるために手間をかける、その心持ち。

「京都の女の人は、手間をかけるべきところはきちんとかけていましたが、始末の料理などは手早くおいしくを心がけていました」

どう調理するか、それぞれの家で頭をつかっていたという。

「なんでこういう知恵が、いまの娘さんになくなってしまったんでしょう。常備菜なんかもそうです。ちりめん山椒なんて、いっぺんにたくさん炊いておけば、他所から何かもらったときのお返しにもできるのに。〝腹巻き〟みたいな服（どうやら、キャミソール）着てると、そうなるんでしょうか。河原町を歩いてる娘さんたちを見ると、そう思ってしまいます」

たくあんのぜいたく煮

材料(4人分)　古漬け沢庵1本、だし雑魚1つかみ、たかのつめ2、3本、醤油カップ1、みりんカップ1/3、酒カップ1/2
作り方　沢庵は5mmほどの輪切りにし、水につけて塩出ししておく。時々、水を替えながら塩出しした沢庵を、ひたひたの水で歯応えを残すくらいにゆで、だし雑魚、種をとったたかのつめ、醤油、みりん、酒を入れる。材料がかぶらないようなら、酒を足す。最初は強火で煮て、沸騰したら火を弱めて汁気がなくなるまで煮る。落とし蓋を使い、かき混ぜないように。

だし雑魚は頭、わた、骨を除いて三枚に。この手間がぜいたくで、味を決める。

ちりめん山椒

材料(4人分)　ちりめん雑魚400g、実山椒カップ1、濃い口醤油150cc、酒250cc、みりん70cc
作り方　鍋にすべての調味料を入れ、中火で煮立てる。煮立ったら、ちりめん雑魚を加えて火を弱め、煮詰めていく。煮詰まる直前に実山椒を加えて火を止めてできあがり。「ちりめん山椒は実山椒を早く入れすぎると色も香りも悪くなります。私は昔から一度にたくさん炊いておき、近所におすそ分けしたり、いただきもののお返しに使ったりしていました」

首藤さんは多めに炊いて袋に入れて保存している。

かつおと昆布のふりかけ

材料(作りやすい分量)　だしをとったあとの昆布、かつお節、砂糖小匙1、醤油小匙1、みりん小匙1、酒小匙1、白ごま小匙1
作り方　昆布は細かく刻み、かつお節と一緒に鍋に入れる。砂糖、醤油、みりん、酒を加えて菜ばしで混ぜながら強火で炒る。菜ばしは3組か4組使うと混ぜやすい。汁気が引いてきたら白ごまを加えて火を止める。白ごまの代わりに松の実でも香ばしい。冷蔵庫に入れれば2週間ほどは保存できる。「私の場合、ここで使うかつお節は、かんなで削ったものです。ほんと、おいしいんです」

ふきの葉の佃煮

材料(作りやすい分量)　ふきの葉3把分、みりん大匙2、酒大匙3、醤油カップ1/2、ちりめん雑魚1つかみ
作り方　ふきの葉は、よく洗っておく。たっぷりの湯で色よくゆでる。あまりゆですぎないうちに引きあげたら、20分ほど水にさらし、かたく絞って水気を切る。このまま、削りかつおと醤油だけでいただいても、ほろ苦くておいしい。佃煮にするときは、調味料をすべて合わせて中火にかけ、ふきの葉とちりめん雑魚を加える。「菜箸を3、4組使い、かき混ぜながら、ぱろぱろになるまで炒めてください」

しじみの炊いたん

材料(4人分)　しじみ(むき身)400g、酒100cc、みりん20cc、砂糖大匙1、醤油大匙3、水50cc、針生姜
作り方　しじみはざるに入れて何度も振り洗いする。砂を充分に抜いたら、水気を切って鍋に入れ、調味料と水を加え、針生姜をのせる。落とし蓋をして中火にかけ、煮立ってきたらアクを取る。アクを取ったら中火のまま煮汁がなくなるまで煮る。「琵琶湖の瀬田産の身が小ぶりのしじみで炊くとおいしいんですが、最近は輸入物に押されてしまって、なかなか……」

落とし蓋はコーヒーフィルター。広げると円形になる。

上・「かつお節はここまで小さくなるまで削って使います。削るのは今でも私がやっています」中・有次の真鍮製の鍋。打ち出しが擦り切れるほど使い込んである。すき焼きに大活躍。下・ごま炒り。「ごまはよく食べますよ。炒るにも擂るにもコツがあります」

暮らしの知恵 052

053 クロワッサン

设计师利用分隔线在杂志的网格系统中又创建了一个网格。排列布局清晰干净，不同的轮廓图形状给组织有序、层次分明的对页带来了动感

● 76.让文化做主导

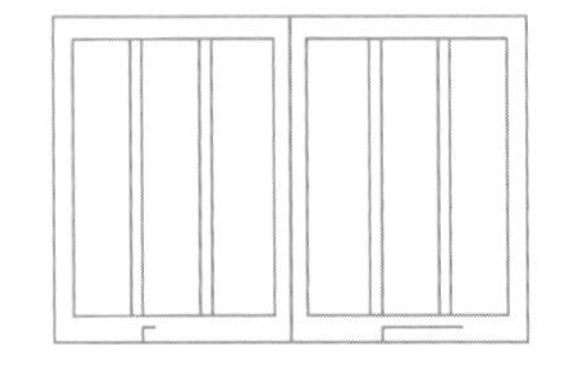

一个好的作品既可以拥有一个强大而清晰的框架和丰富的视觉元素，又能保持不同的内容特点，起到教育观众或读者的目的。设计师可以通过融合文化故事、神话或符号，让一件作品更具感染力。最重要的是让页面变得更丰富，以便读者认识或了解其他文化。

项目
《螺纹》（*Threaded*）杂志

客户
螺纹工作室

设计公司
螺纹工作室

设计团队
凯拉•克拉克（Kyra Clarke），菲奥娜•格里夫（Fiona Grieve），雷根•安德森（Reghan Anderson），菲尔•凯利（Phil Kelly），德斯纳•瓦安加-肖卢姆（Desna Whaanga-Schollum），卡琳•吉本斯（Karyn Gibbons），特拉亚•内瓦（Te Raa Nehua）

图片版权
螺纹媒体有限公司（Threaded Media Limited）

《新开端》（*New Beginnings*）是国际杂志《螺纹》的第20版，由一家总部位于奥克兰的设计机构设计和出版，以介绍毛利艺术及其设计从业人员为主题

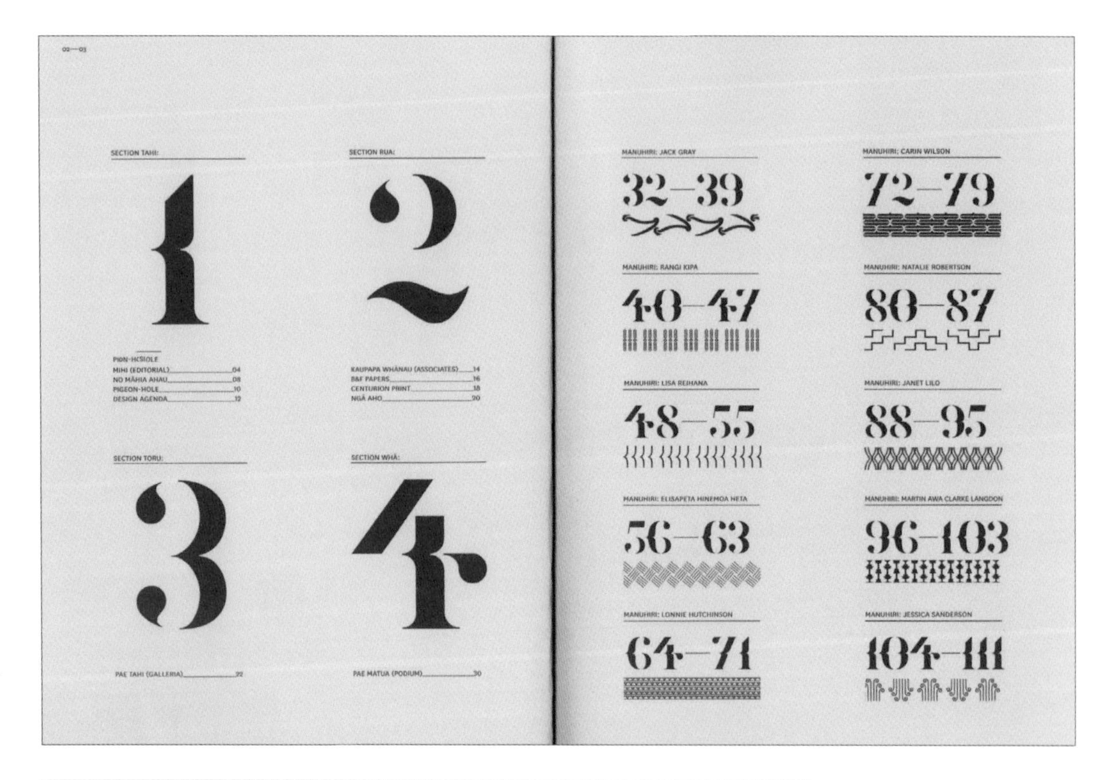

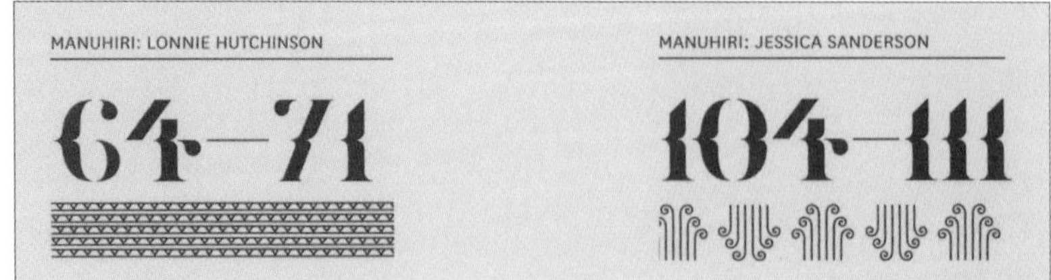

无衬线字体和浪漫风格的数字与来自毛利文化的符号融合。该杂志的编辑专门创作了不同的图案代表杂志中每一位受访者的故事，使杂志中的艺术品成为每一个毛利艺术设计从业者独特的文化符号

下页图： 简单的网格和充足的空间展示了令人惊叹的毛利艺术。装饰图案烘托了艺术文化，也避免了页面过于平淡、刻板。副标题和正文使用了简洁的无衬线字体，与精美的图片形成对比

ON LINE:

I'm acutely aware of how much the power of a line can influence how people read visual material. For me, when I'm working in a sculptural sense I'm analyzing everything by the mana of those lines; if they're transported into low relief or 3D we're talking about edges. Edges are everything. You create a deep or powerful sense of space, direction and form with something that's relatively shallow. The edge of the line enables you to use light to give the impression of depth. I'm aware of it and I just try to exploit it I suppose. There's a beauty in line that's difficult to explain, but I get seduced by the ability to reduce down to linear forms and play with it, there's so much you can do it's really just up to your imagination and over a period of time you get to a point where you can master it and then people, they follow it, they get it. They're not necessarily able to interpret it or explain it but they get it.

The energy contained in the line is no different than the principle of physics. It's the same as how you use your arm to develop a centrifugal force to throw something. The line can do that as well. You can use that energy to influence and to give the impression (of that force) in the same way whether in 2D and 3D.

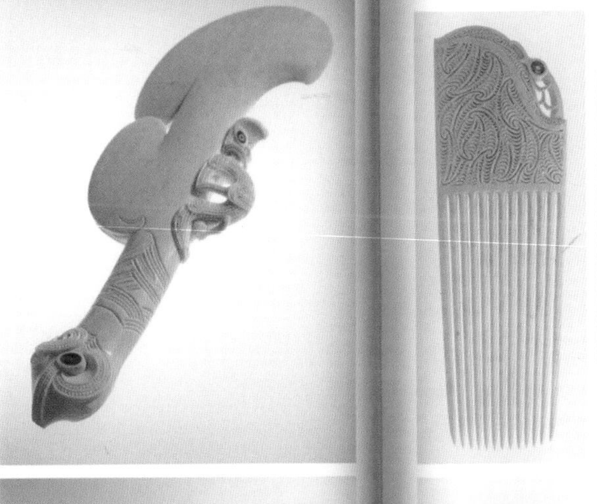

TOP
—
Wahaika, whalebone
•
BOTTOM
—
Kotiate, whalebone

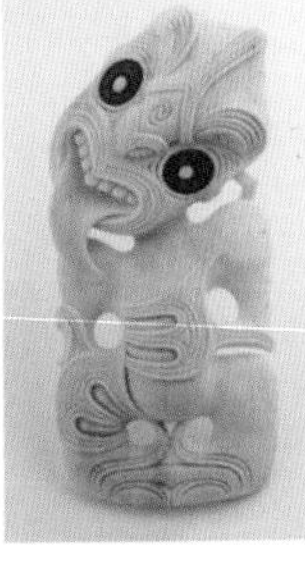

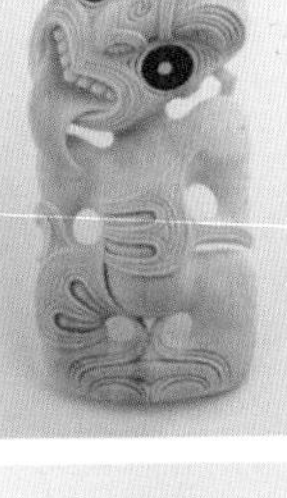

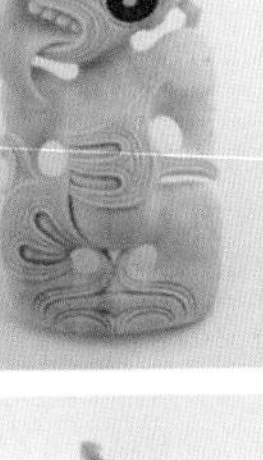

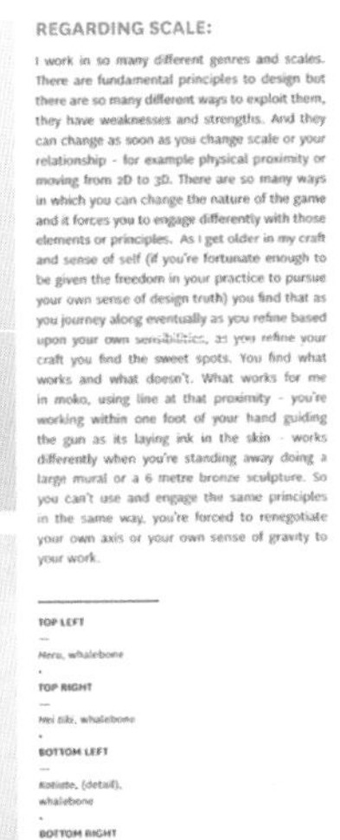

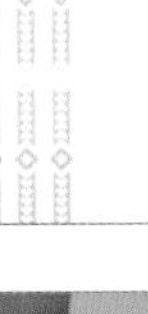

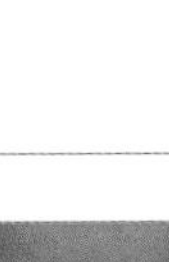

REGARDING SCALE:

I work in so many different genres and scales. There are fundamental principles to design but there are so many different ways to exploit them, they have weaknesses and strengths. And they can change as soon as you change scale or your relationship - for example physical proximity or moving from 2D to 3D. There are so many ways in which you can change the nature of the game and it forces you to engage differently with those elements or principles. As I get older in my craft and sense of self (if you're fortunate enough to be given the freedom in your practice to pursue your own sense of design truth) you find that as you journey along eventually as you refine based upon your own sensibilities, as you refine your craft you find the sweet spots. You find what works and what doesn't. What works for me in moko, using line at that proximity - you're working within one foot of your hand guiding the gun as its laying ink in the skin - works differently when you're standing away doing a large mural or a 6 metre bronze sculpture. So you can't use and engage the same principles in the same way, you're forced to renegotiate your own axis or your own sense of gravity to your work.

TOP LEFT
—
Heru, whalebone
•
TOP RIGHT
—
Hei tiki, whalebone
•
BOTTOM LEFT
—
Kotiate, (detail), whalebone
•
BOTTOM RIGHT
—
Madonna and Child, whaletooth

CONNECTIONS AND PATHWAYS:

My artistic practice is strongly influenced by my political belief that we need to be relevant and that we have a role to bridge the past and the future. I'm lucky enough to grow up in my tribal area. I've had strong cultural connections to my community so I have a sense of allegiance to my culture and community that manifests itself in the work. But I can't change the past. I'm trying to visualize a pathway in the future and trying to use my art as a tool to help lay down some of that pathway. I'm trained as a social scientist as well so if you connect that to my cultural background it's part of my imperative to drive my art to always be forward focused. I believe we have a role to visualize the future and make it happen. If you look over my practice over the past 20-30 years, moko was like that, taonga puoro - Māori musical instruments - was like that, my role in waka in Taranaki was like that. So by kicking those things off and by continuing to push them - the same with taonga whakarakai (adornment arts) - restoring those art forms but restoring them in a way that continues to have relevance not only for now and being present in your work but also into the future. I'm trying to push my work so far ahead that it actually looks futuristic, that people can go 'wow - I really like that'. They can see the footprint of our old world in it but it's also out there tugging on them so the materiality, the aesthetics and the cultural imperative that's subtly locked in there pulls at them.

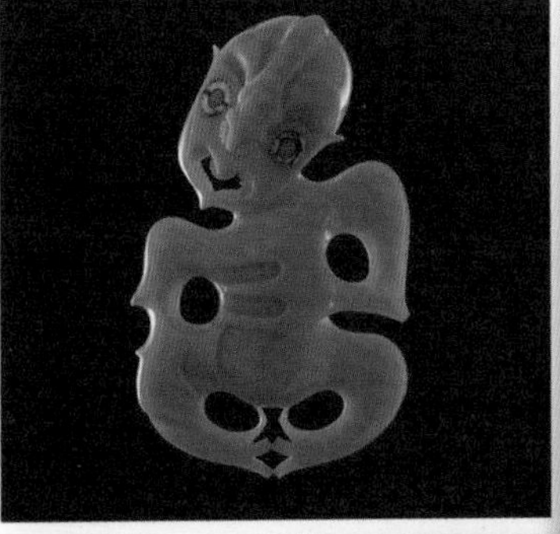

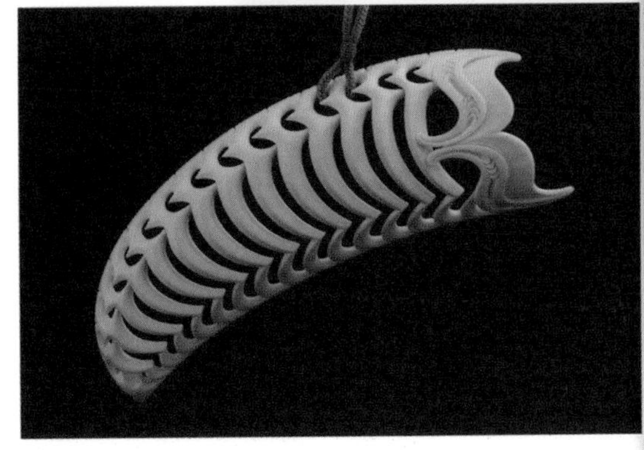

NEW MATERIALS:

The interesting thing was about seeing how the Māori community would respond to media that wasn't seen as valuable, that didn't have a valuable (cultural) attribution, so creating beautiful stuff out of it, it enables itself to be relevant. The beautiful thing about Corian is you can get a great sense of colour - and if you're really good at finishing the work - it's as seductive as what whale tooth, whale bone or pounamu can be. That was the beauty of that exercise, seeing that material being adopted as a taonga. These approaches to materials are interesting journeys, they're not necessarily answers to questions but they're part of the journey.

> "... you're working within one foot of your hand guiding the gun as its laying ink in the skin..."

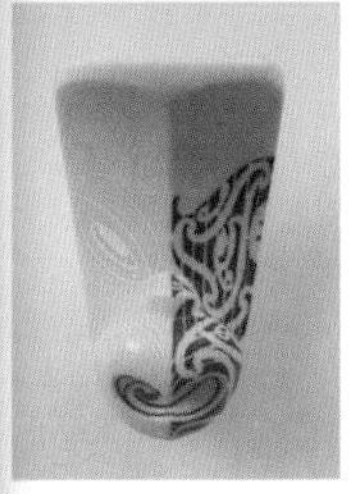

TOP
—
Moko peha
•
BOTTOM
—
Porota, Corian

OPPOSITE PAGE

TOP
—
Hei tiki, Corian
•
BOTTOM
—
Moarei, whaletooth

77.设计一套通用系统

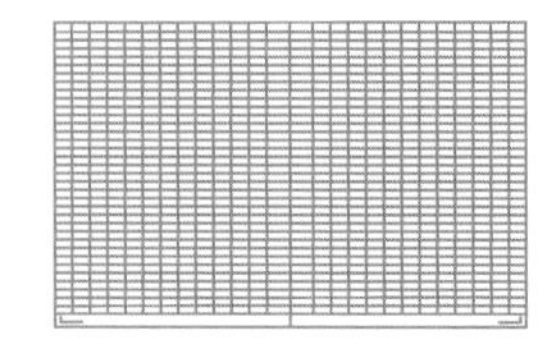

一套通用的系统可以让不同大小、形状的元素和不同的信息形成不同的构图。

先驱

艾伦•卢普顿指出，瑞士网格先驱约瑟夫•穆勒-布罗克曼和卡尔•格斯特纳（Karl Gerstner）将设计程序定义为一组用于构建可视化方案的规则。卢普顿指出了瑞士网格设计规则的关键方面——利用重复的结构制造变化。创建一套通用系统可以让一个项目同时包含紧凑和宽松的页面。

josef müller-brockmann, certitudes visuelles

Les affiches de concert

这个系统网格可以把页面分成两部分、三部分或四部分，也可以将其水平划分

这个强大的网格限制了图片的大小，但可以让页面多样化

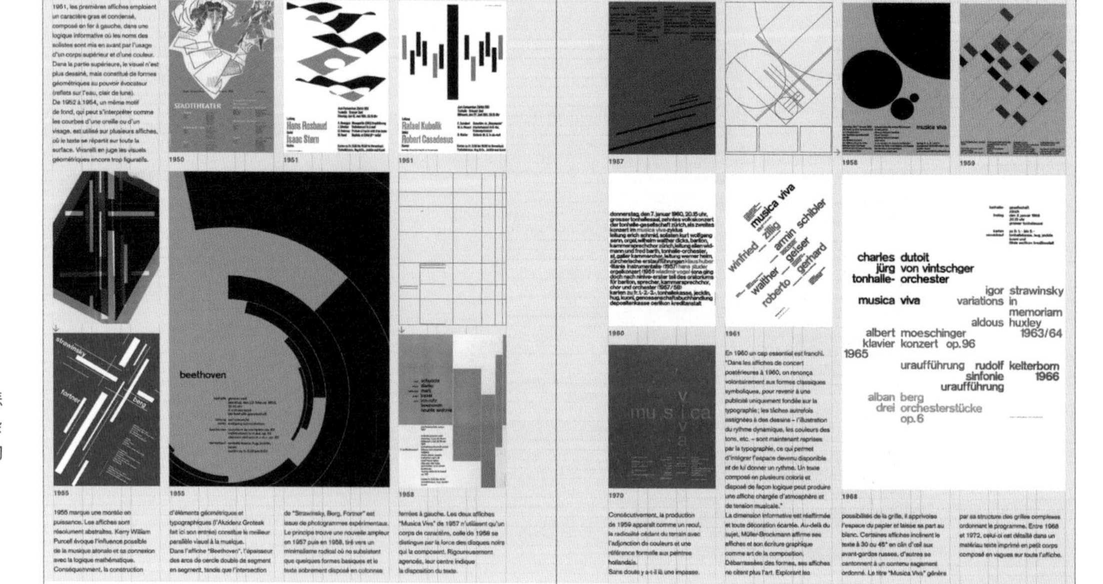

Les affiches de concert

项目
《阶段》杂志

客户
金字塔/《阶段》杂志

设计师
安娜•图尼克

这是一篇评论网格大师约瑟夫•穆勒-布罗克曼作品的杂志文章，文章采用了灵活的网格系统

严谨的网格也可以创造出令人兴奋的设计，精彩的图片和有节奏的布局创造了变化和惊喜

Überholen...?
Im Zweifel nie!

protegez l'enfant !

«En une fraction de seconde, l'affiche doit agir sur la pensée des passants, les contraignant à recevoir le message, à se laisser fasciner avant que la raison n'intervienne et ne réagisse. Somme toute, une agression discrète, mais soigneusement préparée.»

这组页面显示了网格如何轻松地容纳侧边栏，并说明了在网格中如何设计有足够留白的页面

Philosophie de la grille et du design

L'usage de la grille comme système d'organisation est l'expression d'une certaine attitude en ce sens qu'il démontre que le graphiste conçoit son travail dans des termes constructifs et orientés vers l'avenir.

C'est là l'expression d'une éthique professionnelle, le travail du designer doit avoir l'évidence, objective et esthétique qualité du raisonnement mathématique.

Son travail doit être la contribution à la culture générale dont il constitue lui-même une partie.

Le design constructiviste qui est capable d'analyse et de reproduction peut influencer et rehausser le goût d'une société et la façon dont elle conçoit les formes et les couleurs.

Un design qui est objectif, engagé pour le bien-être collectif, bien composé et raffiné constitue la base d'un comportement démocratique. Un design constructif signifie la conversion des lois du design en solutions pratiques. Un travail accompli de façon systématique, en accord avec de stricts principes formels, permet ces exigences de droiture d'intelligibilité et l'intégration de tous les facteurs aussi vitaux pour la vie sociopolitique. Travailler avec un système de grille implique la soumission à des lois valides universellement.

L'usage du système de grille implique la volonté de systématiser, de clarifier ; la volonté de pénétrer à l'essentiel, de concentrer ; la volonté de cultiver l'objectivité au lieu de la subjectivité ; la volonté de rationaliser les modes de production créatifs et techniques ; la volonté d'intégrer des éléments de couleur, de forme et de matière ; la volonté d'accomplir la domination de l'architecture sur l'espace et la surface ; la volonté d'adopter une attitude positive et visionnaire ; la reconnaissance de l'importance de l'éducation et les effets du travail conçu dans un esprit constructif et créatif.

Tout travail de création visuelle est une manifestation de la personnalité du designer. Il est marqué de son savoir, de son habileté et de sa mentalité. – Josef Müller-Brockmann

the architectonic in graphic design the concert poster series of josef müller-brockmann

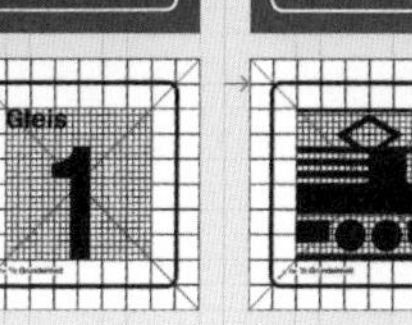

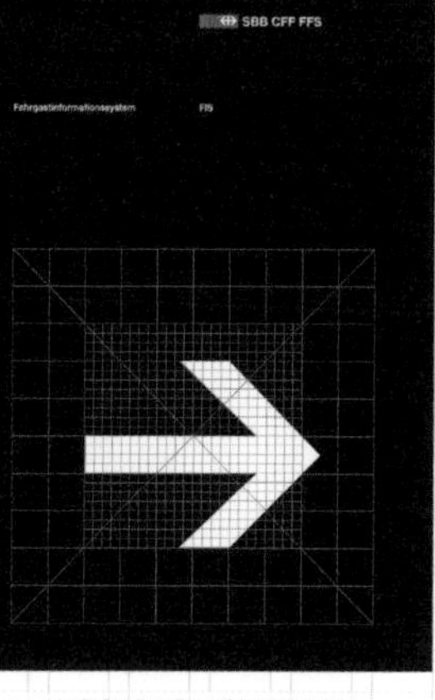

Bern Wylerfeld
CFF Cargo
SBB CFF FFS

● 78.展示字体的粗细和页面层次

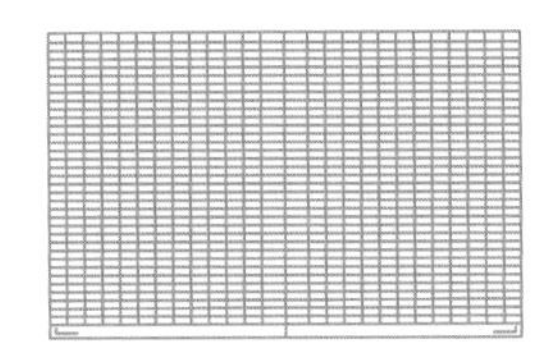

一个基于瑞士网格基础的作品可以让文本更有趣。这个系统利用视觉化手段传达信息，所以内容清晰易读。多栏网格可以容纳多样的信息，也能让图片和彩色方框与各部分信息互相搭配。这样的系统有丰富的变化，可以让素材得到优化。

7 GREAT SERIES. 7 GREAT EXPERIENCES!

COLTRANE
SEPTEMBER 14, 15 & 16, 2006
Blue tranes run deeper. Ecstatic and somber, secular and sacred, John Coltrane's musical sermons transform Rose Theater into a place of healing and celebration with orchestrations of his small group masterpieces "My Favorite Things," "Giant Steps," "Naima," and more. Join us as the **LCJO** with **Wynton Marsalis** marks the 80th year since the birth of one of

WYNTON AND THE HOT FIVES
SEPTEMBER 28, 29 & 30, 2006
Hearts beat faster. It's that moment of pure joy when a single, powerful voice rises up from sweet polyphony. Louis Armstrong's Hot Five masterpieces—"West End Blues," "Cornet Chop Suey," and others—quicken the pulse with irresistibly modern sounds. **Wynton Marsalis, Victor Goines, Don Vappie, Wycliffe Gordon**, and others re-imagine the recordings that defined jazz, and then bring that pure joy to the debut of equally timeless new music inspired by the original.

RED HOT HOLIDAY STOMP
DECEMBER 14, 15 & 16, 2006
Tradition gets fresher. When Santa and the Mrs. get to dancin' the "New Orleans Bump," you know you're walking in a *Wynton Wonderland*—a place where joyous music meets comic storytelling. **Wynton Marsalis, Herlin Riley, Dan Nimmer, Wycliffe Gordon, Don Vappie**, and others rattle the rafters with holiday classics swung with Crescent City style. *Bells, baby. Bells.*

THE LEGENDS OF BLUE NOTE
APRIL 26, 27 & 28, 2007
Bop gets harder. The music is some of the best ever made—Lee Morgan's *Cornbread*, Horace Silver's *Song for My Father*, Herbie Hancock's *Maiden Voyage*—all wrapped up in album cover art as bold and legendary as the music inside. The **LCJO** with **Wynton Marsalis** debuts exciting and long-overdue big band arrangements of the best of Blue Note, complete with trademark cracklin' trumpets, insistent drums, and all manner of blues.

IN THIS HOUSE, ON THIS MORNING
MAY 24, 25 & 26, 2007
Tambourines testify. It's that sweet embrace of life—sometimes celebratory, sometimes solemn—rising from so many houses on so many Sundays. We mark the 15th anniversary of Wynton's first in-house commission, a sacred convergence of gospel and jazz that

FUSION REVOLUTION: JOE ZAWINUL
OCTOBER 27 & 28, 2006
Grooves ask for mercy, mercy, mercy. Schooled in the subtleties of swing by Dinah Washington, keyboardist **Joe Zawinul** brought the fundamentals of funk to Cannonball Adderley, the essentials of the electric to Miles Davis, and carried soul jazz into the electric age with his band Weather Report. Now the **Zawinul Syndicate** takes us on a hybrid adventure of sophisticated harmonies, world music rhythms, and deeply funky grooves. *Mercy.*

BEBOP LIVES!
JANUARY 26 & 27, 2007
Feet tangle and neurons dance. Fakers recoil, goatees sprout, and virtuosos take up their horns. Charlie Parker and Dizzy Gillespie set the bebop revolution in motion, their twisting, syncopated lines igniting the rhythms of jazz. Latter day fakers beware as the legendary **James Moody** and **Charles McPherson**, the alto sax voice of Charlie Parker in Clint Eastwood's *Bird*, raise battle axes and *swing*.

CECIL TAYLOR & JOHN ZORN
MARCH 9 & 10, 2007
Souls get freer. Embark on a sonic voyage as the peerless **Cecil Taylor** navigates us through dense forests of sound—percussive and poetic. He is, as Nat Hentoff proclaimed, "a genuine creator." The voyage banks toward the avant-garde as **John Zorn's Masada** with **Dave Douglas** explores sacred and secular Jewish music and the "anguish and ecstacy of klezmer." Musical wanderlust *will* be satisfied.

THE MANY MOODS OF MILES DAVIS
MAY 11 (Kisor/blanchard) &
MAY 12 (Payton/Miller), 2007
Change gets urgent. "I have to change," Miles said, "It's like a curse." And so his trumpet voice—tender, yet with that *edge*—was bound up in five major movements in jazz. The LCJO's **Ryan Kisor** opens with bebop and the birth of the cool. GRAMMY®-winner **Terence Blanchard** interprets hard bop and

上图和下页图： 这个宣传册的版面在文字粗细、行距、标签标题和标语上呈现出了很多变化，层级清晰明了。带有色彩的模块标出了七个不同的内容系列。每个彩色模块内的版式都简洁而均衡，不同字号和粗细的字体清晰地展示了信息。彩色模块的设置非常成功，是这个宣传册总体布局中的次级布局，优雅的字体和对齐方式使彩色模块犹如一个个小标语

项目
订阅宣传册

客户
林肯中心爵士乐团

设计师
博比·马丁

排版设计将页面编排得非常丰富，也让页面可读性更强

JAZZ AT LINCOLN CENTER'S 06-07 SEASON

From Satchmo's first exuberant solo shouts to Coltrane's transcendent ascent, we celebrate the emotional sweep of the music we love by tracing the course of its major innovations. Expression unfolds in a parade of joyous New Orleans syncopators, buoyant big band swingers, seriously fun beboppers, cool cats romantic and lyrical, blues-mongering hard boppers, and free and fusion adventurers. From all the bird flights, milestones, and shapes of jazz that came, year three in the House of Swing is a journey as varied as the human song itself, and the perfect season to find your jazz voice.

4 — ALJO SERIES

Afro-Latin Jazz Orchestra with Arturo O' Farrill
3 Concerts
Rose Theater, 8pm

BEBO VALDES
OCTOBER 13 & 14, 2006

Mambo migrates. Bebo Valdes is a true legend. *The Los Angeles Times* calls him "elegant and restrained, a product of the golden era of Cuban music, when delicate danzones and tasty cha-cha-chas reigned supreme." When Bebo first lit up the Tropicana with his irresistible mambo beat, a musical blaze spread from Havana to the dance halls of New York City. Today, after over 40 years of exile in the cooler climes of Sweden, Bebo still brings the heat of his native island like no one else. Joined by **Arturo O'Farrill** and the **Afro-Latin Jazz Orchestra**, the 87-years-young Afro-Cuban pioneer serves up *el ritmo caliente* with the United States premiere of his GRAMMY®-winning *Suite Cubana, ¡Viva Bebo!*

5 — SM SERIES

Singers Over Manhattan
4 Concerts
The Allen Room
7:30pm & 9:30pm

STEPHANIE JORDAN & THE WESS ANDERSON QUARTET
OCTOBER 20 & 21, 2006

Standards get fresher. Every so often a new voice stands up and proclaims itself, but few do so with such supreme depth and understated soul. Emerging from the New Orleans jazz family Jordan, **Stephanie Jordan** was last year's *Higher Ground Concert's* "real discovery" (*JazzTimes*). She's joined in sweet counterpoint by **Wess "Warmdaddy" Anderson**, whose buoyant saxophone voice continues to proclaim itself with unmatched joy and warmth.

WILLIE NELSON SINGS THE BLUES
JANUARY 12 & 13, 2007

Blues get democratic. He's got the right to sing the blues. As folk legend **Willie Nelson** told the great B.B. King and the late, great Ray Charles, "Gentlemen, I think I'm the only one here who actually had to pick cotton." Country outlaw Willie Nelson and Crescent City son **Wynton Marsalis** come together on two stellar evenings in The Allen Room to demonstrate soulfully why the blues should be our national anthem.

6 — SS SERIES

Singin' & Swingin'
3 Concerts
The Allen Room
7:30pm & 9:30pm

COLTRANE/HARTMAN
SEPTEMBER 15 & 16, 2006

Life gets lusher. Between the grooves of a deeply romantic recording that stands at the pinnacle of vocal and instrumental harmony, John Coltrane and Johnny Hartman were relaxed and electric, muscular and gentle. Tenor saxophonist **Todd Williams** with his "commanding tone" (*The New York Times*) and vocalist **Kevin Mahogany**, who makes "you want to just sit down and listen" (*USA Today*), recast the sophisticated balladry of a recording that flows like an "April breeze on the wings of spring..." All against a backdrop of Coltrane's most beloved instrumental ballads.

PAQUITO D'RIVERA
NOVEMBER 17 & 18, 2006

Streams converge. The jazz-meets-classical clarinet tradition of Benny Goodman is carried into a new century by the seven-time GRAMMY®-winner and 2005 National Medal of Arts recipient **Paquito D'Rivera**. "A formidable musician... with lovely, clear low registers," (*The New York Times*) Paquito and company explore the dynamic Third Stream convergence of classical and jazz with interpretations of the music of Ravel, Bartók, and Stravinsky.

THE BIRTH OF COOL
MARCH 30 & 31, 2007

Whispers shout louder. Cool. Lester Young invented the word as his saxophone angled toward the heavens—*original cool*. Count Basie set Kansas City ablaze without breaking a sweat—*slow hand cool*. Billie Holiday relaxed each syllable and clichés wilted—*tragic cool*. And Miles Davis caressed melody like a pixie with a blowtorch—*blue flame cool*. Pianist **Bill Charlap**, who "approaches a song the way a lover approaches his beloved" (*TIME* magazine), leads an ensemble in a celebration of the classics of cool, and as the last note drops, we'll all be cool.

7 GREAT SERIES. 7 GREAT EXPERIENCES!

2 — JJ SERIES

Jazz Jam
4 Concerts
Rose Theater, 8pm

WYNTON AND THE HOT FIVES
SEPTEMBER 28, 29 & 30, 2006

Hearts beat faster. It's that moment of pure joy when a single, powerful voice rises up from sweet polyphony. Louis Armstrong's Hot Five masterpieces—"West End Blues," "Cornet Chop Suey," and others—quicken the pulse with irresistibly modern sounds. **Wynton Marsalis, Victor Goines, Don Vappie, Wycliffe Gordon**, and others re-imagine the recordings that defined jazz, and then bring that pure joy to the debut of equally timeless new music inspired by the original.

RED HOT HOLIDAY STOMP
DECEMBER 14, 15 & 16, 2006

Tradition gets fresher. When Santa and the Mrs. got to dancin' the "New Orleans Bump," you know you're walking in a *Wynton Wonderland*—a place where joyous music meets comic storytelling. **Wynton Marsalis, Herlin Riley, Dan Nimmer, Wycliffe Gordon, Don Vappie**, and others rattle the rafters with holiday classics swung with Crescent City style. *Bells, baby. Bells.*

THE LEGENDS OF BLUE NOTE
APRIL 26, 27 & 28, 2007

Bop gets harder. The music is some of the best ever made—Lee Morgan's *Cornbread*, Horace Silver's *Song for My Father*, Herbie Hancock's *Maiden Voyage*—all wrapped up in album cover art as bold and legendary as the music inside. The **LCJO** with **Wynton Marsalis** debuts exciting and long-overdue big band arrangements of the best of Blue Note, complete with trademark crackin' trumpets, insistent drums, and all manner of blues.

IN THIS HOUSE, ON THIS MORNING
MAY 24, 25 & 26, 2007

Tambourines testify. It's that sweet embrace of life—sometimes celebratory, sometimes solemn—rising from so many houses on so many Sundays. We mark the 15th anniversary of Wynton's first in-house commission, a sacred convergence of gospel and jazz that gave rise to a new sound, at once modern and familiar. When **Marsalis, Riley, Gordon, "Warmdaddy,"** and **Williams** testify, the House of Swing will shake down to its drum skin floors.

3 — MM SERIES

Music of the Masters
4 Concerts
Rose Theater, 8pm

FUSION REVOLUTION: JOE ZAWINUL
OCTOBER 27 & 28, 2006

Grooves ask for mercy, mercy, mercy. Schooled in the subtleties of swing by Dinah Washington, keyboardist **Joe Zawinul** brought the fundamentals of funk to Cannonball Adderley, the essentials of the electric to Miles Davis, and carried soul jazz into the electric age with his band Weather Report. Now the **Zawinul Syndicate** takes us on a hybrid adventure of sophisticated harmonies, world music rhythms, and deeply funky grooves. *Mercy.*

BEBOP LIVES!
JANUARY 26 & 27, 2007

Feet tangle and neurons dance. Fakers recoil, goatees sprout, and virtuosos take up their horns. Charlie Parker and Dizzy Gillespie set the bebop revolution in motion, their twisting, syncopated lines igniting the rhythms of jazz. Latter day fakers beware as the legendary **James Moody** and **Charles McPherson**, the alto sax voice of Charlie Parker in Clint Eastwood's *Bird*, raise battle axes and *swing*.

CECIL TAYLOR & JOHN ZORN
MARCH 9 & 10, 2007

Souls get freer. Embark on a sonic voyage as the peerless **Cecil Taylor** navigates us through dense forests of sound—percussive and poetic. He is, as Nat Hentoff proclaimed, "a genuine creator." The voyage banks toward the avant-garde as **John Zorn's Masada** with **Dave Douglas** explores sacred and secular Jewish music and the "anguish and ecstacy of klezmer." Musical wanderlust *will* be satisfied.

THE MANY MOODS OF MILES DAVIS
MAY 11 (Kisor/Blanchard) & MAY 12 (Payton/Miller), 2007

Change gets urgent. "I have to change," Miles said. "It's like a curse." And so his trumpet voice—tender, yet with that edge—was bound up in five major movements in jazz. The LCJO's **Ryan Kisor** opens with bebop and the birth of the cool. GRAMMY®-winner **Terence Blanchard** interprets hard bop and the modal *Kind of Blue*. New Orleans adventurer **Nicholas Payton** conjures the great 60s quintet. And Miles vet **Marcus Miller** electrifies with fusion.

CUBANA BE CUBANA BOP
JANUARY 12 & 13, 2007

Cultures collide and rhythm explodes. "Dizzy no peaky pani," said conguero Chano Pozo, "I no peaky engly, but boff peak African." When they wed Afro-Cuban grooves to the frenetic language of bebop, their music—cubop—burst on the scene with the heat of an inferno. The **Afro-Latin Jazz Orchestra** with **Arturo O'Farrill** revisits Machito and Dizzy Gillespie masterpieces—"Panga," "Manteca," "Tin Tin Deo," "A Night in Tunisia"—along with new works that demonstrate that when jazz gets tropical, everybody moves.

TODO TANGO
APRIL 13 & 14, 2007

Swing gets sultry. Dancers step closer when jazz treads on South American shores. Tango crusader, composer, bassist, and arranger **Pablo Aslan** joins the **Afro-Latin Jazz Orchestra** with **Arturo O'Farrill** as they explore the dynamic intersection where jazz and tango converge. Join us as we celebrate the living tradition of tango with performances including the legendary music of Astor Piazzola, along with new works that pay tribute to the rich musical culture of Argentina.

DIANNE REEVES
APRIL 20 & 21, 2007

Satin shimmers divinely. At once shimmering and sultry, four-time GRAMMY® winner—most recently for the soundtrack to the Oscar®-nominated *Good Night, and Good Luck*—**Dianne Reeves** is in a class by herself—powerful when soft, intimate at a fever pitch, and agile at any tempo. With her trio, classics are rendered with timeless grace and style.

DARIN ATWATER GOSPEL
MAY 25 & 26, 2007

Spirits run deeper. The rhythms of the sanctified church will shatter The Allen Room windows. Masterful young pianist **Darin Atwater**, composer-in-residence of the Baltimore Symphony Orchestra and artistic director of the Soulful Symphony is joined by **Kim Burrell**, a talent "graced with the thunder of a gospel shouter and the sophistication of a classy jazz chanteuse" (*Billboard*). With guest singers—full-throated and gossamer—they raise voices in a divine confluence of jazz, classical, and gospel.

7 — JAYP SERIES

Jazz for Young People℠
3 Concerts
Rose Theater, 12pm & 2pm

What an amazing show!...every time I go to a Young People's concert they get better and better. The lessons were great, the pacing was perfect and the singing along was such a good way to get the kids involved...beautiful!"
—Maria Afreshouer, Vocal & General Music Hutchison Colonial Elementary Schools

WHAT IS AN ARRANGER?
DECEMBER 2, 2006

Brass and reeds reconcile. How do 15 strong-willed musicians come together in perfect harmony? How does the standard become fresh again? Thank the arranger, the unsung hero who musically choreographs every show, bringing order and imagination to the bandstand. The **LCJO** with **Wynton Marsalis** explore the techniques arrangers use to help everyone get along and sound good, good, good.

WHAT IS LATIN JAZZ?
MARCH 3, 2007

Rhythm becomes everything. What happens when you put a little Latin in your jazz? As Latin jazz pioneer Mario Bauzá explained, "You walk with rhythm, you talk with rhythm, you eat your food with rhythm." **Arturo O'Farrill** and the **Afro-Latin Jazz Orchestra** show us how Dizzy, Machito, Bauzá, and others infected the lines of jazz with Latin grooves and started a rhythm epidemic. Watch out! Your knees will become bongos, your kid sister will turn into a timbale!

HOW DO WE CREATE JAZZ MOODS?
MAY 19, 2007

Moods mingle. Jazz is more than a mood indigo. In the beat of a heart it can take you from April in Paris to autumn in New York, from a stormy Monday to the sunny side of the street. Host **Wynton Marsalis** and friends show us how jazz's changing tones, colors, rhythms, and tempos take us from moanin' to groovin' high!

1 — LCJO SERIES

Lincoln Center Jazz Orchestra with Wynton Marsalis
4 Concerts
Rose Theater, 8pm

COLTRANE
SEPTEMBER 14, 15 & 16, 2006

Blue tranes run deeper. Ecstatic and somber, secular and sacred, John Coltrane's musical sermons transform Rose Theater into a place of healing and celebration with orchestrations of his small group masterpieces "My Favorite Things," "Giant Steps," "Naima," and more. Join us as the **LCJO** with **Wynton Marsalis** marks the 80th year since the birth of one of the most admired, influential, and adventurous artists in the history of jazz.

GERSHWIN
NOVEMBER 16, 17 & 18, 2006

Rhapsodies get bluer. "Composers have been walking around jazz like a cat around a plate of soup," said legendary conductor Walter Damrosch, "waiting for it to cool so that they could enjoy it without burning their tongues." George Gershwin, however, had *chutzpah*. The **LCJO** with **Wynton Marsalis**, special guest **Marcus Roberts**, and the **American Composers Orchestra** perform Gershwin's groundbreaking *Rhapsody in Blue*, and then underscore vocal performances of Nelson Riddle's finest arrangements from the Gershwin songbook. The evening closes as composer **Derek Bermel** brings his innovative vision to an exciting new work.

JAZZ AND ART
FEBRUARY 22, 23 & 24, 2007

Sound bleeds color. The lipstick reds of a bebop fanfare and midnight blues of a low-down dirge syncopate our musical canvas. Inspired by the **Museum of Modern Art's** collection, the **LCJO** with **Wynton Marsalis** performs the music that moved Mondrian, Bearden, Pollock, and others to saturate their art with the rhythmic energy of jazz. Then in a debut work, *Down Beat* magazine 2005 "Rising Star" **Ted Nash** interprets in song the masterworks of Picasso, Chagall, and other twentieth century masters.

THE SONGS WE LOVE
MARCH 29, 30 & 31, 2007

Perfection endures. They are arranged to perfection—"April in Paris" arranged by Wild Bill Davis, "Summertime" by Gil Evans, and many more—and they're our life soundtracks, elevating the everyday, making the mundane magical. The **LCJO** with **Wynton Marsalis** plays some of the greatest arrangements of our favorite songs—swinging and supple, sophisticated and spirited—and remind us all over again how great music becomes legendary.

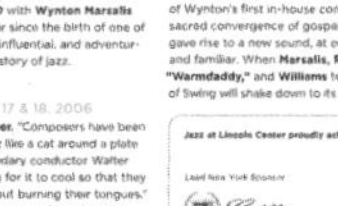

ROSE THEATER

THE ALLEN ROOM

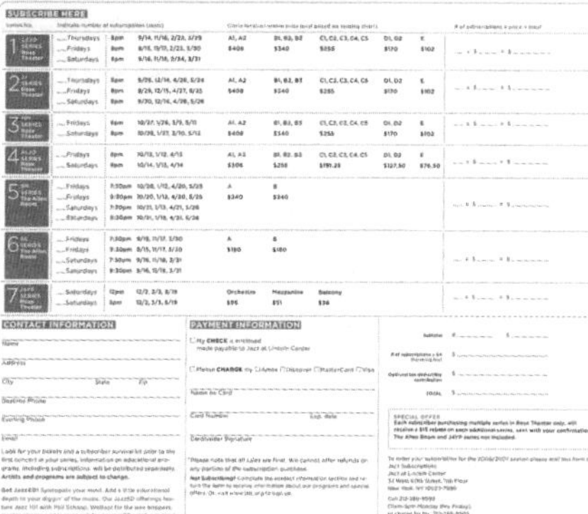

SUBSCRIBE HERE

CONTACT INFORMATION

PAYMENT INFORMATION

● 79.使用Helvetica字体

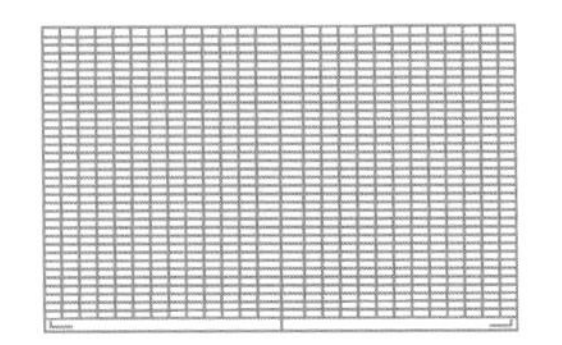

2007年，Helvetica字体的五十周年纪念活动使这款经典而简洁的无衬线字体成为焦点。为什么Helvetica字体会与瑞士网格紧紧联系在一起？Helvetica源于拉丁语“Helvetia”，即“瑞士”的意思。该字体最初被命名为“Neue Haas Grotesk”，在20世纪50年代，它开始与有序的网格一起使用，定义了现代主义的设计风格。

细长而优雅的Helvetica字体看起来平和而精致

项目
Helvetica字体展示

客户
设计卡片工作室，威曼•德鲁克斯（Veenman Drukkers），康斯莱设计工作室（Kunstvlaai）

摄影师
贝丝•托恩德罗

Helvetica字体可以有多种粗细和大小的变化。中等大小和粗体的字体通常用于表达严肃的观点。较小的字号能带来简约、奢华和禅意的感觉

K_nst
VI__.

Art Pie
International
A.P.I.

Een boek navertellen
op video in precies
één minuut of kom

Win 1000 euro

naar de Kunstvlaai A.P.I.
bij de stand van The One Minutes en
maak hier jouw boek in één minuut.
Van 10–18 mei 2008

Westergasfabriek
Haarlemmerweg 6-8
Amsterdam
www.kunstvlaai.nl

Helvetica字体具有可读性强的特点，对印刷来说，这一特点非常重要

● 80.改变分隔线的粗细

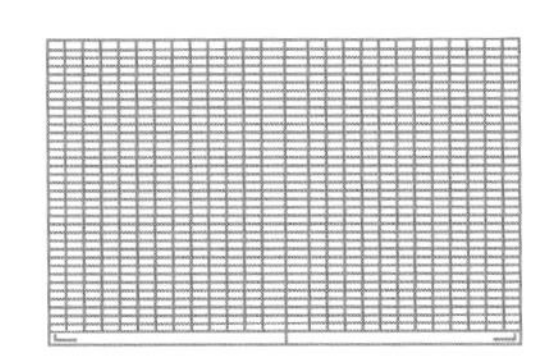

分隔线有很多作用，它可以用作导航栏、标题栏、图片的基准线、分区工具和版头。

Home Recent News Clients Awards Contact

United States
National Parks Service
Washington, DC
Publications Program
1977

In order to achieve better identification and financial savings through standardization of every aspect of the publications program, we designed a modular system that determined everything from the paper size to graphics to cartography and illustration.

Corporate Identity 1 of 3 ▶

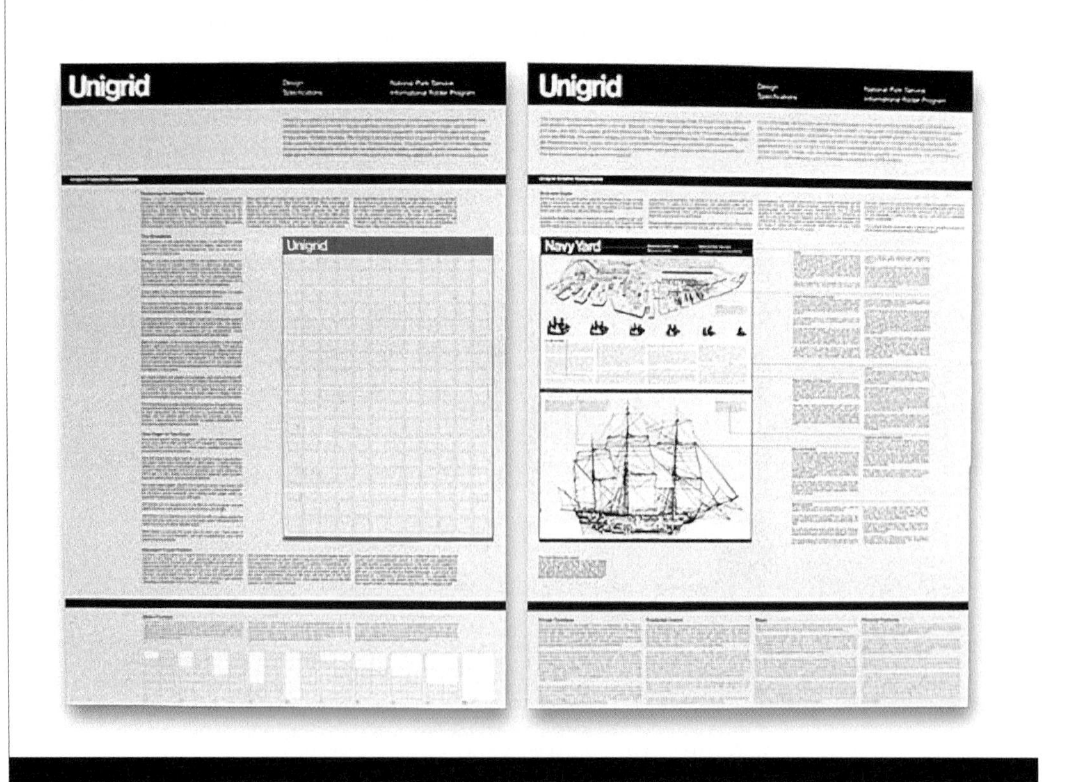

项目
维格尼利联合公司（Vignelli Associates）网站

客户
维格尼利联合公司

设计师
达尼·皮德曼（Dani Piderman）

设计总监
马西莫·维格尼利（Massimo Vignelli）

马西莫·维格尼利是一名网格设计大师，他在网站上展示他的品牌。这组页面是本书对维格尼利及其同事的致敬

上页图（上）：维格尼利联合公司延续了一贯的风格，网页的设计井然有序

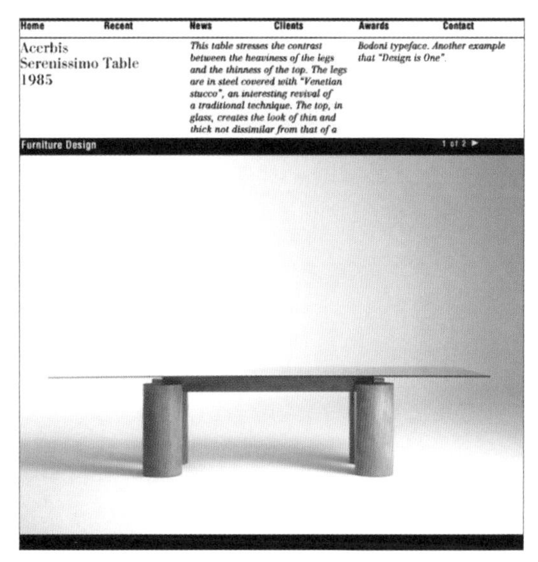

不同粗细的分隔线既可以分隔信息，又能包含信息

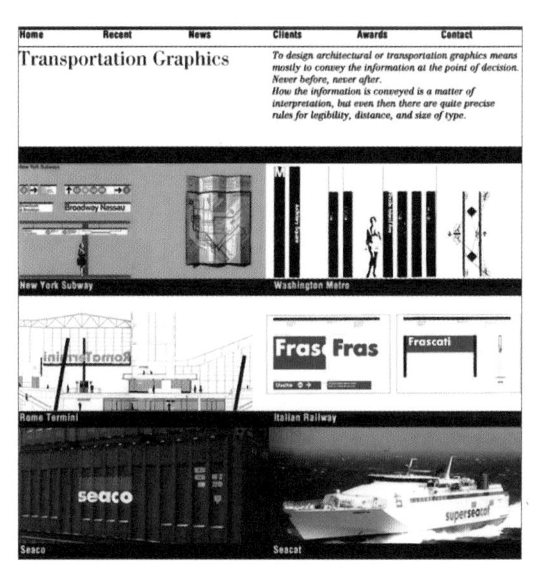

上页图（下）：标题采用了Franklin Gothic粗体，与Bodoni正体及Bodoni斜体形成对比与互补，为瑞士风格的设计增添了一些意大利风情

● 81.混用垂直和水平的层次

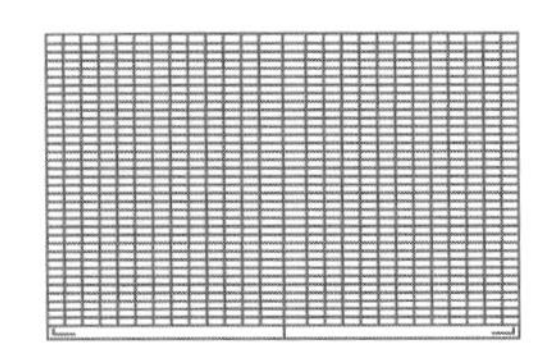

将页面划分成轮廓清晰的区域可以使信笺、表格和收据美观实用。垂直和水平的网格可以同时使用，以不太常用但又能容纳所有必要元素的方式组织各信息单元。

IS

INDUSTRIES stationery

91 Crosby Street
New York, NY 10012
212.334.4447
www.industriesstationery.com

ITEM NUMBER	DESCRIPTION	QUANTITY	PRICE	EXTENSION
11.150.3	Small Spiral Pads with Black cover/Colorfest pages-set of 3	1	16.50	16.50
71.120.2	SpinlSquare Notebook PopPrints Khaki	1	6.50	6.50
71.120.1	SpinlSquare Notebook PopPrints Blue	1	6.50	6.50

SALES RECEIPT

DATE
4/8/2008

REFERENCE NUMBER
80901

SALESPERSON
CE

SOLD TO

SHIP TO

RETURN POLICY
Merchandise may be returned for exchange or store credit within 14 days of purchase with the store receipt. Sale merchandise is non-returnable. All returns must be in saleable condition.

STORE HOURS
Monday-Saturday 11:00-7:00
Sunday Noon-6:00

MERCHANDISE TOTAL	SHIPPING	OTHER CHARGES	DISCOUNT	TAXABLE SUBTOTAL	SALES TAX	NON TAX SALES	TOTAL	AMOUNT PAID	BALANCE DUE
29 50				29.50	2.47		**31.97**	31.97	

项目
发票收据

客户
INDUSTRIES文具公司

设计师
德鲁•苏扎（Drew Souza）

这张收据的设计让人想起赫伯特•拜耶（Herbert Bayer）的设计方法，将整个页面视为一个可分割的平面

本页和上页图：这个收据同时使用了水平和垂直的层级，为各部分数据打造了一个可以被清晰分割的“容器”。如果没有销售信息，收据只是一个漂亮的抽象作品，而有了这些基本信息，该收据就是一个实用的系统

IS

INDUSTRIES stationery

91 Crosby Street
New York, NY 10012
212.334.4447

www.industriesstationery.com

SALES RECEIPT

DATE

REFERENCE NUMBER

SALESPERSON

SOLD TO

SHIP TO

RETURN POLICY

Merchandise may be returned for exchange or store credit within 14 days of purchase with the store receipt. Sale merchandise is non-returnable. All returns must be in saleable condition.

STORE HOURS

Monday-Saturday 11:00-7:00
Sunday Noon-6:00

ITEM NUMBER | DESCRIPTION | QUANTITY | PRICE | EXTENSION

MERCHANDISE TOTAL | SHIPPING | OTHER CHARGES | DISCOUNT | TAXABLE SUBTOTAL | SALES TAX | NON TAX SALES | TOTAL | AMOUNT PAID | BALANCE DUE

SALES DRAFT

DATE

REFERENCE NUMBER

SALESPERSON

SOLD TO

DISCOUNT

MERCHANDISE TOTAL

SHIPPING

OTHER CHARGES

TAXABLE SUBTOTAL

SALES TAX

NON TAX SALES

TOTAL

AMOUNT PAID

BALANCE DUE

PAID BY

PAID BY

82.制造惊喜的效果

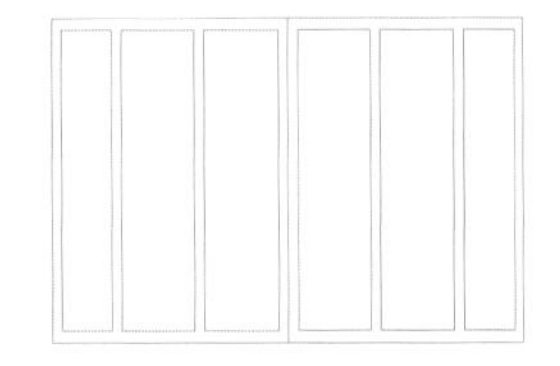

严谨的网格系统可以为一本杂志的设计提供一个总体规划，但是过于复杂的结构会使页面变得单调乏味。出乎意料地插入某个元素，打破网格，或适当的留白可以吸引读者的注意力。

项目
《没有男性的国度》

客户
The Wing公司

设计公司
Pentagram公司

创意总监
艾米莉·欧本曼

合伙人
艾米莉·欧本曼

高级设计师
克里斯汀娜·霍根

设计师
伊丽莎白·古德斯皮德，乔伊·崔罗

项目经理
安娜·梅克斯勒

设计师设计了一个强大的整体网格，但特别插入了海报、贴纸和视觉双关语来打破网格系统

Photograph from California Department of Transportation

Twelve miles from downtown and five miles from the sea, I-10 arrives to cross the I-405 at the most beautiful freeway interchange in L.A.—the Santa Monica-San Diego interchange. On a satellite map, these intertwined roads look something like a game of cat's cradle between fingers. The scale, from head-on, is vast and monumental, as striking as the Gateway Arch in repose. In a more pastoral setting, it might have been called land art. In a city defined by its lack of a center, it seems to both enable and pay homage to the sprawl.

72

Feature

Three stacked ramps spring forth from small hills, scattering cars in all four directions. This is the junction between normal L.A. and the parts of L.A. that we always see in movies—the beaches and hills and the very big houses. No doubt my fondness for its ramps has something to do with how they often lead to fun.

This past spring, driving on the 10, preparing to change to the 405, I noticed a light green highway sign with the junction's formal municipal name: the Marilyn Jorgenson Reece Memorial Interchange. Sitting in traffic, I typed the honorific into Google, hoping to learn what a woman must do in order to win a junction for herself. The search results were few—an obituary, some blog posts—and all seemed to draw from the same pool of facts: the first licensed female engineer in California, designed the Santa Monica-San Diego interchange, died at 77 in 2004. Reece was a readymade Woman of the Month, a frequent subject of those bland but nice blurbs that often say more about corporate obligation than the actual women they seek to reward.

Her image results were far more compelling, or at least lent some inadvertent texture to her life. Reece, it seems, was a snappy dresser—a fan of polka-dot blouses and smocks before the era of easy office separates. In one photo of the interchange under construction, she wears a linen skirt suit with a Peter Pan collar. Cranes and pylons rise from the dirt, casting her ballet flats as comically flimsy. It's hard to overstate the strangeness of the photo. The scene looks cribbed from a B-movie—a woman swept up from the pew of her church and set down unharmed at the helm of a job site. Archives are full of such photos of men, surveying big holes in their hard hats and suits. Reece looks more focused than photo-op ready. A gentle breeze ruffles her jacket but scarcely moves her hair-sprayed hair.

Back at my house, I tried to learn more. As the daughter of a working mother and the Spice Girls, I still feel a surge of neoliberal pep at the thought of a female-designed highway junction (or, if you prefer, a *she*-way int-*her*-change). Reece, like most private citizens of her time, barely makes a blip in the searchable

"The worst-designed freeways make you wrestle for this thrill; the best, like Reece's, induce it without effort."

news, despite the relative high profile of her work. The day-to-day details of her life and career are less readily available than the same information for any given ex-boyfriend's new girlfriend. I tracked down a colleague from her time at CalTrans who never responded to my interview request. I tweeted at the former Secretary of State who sponsored the light green freeway sign. Deep in the text of a press release, I learned of a $1,000 prize in her name, awarded to a Pomona middle school student for her "Empirical Analysis of Wooden Structures."

It's nice how women cross paths across time but not especially interesting. (I suppose this was the inadvertent message of *The Hours*.) Getting more serious, I filed a public records request. Dredging up some nouns that might pertain to engineering, I wrote: "Any correspondence, designs, photos, or proposals would be useful." The CalTrans librarian emailed me back, asking me to clarify the nature of my interest. The records form acknowledged litigation research but didn't have a box for vague curiosity relating to the life of a former employee. The librarian said she'd see what she could do, but she sounded overworked and confused by my request. I didn't expect to hear from her again.

73

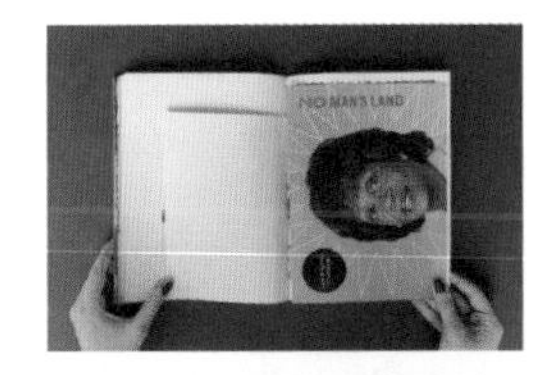

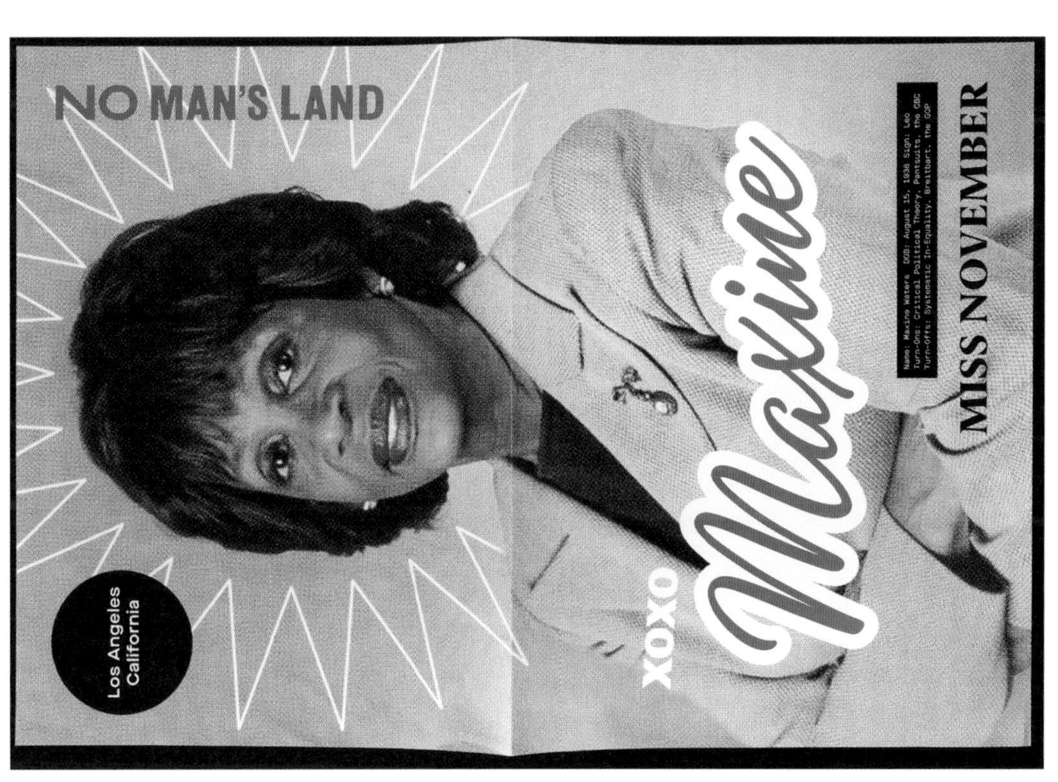

为了应对数字化的浪潮，与读者（用户、浏览器）建立联系，出版物变得越来越多样化。在数字时代，出版纸质杂志是有风险的，这本杂志旨在为有相同价值观的女性群体建立一个社区。杂志中加入了一点儿能激发触觉的元素——贴纸和折叠海报，海报的主角是一位颇有成就的政治家。这改变了其他出版物在插页上使用“日历女郎”的惯例

上页图及本页左图：以“加利福尼亚州首位女性立交桥设计师用人行道创作诗歌”为标题的专题报道，利用一句简短的具象诗打破了网格，并在排版上呼应了立交桥的形状

● 83.改变大小

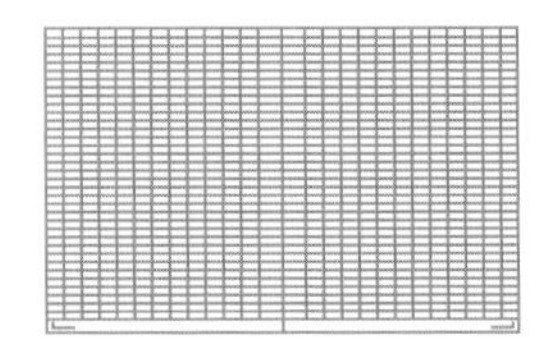

一旦确定了整体网格，就要开始考虑尺寸、空间、大小和版式等问题。从文本的内容和重要性出发，图片和文本的大小可以是不同的，也可以是变化的，这取决于素材所需的空间大小。

这张封面图片清楚地展示了排版可以极其简洁

项目
《绿色是什么？》（*What Is Green?*）

客户
触手可及设计公司（Design within Reach）

设计公司
触手可及设计公司

创意总监
詹妮弗·莫拉（Jennifer Morla）

艺术总监
迈克尔·塞纳托（Michael Sainato）

设计师
詹妮弗·莫拉，蒂姆·袁（Tim Yuan）

文案
格温多林·霍顿（Gwendolyn Horton）

绿色和可持续性是许多公司关注的全球性热点话题，包括一直具有生态意识的触手可及设计公司。这个作品的前13页包含了与生态相关的内容，采用了不同的版式布局，具有一种流畅感

As if it wasn't challenging enough to choose between one color and another, now there's green, which comes loaded with its friends: sustainable, eco-friendly, cradle-to-cradle, recycled, recyclable, small footprint, low-VOC, Greenguard, LEED and FSC-certified. Being a design company, we're encouraged by the increasing number of smart solutions to improve the planet. But we know that not all items fit into every category of ecological perfection. At DWR, we believe in honestly presenting our assortment so you can choose what's best for you. We also believe in selling products that last. We're all doing our part, and we welcome your response when we ask, "What is green?"
DESIGN
WITHIN
REACH

在第一页，一段很长的文本填满了整个网格

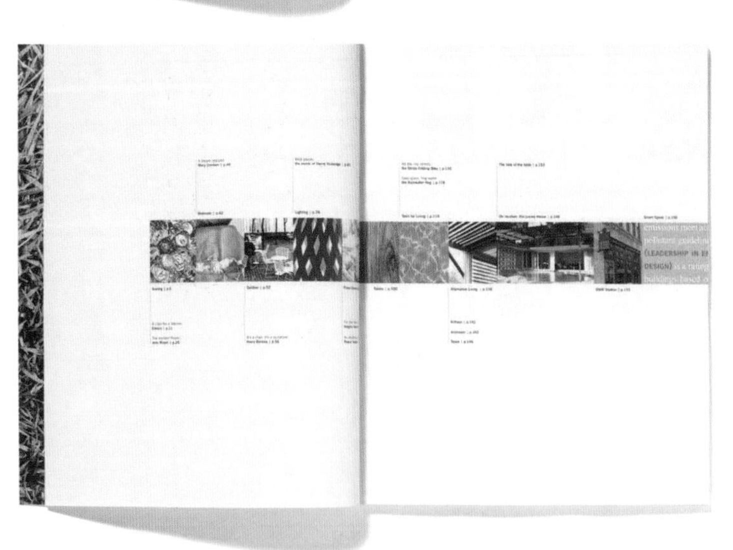

由于素材体量发生了巨大变化，目录页面采用了水平设置，增加了版面的流动性。引线引导读者阅读目录，缩略图起到引导读者快速浏览内容的作用

Green is up-cycling cans into a chair that lasts 150 years.

The hand-brushing department at Emeco, U.S.A.
At Emeco, all aluminum waste is recycled, even the aluminum dust that's filtered out of the air.

The upside of up-cycling aluminum: chairs for a lifetime or two.
When Emeco started making its aluminum chairs in 1944, you can be darn sure there wasn't a marketing brief that said, "Make it attractive to the eco-conscious community." Emeco had other things on its mind, namely how to make a chair withstand a torpedo blast. The irony is that Emeco chairs have become an outstanding example of what's commonly referred to as "green." To create the 1006 Navy Chair (1944), Emeco invented a 77-step process to satisfy the military's need for lightweight, corrosion-resistant chairs for destroyers and submarines. In the process, the company invented a method to make aluminum three times stronger than steel, and a chair so durable that it has an estimated lifespan of 150 years. Legend has it that Wilton Dinges, who founded Emeco in 1944, actually tossed a 1006 Navy Chair out the window of a six-story building. The people on the sidewalk below were a bit surprised, but the chair was fine, with the exception of a few scratches. Today, everything Emeco makes is still manufactured by hand using the same 77-step patented process. Emeco chairs and tables all begin with 80% recycled aluminum, which requires only 5% of the energy needed to produce virgin aluminum, and they're all made in Pennsylvania, U.S.A. Emeco's all-aluminum chairs and stools are built to last, and generations from now, when your great-great-grandchildren finally manage to wear out a chair that's tested to withstand 1,700 pounds of weight (big kids), the aluminum can be 100% recycled and made into something else. In recent years, Emeco has partnered with Philippe Starck, Norman Foster and others to create classic designs for a new century, and these collections are made in the same facility, using the same processes and by the same people who make everything else at Emeco. Perhaps Philippe Starck said it best when he explained that "working with Emeco has allowed me to use a recycled material and transform it into something that never needs to be discarded – a tireless and unbreakable chair to enjoy for a lifetime. It is a chair you never own, you just use it for a while until it is the next person's turn." On the next page you'll find Emeco chairs and stools, all of which contribute to LEED™ credit #4.2 Recycled Content (and credit #5.1 if shipped within 500 miles of Hanover, Pennsylvania). For the entire **Emeco Collection**, visit dwr.com.

DESIGN WITHIN REACH: APRIL 2008 | 11

这些布局显示了文本大小的变化，其中一个页面的文本框非常宽。通常，这种方式并不合适，然而在这个案例中，设计风格和内容信息都超出了常规的设计原则。在这个页面中，读者会了解到关于可回收铝制椅子的情况，其中关于椅子的描述非常简练，是该页版式的一大亮点

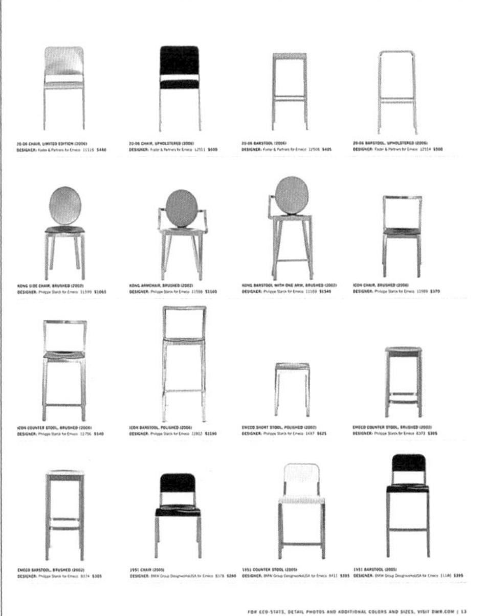

84.问问自己什么可以舍弃

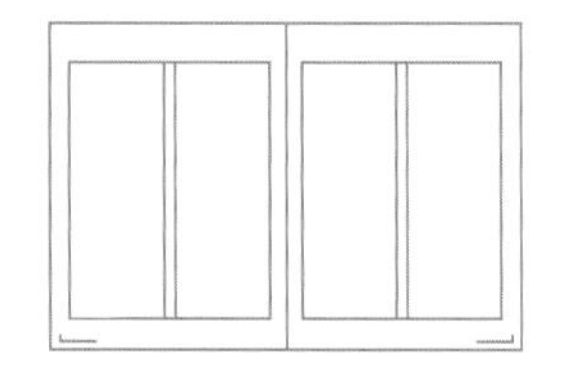

当照片很精彩时，不要破坏它。有时候，最好的方式就是把照片尽可能放大，尽可能少裁剪，或者完全不裁剪，并且让照片上没有多余的字或图案。换言之，让照片与网格结合，但也要让照片保留自己的精髓。

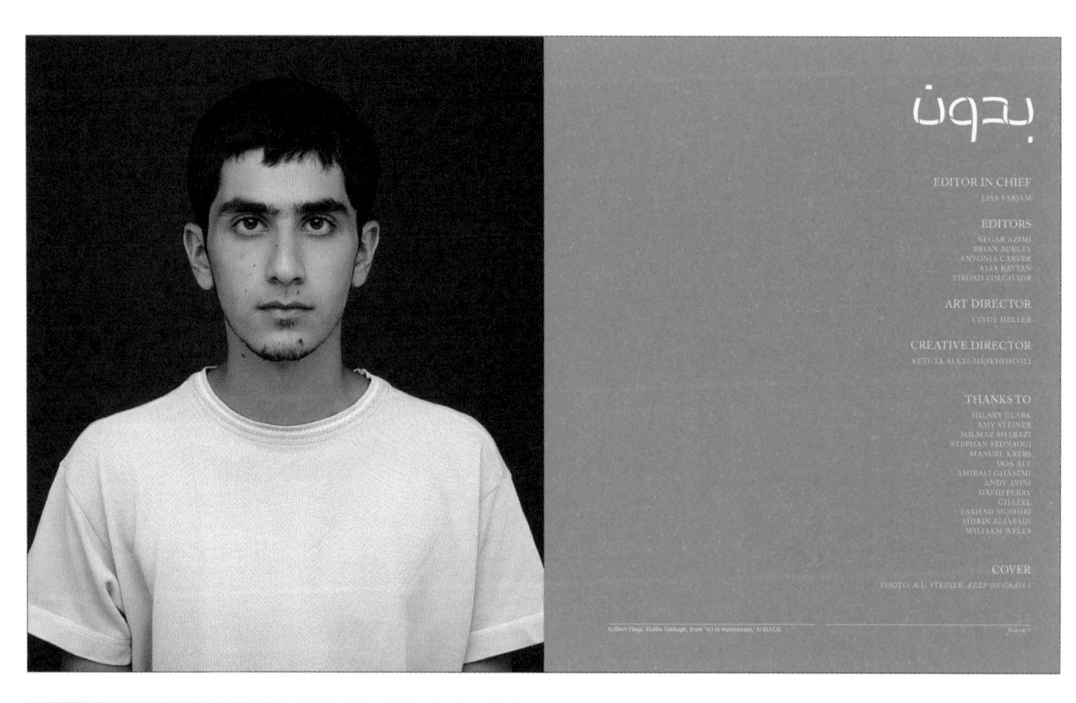

本页图和下页图：没有必要对这些照片做任何处理，因为这些照片本身就很有意义

项目
杂志

客户
比杜恩（Bidoun）

创意总监
凯图塔-阿列克西•梅斯基什维利（Ketuta-Alexi Meskhishvili）

设计师
辛迪•海勒（Cindy Heller）

摄影师
吉尔伯特•哈格（Gilbert Hage）（本页肖像照），西莉亚•彼得森（Celia Peterson）（下页工人照）

82 Tourism

Cautious Radicals

Art and the invisible majority

By Antonia Carver

At the 2005 Sharjah Biennial, artist Peter Stoffel attempted to get himself banned. Taking inspiration from the notices placed by employers in local newspapers, featuring the names, nationalities, passport numbers and mug shots of ex-employees, Stoffel requested that the biennial's organizing body fire him and announce his occupational demise in the same way. Other potential employers—presumably those organizing another biennial in the UAE—would be hiring him "at their own risk and responsibility." At the same time, the biennial would write Stoffel a recommendation letter "acknowledging his reliable services as an artist," which would be freely available to visitors to the biennial.

The artist's conceit turned out to be more potent than the proposed work itself. In keeping with the generally taboo nature of discussion surrounding the rights of the Gulf's underclass of foreign maids and laborers, the biennial organizers declined to go along with Stoffel's ruse. During the exhibition, he showed two panels of text—one a narrative explaining his concept and the outcome, the other a page from a local newspaper with advertisements placed by "sponsors" of Sri Lankans and Pakistanis who had "absconded from duty" and were therefore now outside the employer's responsibility.

For Gulf-based biennial visitors, Stoffel's project was audacious in its attempt to query the region's strict racial and financial hierarchy of workers' rights. (Since the biennial, new legislation has begun to address both the rights of the employee in the transferral of sponsorship and the prerogative of sponsors to impose the customary six-month ban—from the country, and/or from working for a competitor company—on some employees.)

As he describes it, Stoffel attempted to establish a connection between the smallest minority in the UAE, that of the immigrant artist, and the largest, the immigrant laborer. (About two-thirds of the UAE's work force comes from abroad, and about a quarter of all expats work as unskilled laborers for construction companies.) Stoffel concluded that the "two parallel lines of the biennial artist and the Pakistani worker never cross, and that is the paradox of the paradox: that even at an imaginary point, within an artwork, it's impossible to establish a connection."

Despite being the largest segment within the UAE population, the foreign working class remains by and large a faceless majority, known only to the wealthy minority through increasingly ballsy local media stories. Every week, the usually self-censoring UAE newspapers detail gory tales of trafficking, suicide, and rape; of false promises made by dubious foreign employment agencies and mounting debts; of dehydration while working in extreme summertime heat and humidity; of industrial accidents and loan sharks; of depressed, desolate labor camps. The Indian Embassy's official list of its functions includes such grisly tasks as "processing applications received for providing free air tickets by Air India/Indian Airlines for transportation of dead bodies of destitute/stranded/absconded Indian nationals."

In many ways, the situation faced by the Gulf's legions of indentured laborers is mirrored worldwide, from Chinese cockle-pickers in the UK to Mexican meatpackers in US abattoirs. But the particular state of affairs in Dubai, with its rapid growth and surface profligacy, takes a microscope to what's vaguely termed globalization.

Tourism 83

Phoos of laborers in Dubai by Celia Peterson, 2005, courtesy of Celia Peterson and arabianEye

84 Tourism

Tourism 85

● 85.善用侧边栏

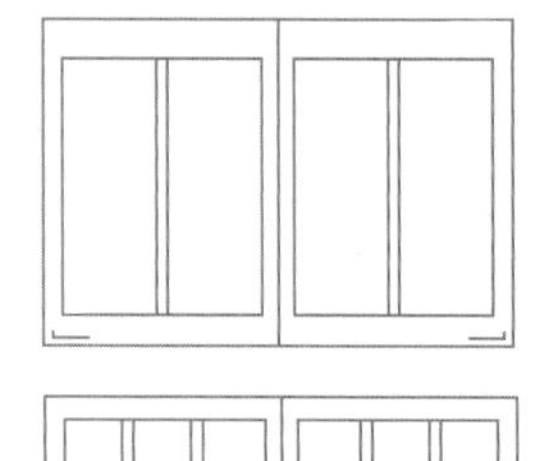

侧边栏一般用于容纳由正文延伸出的扩展内容，这些内容通常与主文本相关，但又需要与主文本分离。侧边栏可以放在网格内，它们的功能是容纳附加信息，而不是打破网格。

COST&PRICE／コスト&プライス
［2007年3月期］

95%の建設会社で鉄鋼コスト増
資材・労務単価の上昇は止まらず

資材価格が高騰している。本誌の調査では、建設会社の9割以上が鉄鋼や金属製品が値上がりしたと回答。資材コスト上昇分を建設単価に反映させていない建設会社は1割にも満たない。労務費も今後の建築工事需要を見込んで上昇基調にあり、建設コストの上昇はまだ続きそうだ。 ※価格はすべて税抜き

●資材コスト上昇の建設単価への反映

資材費動向

［鉄鋼・金属類］
市況は上伸して
さらに高値を目指す

［木材］
国際情勢の変化により
木材価格は上昇

［セメント］
大口需要家の抵抗で
値上げ交渉は難航

●資材単価の推移（東京地区）

96 NIKKEI ARCHITECTURE 2007-5-14

2007-5-14 NIKKEI ARCHITECTURE 97

一个组织严密的网格通常会提供不同规格（多栏、双栏或单栏）的侧边栏或文本框

项目
《日经建筑》（*Nikkei Architecture*）杂志

客户
《日经建筑》杂志

设计公司
ar

在这本建筑杂志中，技术性的信息都以方框和图表的形式呈现

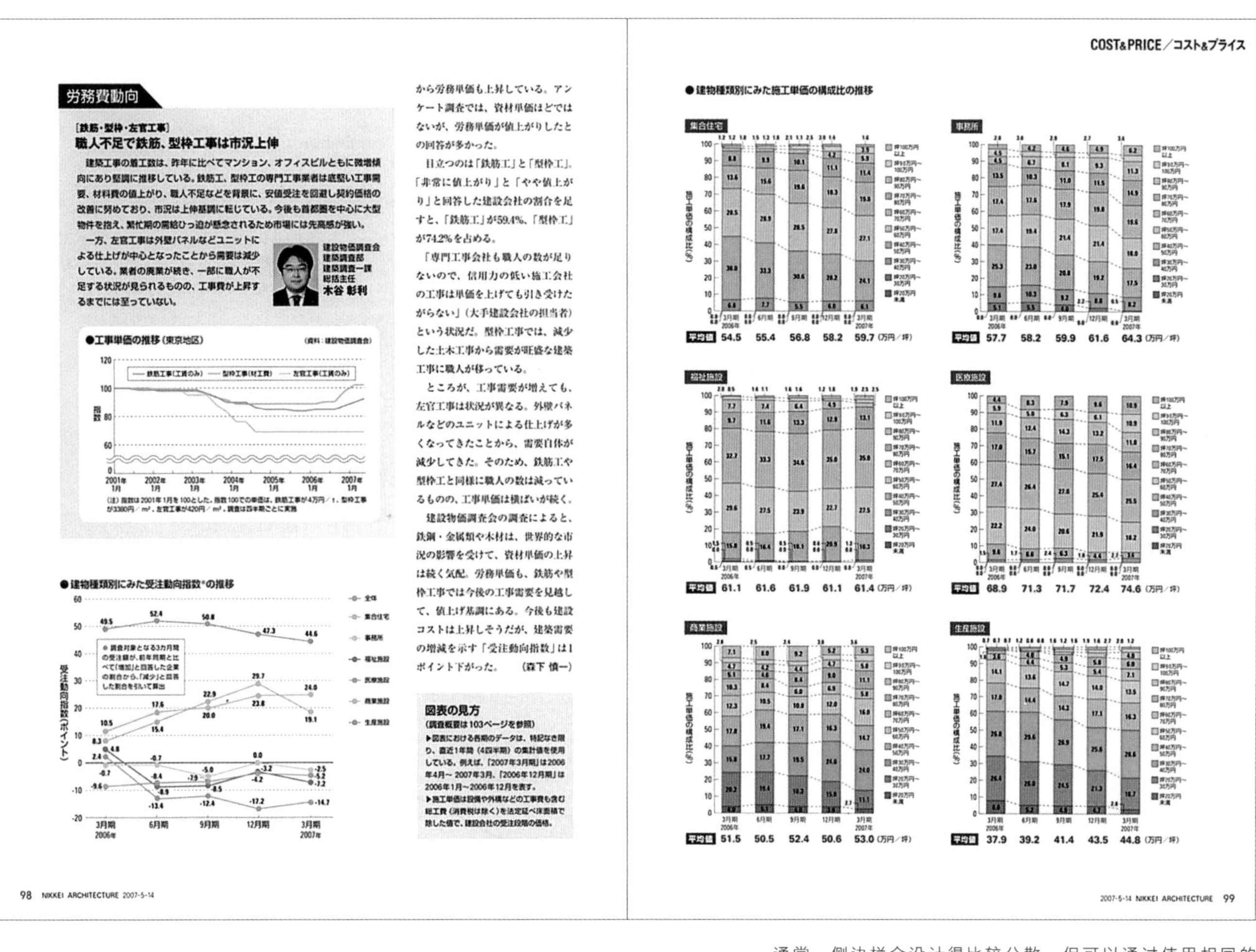

COST&PRICE／コスト&プライス

労務費動向

[鉄筋・型枠・左官工事]

職人不足で鉄筋、型枠工事は市況上伸

建築工事の着工数は、昨年に比べてマンション、オフィスビルともに微増傾向にあり堅調に推移している。鉄筋工、型枠工の専門工事業者は底堅い工事需要、材料費の値上がり、職人不足などを背景に、安値受注を回避し契約価格の改善に努めており、市況は上伸基調に転じている。今後も首都圏を中心に大型物件を抱え、繁忙期の需給ひっ迫が懸念されるため市場には先高感が強い。

一方、左官工事は外壁パネルなどユニットによる仕上げが中心となったことから需要は減少している。業者の廃業が続き、一部に職人が不足する状況が見られるものの、工事費が上昇するまでには至っていない。

建設物価調査会
建築調査部
建築調査一課
総括主任
木谷 彰利

●工事単価の推移（東京地区）（資料：建設物価調査会）

●建物種類別にみた受注動向指数*の推移

から労務単価も上昇している。アンケート調査では、資材単価ほどではないが、労務単価が値上がりしたとの回答が多かった。

目立つのは「鉄筋工」と「型枠工」。「非常に値上がり」と「やや値上がり」と回答した建設会社の割合を足すと、「鉄筋工」が59.4%、「型枠工」が74.2%を占める。

「専門工事会社も職人の数が足りないので、信用力の低い施工会社の工事は単価を上げても引き受けたがらない」（大手建設会社の担当者）という状況だ。型枠工事では、減少した土木工事から需要が旺盛な建築工事に職人が移っている。

ところが、工事需要が増えても、左官工事は状況が異なる。外壁パネルなどのユニットによる仕上げが多くなってきたことから、需要自体が減少してきた。そのため、鉄筋工や型枠工と同様に職人の数は減っているものの、工事単価は横ばいが続く。

建設物価調査会の調査によると、鉄鋼・金属類や木材は、世界的な市況の影響を受けて、資材単価の上昇は続く気配。労務単価も、鉄筋や型枠工事では今後の工事需要を見越して、値上げ基調にある。今後も建設コストは上昇しそうだが、建築需要の増減を示す「受注動向指数」は1ポイント下がった。（森下 慎一）

図表の見方

（調査概要は103ページを参照）

▶図表における各期のデータは、特記なき限り、直近1年間（4四半期）の集計値を使用している。例えば、「2007年3月期」は2006年4月～2007年3月、「2006年12月期」は2006年1月～2006年12月を表す。

▶施工単価は設備や外構などの工事費も含む総工費（消費税は除く）を法定延べ床面積で除した値で、建設会社の受注段階の価格。

●建物種類別にみた施工単価の構成比の推移

集合住宅　平均値 54.5　55.4　56.8　58.2　59.7（万円／坪）

事務所　平均値 57.7　58.2　59.9　61.6　64.3（万円／坪）

福祉施設　平均値 61.1　61.6　61.9　61.1　61.4（万円／坪）

医療施設　平均値 68.9　71.3　71.7　72.4　74.6（万円／坪）

商業施設　平均値 51.5　50.5　52.4　50.6　53.0（万円／坪）

生産施設　平均値 37.9　39.2　41.4　43.5　44.8（万円／坪）

98 NIKKEI ARCHITECTURE 2007-5-14

2007-5-14 NIKKEI ARCHITECTURE 99

通常，侧边栏会设计得比较分散，但可以通过使用相同的颜色、字体或分隔线，以形象的方式让侧边栏与主要内容区域产生关联

● 86.研究大师的作品

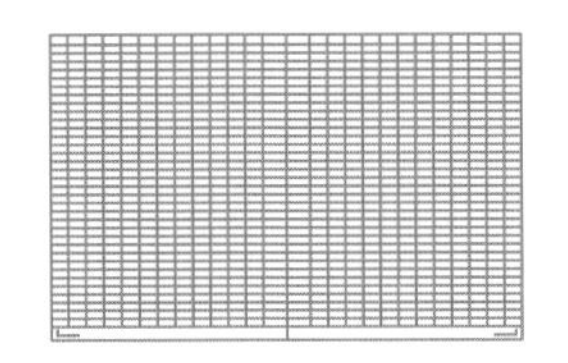

研究平面设计大师的作品可以学习大师的设计风格，同时又能形成自己对网格系统的新的理解。

设计首页版面时，可以效仿瑞士原创设计大师的作品，先对整体方案做深入了解，这样会产生新的灵感，而不是一味模仿某个元素。

1914 Naissance de Josef Müller le 9 mai à Rapperswil (Suisse).

1930 Apprentissage à Zurich.

1932 – 1935 Suit les cours de Ernst Keller et Alfred Willimann à l'école des arts et métiers de Zurich.

1934 Débute sa carrière comme illustrateur, scénographe d'exposition et décorateur de théâtre.

1939-1945 Lieutenant dans l'armée suisse.

1943 Épouse Verena Brockmann (violoniste) et adopte le nom de Josef Müller-Brockmann. Courante en suisse, cette pratique le distingue des autres Müller.

1944 Naissance de leur fils Andreas.

1950 Premières commandes du Zürich Tonhalle.

1951 Membre de l'AGI.

1952-1958 Affiches de sécurité pour l'Automobile-Club de Suisse.

1956 Voyages en Amérique : conférences à Aspen et New York, voyage au Mexique avec Armin Hoffman. Songe à s'installer aux États-Unis.

1957 Nommé professeur à l'école des arts et métiers de Zurich.

1958

1960 Quitte l'école de Zurich. Conférence à Tokyo.

1961 Second voyage au Japon.

1962 Contrats de design et de conseil en RFA. Enseigne à Ulm.

1964 Mort de Verena dans un accident de voiture.

1965 Fonde avec Eugen et Kurt Federer la Galerie 58 (puis Galerie Seestrasse) dans la maison familiale. Le lieu deviendra une référence de l'art concret. Conférences et expositions de Johannes Itten, Max Bill, Karl Gerstner… Fermeture en 1990.

1967 Fondation de Müller-Brockmann & Co avec Max Baltis, Ruedi Rüegg et Peter Andermatt et une quinzaine d'employés.

1971 Mariage avec Shizuko Yoshikawa.

1976 Brouille avec Baltis et Rüegg, qui quittent l'agence. Müller-Brockmann dirige seul l'agence jusqu'à sa fin en 1984.

1975 – 1980 Système de signalétique pour les chemins de fer suisses.

1981

1989

1994

1996 Meurt le 30 août.

1989 *Fotoplakate – Von den Anfängen bis zur Gegenwart.*

1994 *Mein Leben : Spielerische Ernst und ernsthaftes Spiel,*

82:

项目
《阶段》杂志

客户
金字塔/《阶段》杂志

设计师
安娜•图尼克

这本杂志刊登了一篇关于设计师约瑟夫•穆勒-布罗克曼的文章，其中收录了他的生平、著作封面和有影响力的图片，其版式设计是一个值得珍藏的网格基础应用通过对约瑟夫·穆勒-布罗克曼作品的敏锐观察，设计师将这些页面设计得非常丰富。这个案例既是一个精彩的作品，也是一个独立的研究成果

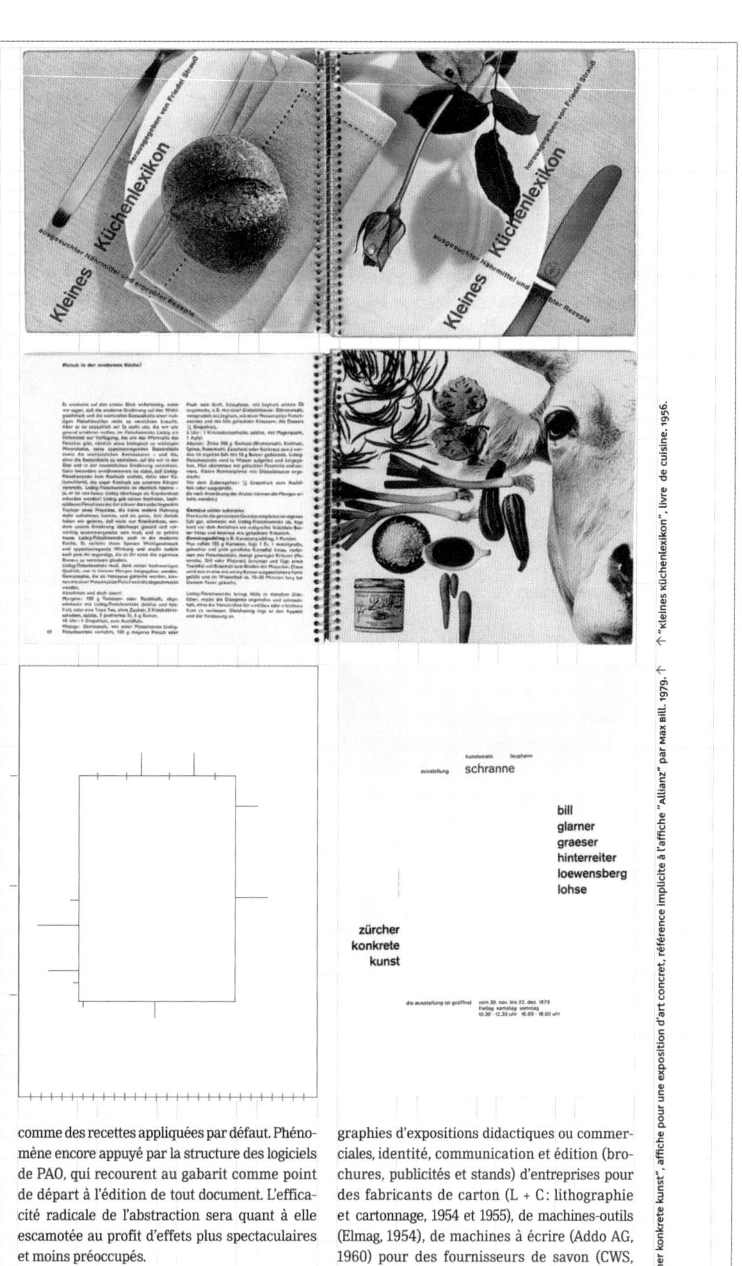

« Plus la composition des éléments visuels est stricte et rigoureuse, sur la surface dont on dispose, plus l'idée du thème peut se manifester avec efficacité. Plus les éléments visuels sont anonymes et objectifs, mieux ils affirment leur authenticité et ont dès lors pour fonction de servir uniquement la réalisation graphique. Cette tendance est conforme à la méthode géométrique. Texte, photo, désignation des objets, sigles, emblèmes et couleurs en sont les instruments accessoires qui se subordonnent d'eux-mêmes au système des éléments, remplissent, dans la surface, elle-même créatrice d'espace, d'image et d'efficacité, leur mission informative. On entend souvent dire, mais c'est là une opinion erronée, que cette méthode empêche l'individualité et la personnalité du créateur de s'exprimer. »

↑ "kleines küchenlexikon", livre de cuisine, 1956. ↑ "zürcher konkrete kunst", affiche pour une exposition d'art concret, référence implicite à l'affiche "Allianz" par Max Bill, 1979.

comme des recettes appliquées par défaut. Phénomène encore appuyé par la structure des logiciels de PAO, qui recourent au gabarit comme point de départ à l'édition de tout document. L'efficacité radicale de l'abstraction sera quant à elle escamotée au profit d'effets plus spectaculaires et moins préoccupés.

ceci dit, au boulot

Depuis ses débuts de scénographe, Müller-Brockmann a réalisé un grand nombre de travaux, seul ou à la tête de son agence (1965-1984): scénographies d'expositions didactiques ou commerciales, identité, communication et édition (brochures, publicités et stands) d'entreprises pour des fabricants de carton (L + C: lithographie et cartonnage, 1954 et 1955), de machines-outils (Elmag, 1954), de machines à écrire (Addo AG, 1960) pour des fournisseurs de savon (CWS, 1958) de produits alimentaires (Nestlé, de 1956 à 1960) ou pour la chaîne de magasins néerlandais Bijenkorf (1960). En 1962, il décroche d'importants contrats auprès d'entreprises allemandes: Max Weishaupt (systèmes de chauffage) et Rosenthal

:83

● 87.裁剪图片，放大细节

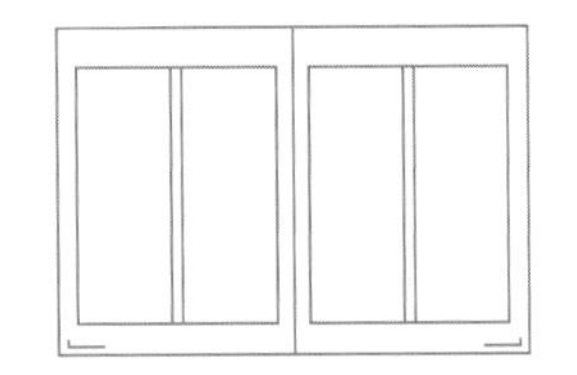

网格可以覆盖整个作品，压倒一切设计构成，也可以是不明显的基础架构。有人说，网格是一种“优雅的、逻辑清晰的、不突兀的布局”。

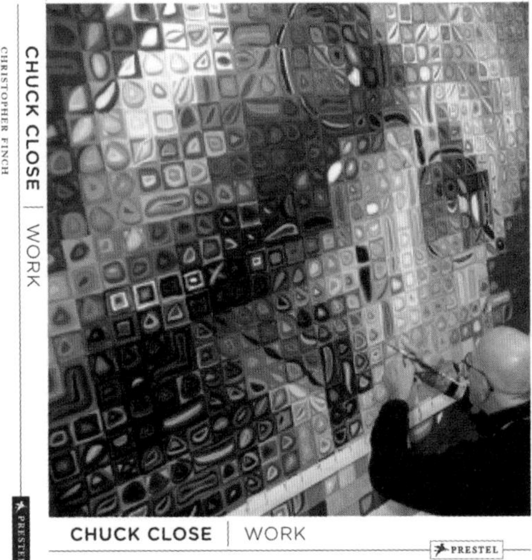

另见10页

这个封面的优点在于它的简洁，以及它对艺术家和艺术家作品的凸显，这是封面在折叠和包装之前的整体版面设计

项目
《查克·克洛斯作品集》

客户
普雷斯特出版社

设计师
马克·迈尔尼克

低调的版式设计优雅地展示了充满个性的绘画

扉页上的图片展示的是艺术家的侧脸

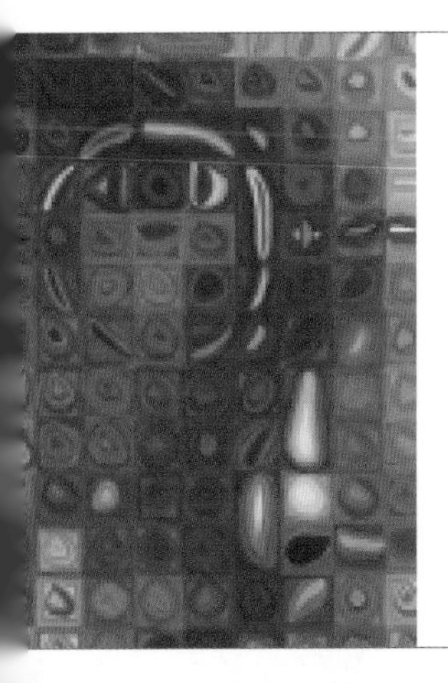

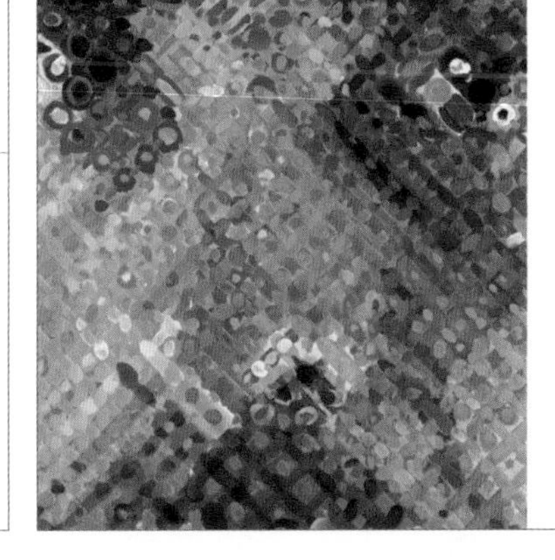

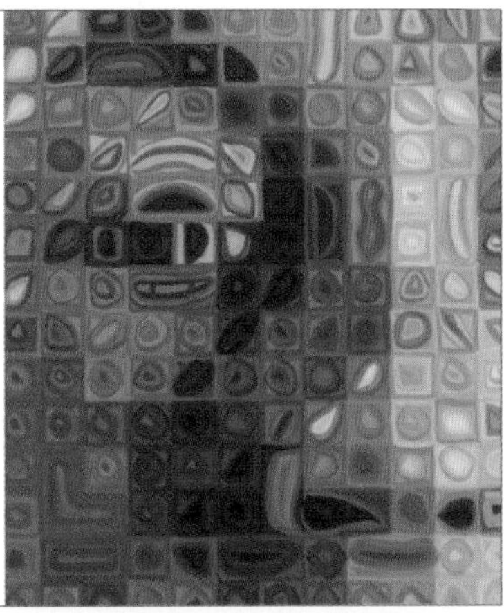
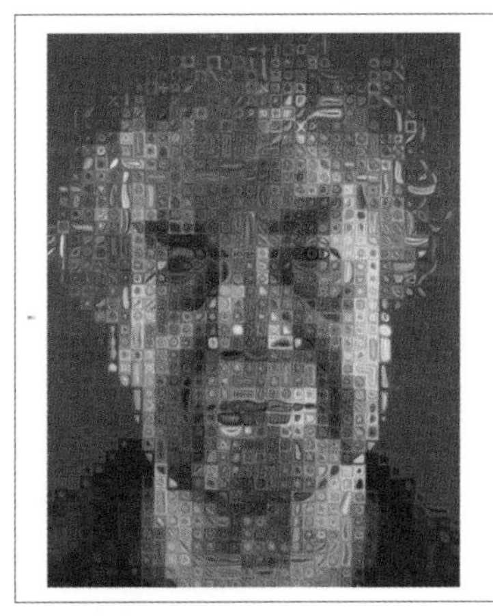

本页左上图：展开的书名页中是放大的眼睛，从中可以捕捉到艺术家的神韵，标题同样非常简洁

本页右上图：在这页里，主要内容及标题明显使用了网格

本页中间的两张图：同样，网格主宰了整个页面

88.改变边界

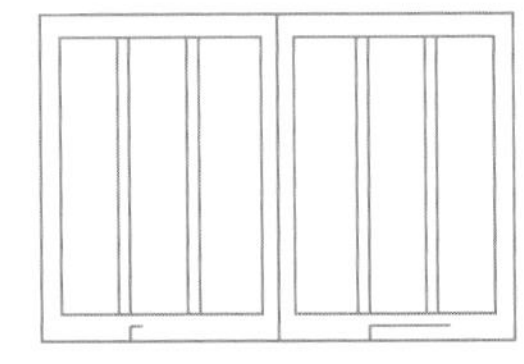

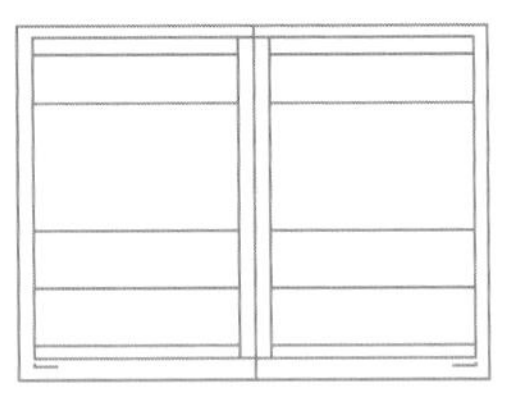

辅助材料也可以设计得像主要文本一样漂亮，从而模糊主要文本和辅助材料之间的边界。辅助材料是指一本书或一个产品名录末尾的内容，如附录、大事记、注释、书目和索引等，这些材料可能很复杂。细节的设计决定了整个项目的设计是否完整，例如，那些常被忽略的页面是否设计得美观、清楚。

项目
"向我展示泰国"（*Show Me Thai*）展览目录

客户
泰国文化部当代艺术与文化办公室（Office of Contemporary Art and Culture, Ministry of Culture, Thailand）

设计公司
实践工作室/泰国（Practical Studio/Thailand）

设计总监
桑提•拉拉查韦（Santi Lawrachawee）

平面设计师
埃卡拉克•潘帕纳瓦特（Ekaluck Peanpanawate），蒙柴•桑提维（Montchai Suntives）

这个展览目录包含许多有用的网格，其中对参与者名录进行了有趣的处理

Appendix C ::
Exhibitions

Year	Exhibition
2006	The 74th Exhibition of the Japan Print Association, 2006 Tokyo Metropolitan Art Museum Ueno, Japan (Group)
2005	Kyoto Hanga 2005 International Print Exhibition Japan and Thailand, Japan
	YOKOHAMA 2005 International Triennale of Contemporary Art: Art Circus, Yokohama, Japan (Group)
	"FT3" The 3rd Fukuoka Asian Art Triennale 2005 : Parallel Realities : Asian Art Now", Fukuoka Asian Art Museum Japan (Group)
	International Print Exhibition, Japan and Thailand, Kyoto City Museum, Japan (Group)
	Fourth Prize : The 6th Kochi International Triennial Exhibition of Prints, Japan (Award)
	The 6th Kochi International Triennial Exhibition of Prints, Japan (Group)
2004	Friendship Through the Test of Time : At Memories of the MONBUSHO Scholarship Alumni and Japanese Friends, The National Gallery, Chao-Fah Road, Bangkok (Group)
2003	Traffic, Fukuoka Art Museum, Kitakyushu Municipal Museum of Art, Fukuoka and Kitakyushu, Japan (Group)
	First Prize : "A solar eclipse 2002" The International Print and Drawing Exhibition Silpakorn University, Bangkok
	Wind of Artist in Residence 2003, Fukuoka Asian Art Museum, Japan
	Print Exhibition Tokyo 2003, at the Art Gallery of the Faculty of Art, Tama Art University, Tokyo, Japan (Group)
	"Happiness : A Survival Guide for Art and Life," Mori Art Museum, Tokyo, Japan (Group)
2002	The 2nd Fukuoka Asian Art Triennale 2002, Theme 'Imagined Workshop' Fukuoka Asian Art Museum, Fukuoka, Japan (Group)
	1st Art Exhibition "Spirit of Thai and Japanese Cultural Exchange 2002"
2002	3rd International Mini Print Triennial 2002, Tama Art University Museum, Tokyo, Japan (Participated Artist)
	5th Kochi International Triennial Exhibition of Prints, Kochi, Japan (Participated Artist)
	15 Tracks : Contemporary Southeast Asian Art, Tama University, Tokyo and Fukuoka Asian Art Museum, Fukuoka, Japan
	"Bangkok in Pink, the Gallery, Yokohama Museum of Art, Japan (Solo)
2001	International Print Triennial in Kanagawa 2001, Japan (Participated Artist)
	Art and Culture Prize, The 12th Fukuoka Asian Culture Prizes 2001, Japan
2000	Memorial Exhibition for Montien Boonma, Fukuoka Asian Art Museum, Fukuoka (Solo)
1999	The 1st Fukuoka Asian Art Triennale 1999, Theme 'Communication: Channels for Hope', Fukuoka Asian Art Museum, Japan (Group)
	Contemporary Art Exchange Exhibition, (Vietnam Thailand Japan), Japan
	4th Kochi International Triennial Exhibition of Print, Kochi, Japan (Participated Artist)
1998	Master Thai Artist in Japan, Art Gallery, the Faculty of Painting, Sculpture and Graphic Arts, Silpakorn University, Bangkok

上页图（上）：一张留白较多的照片与高度网格化排列的照片形成对比

上页图（下）：左侧页面上的文本宽度与右侧页面上两张图片的宽度之和相同。一般不建议把文本排列得过宽，但这个版面的设计效果很好

三栏网格和图表巧妙地产生了一种秩序感

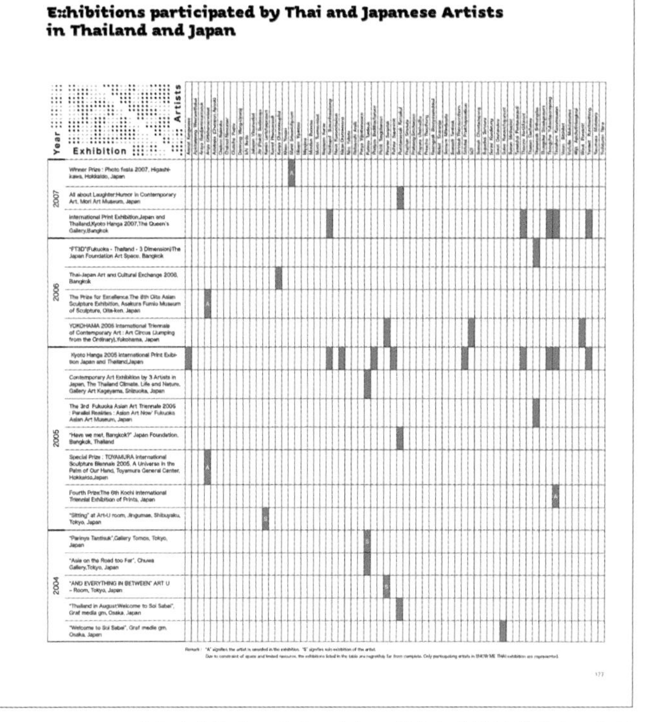

这组表格清晰、美观、有趣，并带有装饰性的图案，为页面增添了质感

● 89.使用模块

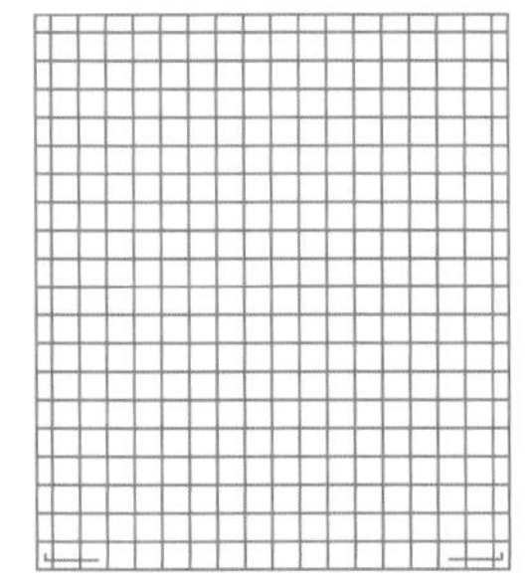

另见21页

如果把设计的步骤分解一下，即使是很难完成的设计也能做得很好。色彩可以创造形状和空间，逐渐减弱的色彩本质上就是一个“负空间”。主色调可以成为前景的一部分，不同部分的叠加可以为整个作品创建另一个维度。尝试使用不同的层次和几何形状，最终的效果可能也会符合黄金比例。

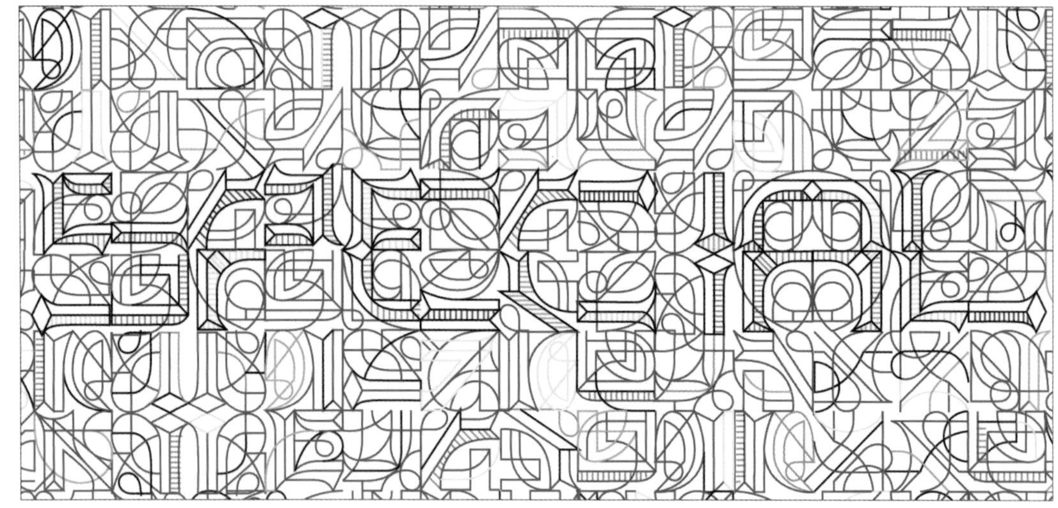

这个网格也是一个字谜，这个字谜经过玛丽安•班杰斯娴熟的设计变得更有深度。玛丽安•班杰斯喜欢“挑战那些我知道的规则，尝试做不熟悉但是效果好的东西”

项目
《卫报》G2（*The Guardian's G2*）字谜特刊的封面

客户
卫报传媒集团（The Guardian Media Group）

设计师
玛丽安•班杰斯

这个封面使用了层层叠叠的线条与方块

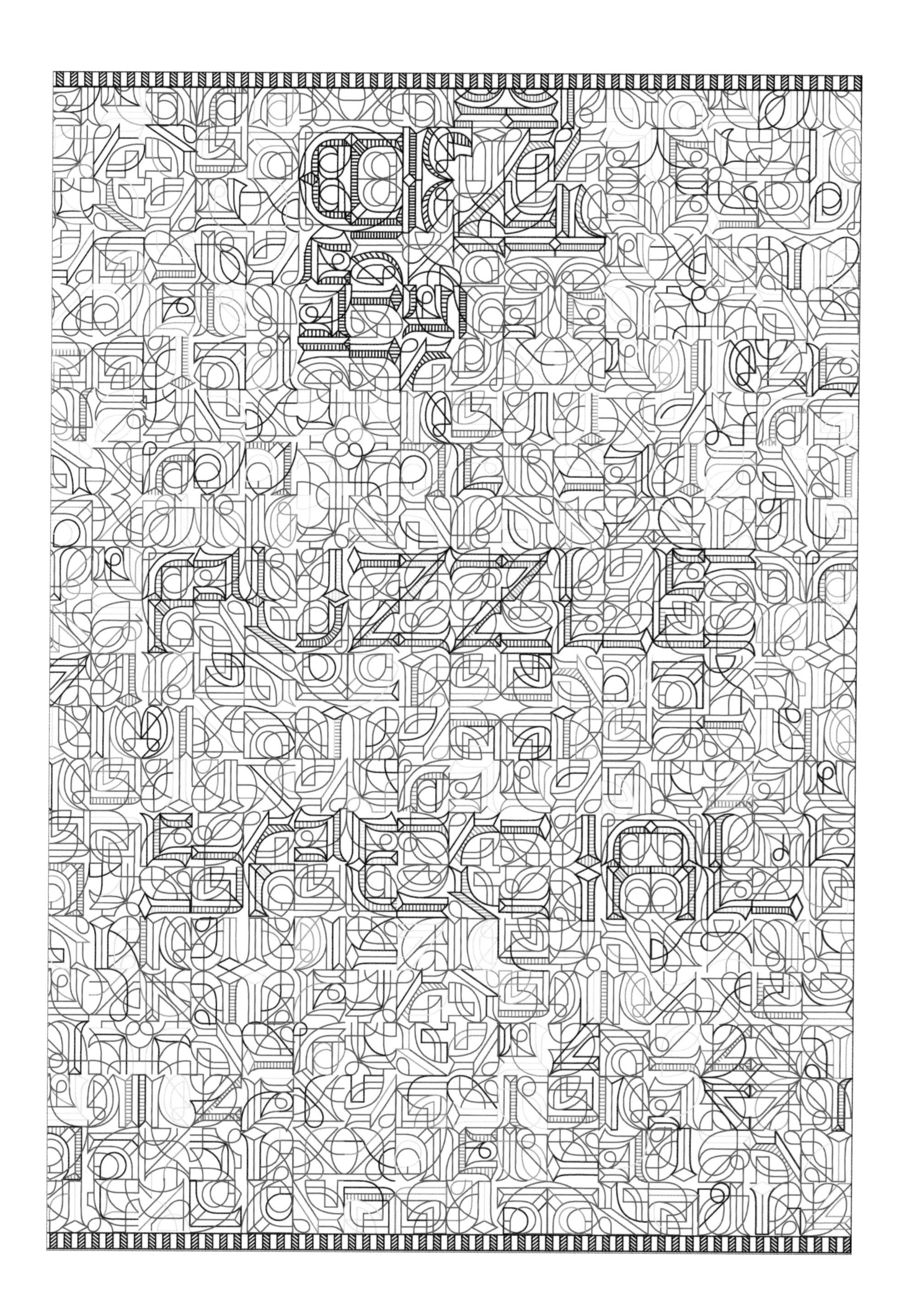

● 90.多维度设计

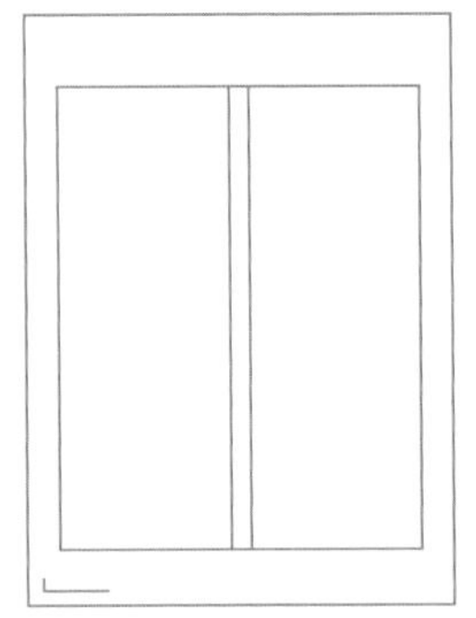

尽管大多数使用网格设计的版式都是平面的，但无论是在印刷的页面上还是在计算机屏幕上，设计师都需要捕捉作品载体的维度。一本宣传册的设计要采用自己的模式，而不是书、菜单或其他模式。以三维的角度构思，在二维的平面上设计，该宣传册采用了手风琴形的折叠形式，从而获得了另外一个维度——深度。

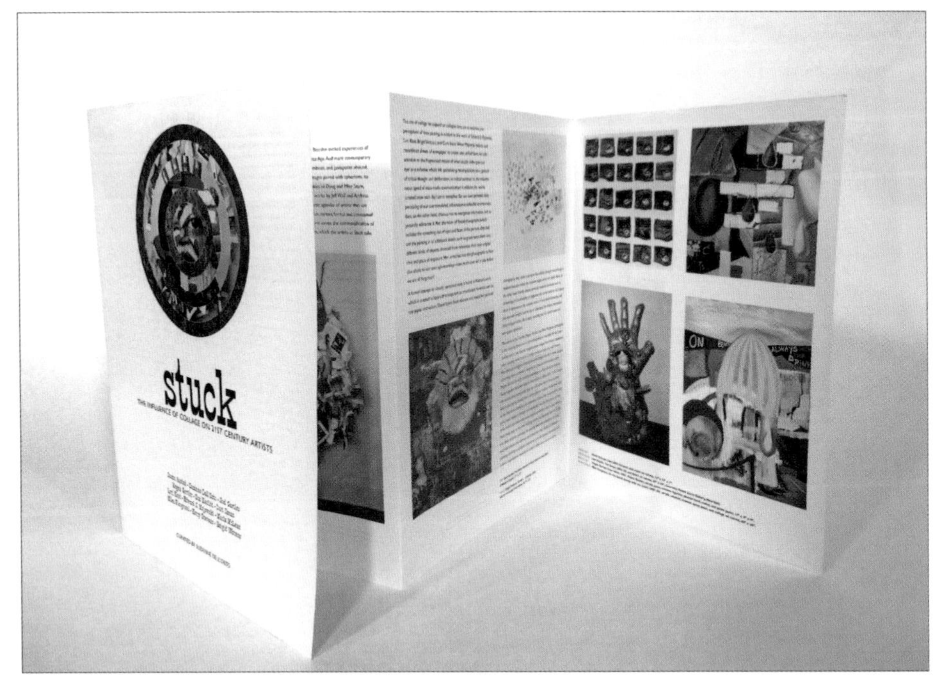

stuck
THE INFLUENCE OF COLLAGE ON 21ST CENTURY ARTISTS

项目
《stuck展览目录》（一个以拼贴画为主题的艺术展览）

客户
莫洛伊学院（Molloy College）

美术馆馆长
约兰德•特林塞博士（Dr. Yolande Trincere）

策展人
苏珊娜•戴尔•奥尔托

设计师
苏珊娜•戴尔•奥尔托

这本拼贴画展览宣传册被巧妙地设计为折叠式，让人想起美术馆展览中一些有趣的艺术品

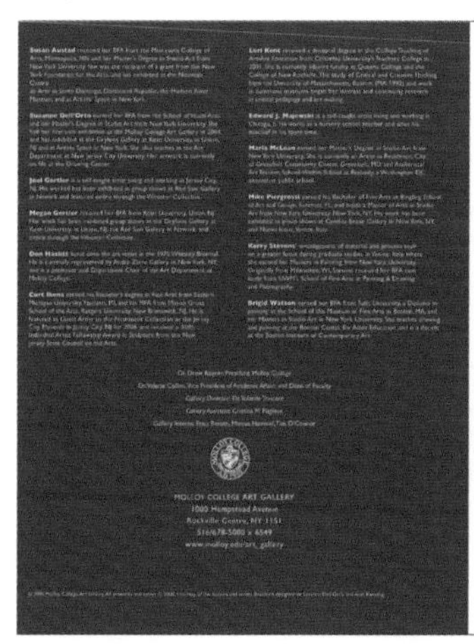
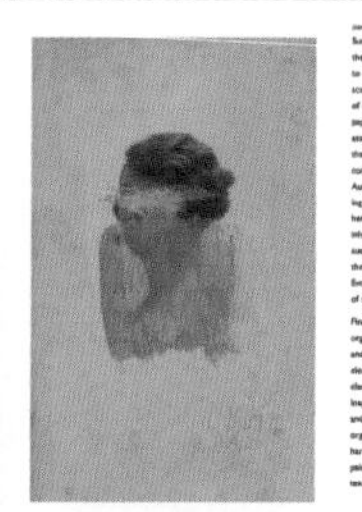

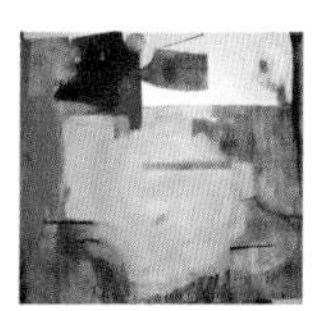

stuck
THE INFLUENCE OF COLLAGE ON 21ST CENTURY ARTISTS

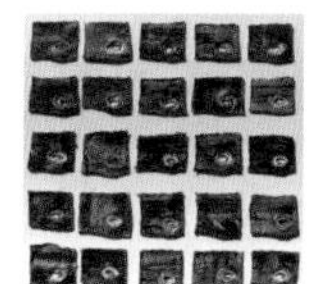
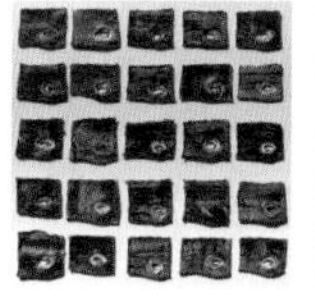
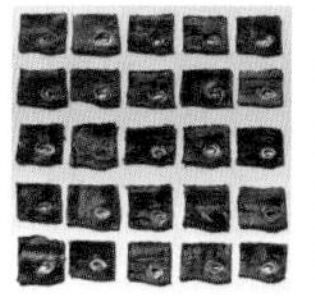

传统的网格为展览中各种各样离奇古怪的拼贴画提供了支撑。一板一眼的字体与精心规划的留白结合，为鲜活的艺术品提供了框架。上图是宣传册的封面和封底，下图是宣传册的内页。这本折叠式小册子采用两面印刷，呈现出一种立体的感觉

上页图： 宣传册共有四个折页的内容，其中一页展示了一本关于解读艺术史的书籍，这部分内容被整齐地放在其中一栏里。庄重的Gill Sans字体和诙谐的P. T. Barnum字体组合在一起，让人想起拼贴画中的并列元素

● 91.着眼全局

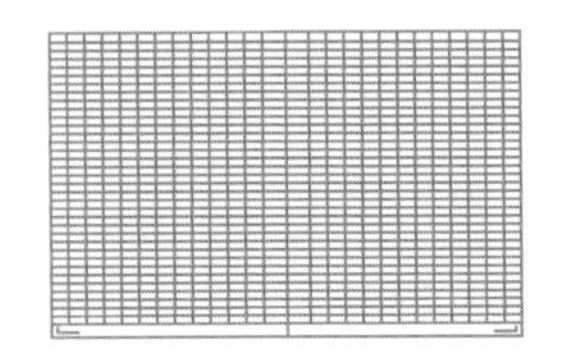

网格的框架可以包含许多叠加的元素。图层叠加时须记住：

◎ 信息类内容的排版需要保证可读性；
◎ 开放的空间对构图来说至关重要；
◎ 没必要填满每个空间。

从表面上看，增加层次可以激发读者的兴趣。在更深层次上，它会帮助你思考如何组合不同的元素。

本页的两张图： 层次增加了版面的立体感，同时又保证了封面和内页的信息很清晰

项目
哥伦比亚大学地球研究所品牌海报

客户
哥伦比亚大学地球研究所

创意总监
马克•英格利斯

设计师
约翰•斯蒂斯洛（John Stislow）

插画师
马克•英格利斯

在这个项目里，富有层次的照片、线条插图和图标增加了作品的深度，并暗示了内容的多层意义，也增加了作品的趣味性

叠加在照片上的元素和透明区域增强了三栏网格的版式效果

在这张讨论复杂健康问题的海报上，设计师只在最上面的图层做了排版设计

● 92.让框架支持所有媒介

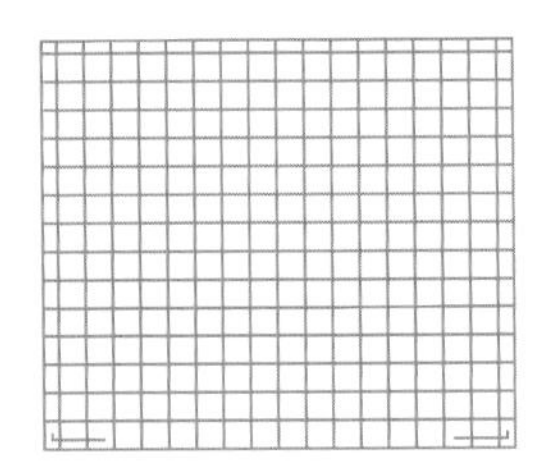

在以字节为单位的空间中，分区和颜色都可以传递信息。有时，使用与客户相关的元素可以帮助我们确定颜色和形式。不同类别的信息可以放在不同的框或相邻的框里，通过导航栏直达网站的各个页面。一个信息密集的网站就像一个大都市，规划整齐但拥挤繁忙，有时，令人头晕目眩的旅程只是开始。

黑色的标题栏和与出租车一样的黄色方框构成了网站的标志性外观

项目
设计出租车（Design Taxi）网页

客户
设计出租车

设计公司
设计出租车

设计总监
亚历克斯•戈（Alex Goh）

来自新加坡的设计出租车公司，其网站设计就像一辆出租车，带用户穿梭在一个个网格之间。高密度的“数字城市”里充斥着框架、分隔线、文本框、导航栏、色彩、阴影、链接和搜索框，但没有星巴克咖啡

由于有大量的素材，所以，网站通过框架区域和不同的灰色阴影部分实现对信息的控制。这段“旅程”有时会有点颠簸，要找到与超链接相对应的标题可能有点困难

版式是为了功能而设计的，而不是为了美观。这个页面的设计主要是以方便更新为目的

● 93.排版可以错落有致

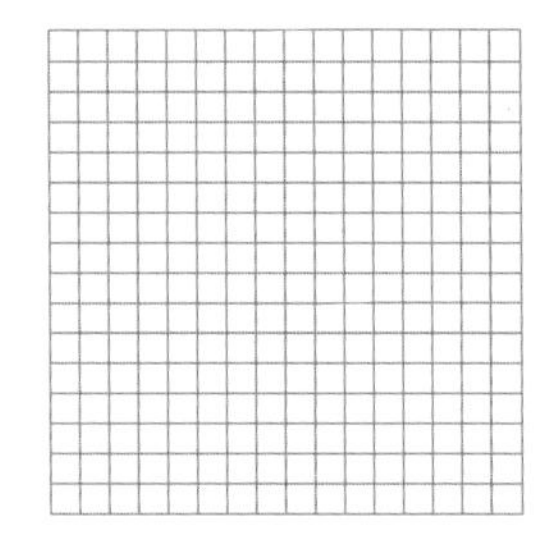

在原有模块的基础上，图形可以被翻转、复制或删除，以保证整个标志系统的新颖和前卫。

项目
圣欧拉利娅（Santa Eula-lia）品牌设计

客户
圣欧拉利娅

设计公司
马里奥·埃斯肯纳兹工作室

设计师
马里奥·埃斯肯纳兹

这是为巴塞罗那的奢侈品店圣欧拉利娅设计的标志，设计源于“X”这个图案（X是代表圣欧拉利娅的符号）

自2006年首次设计该标志以来，马里奥·埃斯肯纳兹工作室一直在原始图案上添加不同元素（数字和符号）

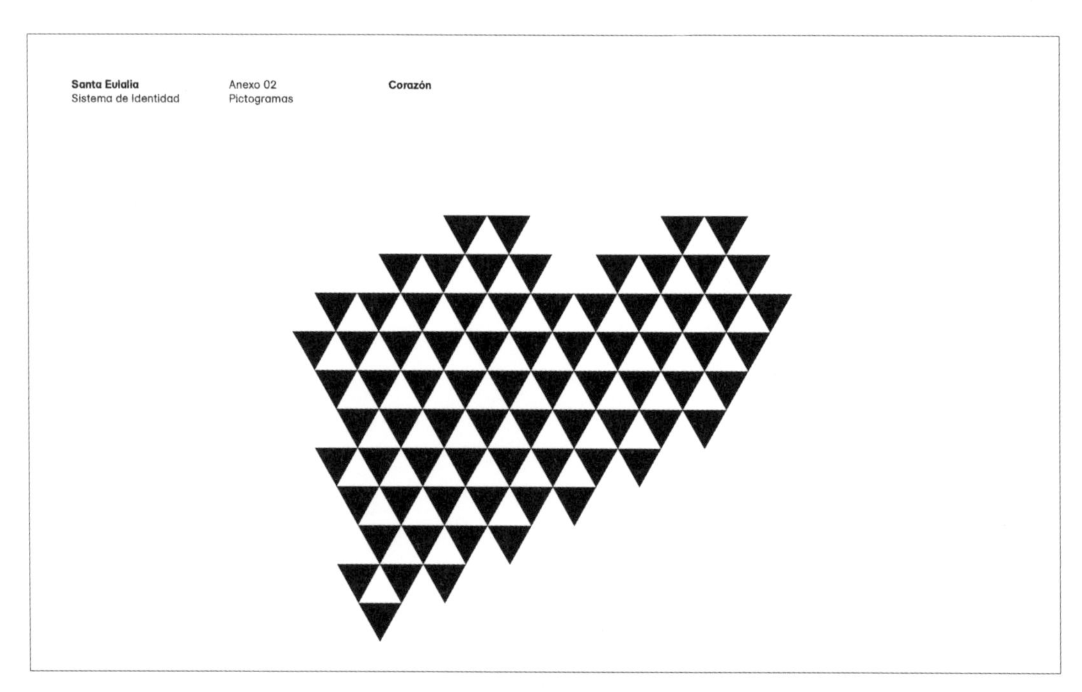

上页两张图：品牌名称和年份使用的无衬线字体和干净的网格与圣欧拉利娅风格化的“X”图案相辅相成

本页图：品牌的基础图案被拓展成了一个更加复杂的网格系统，但是仍然暗含着“X”图案

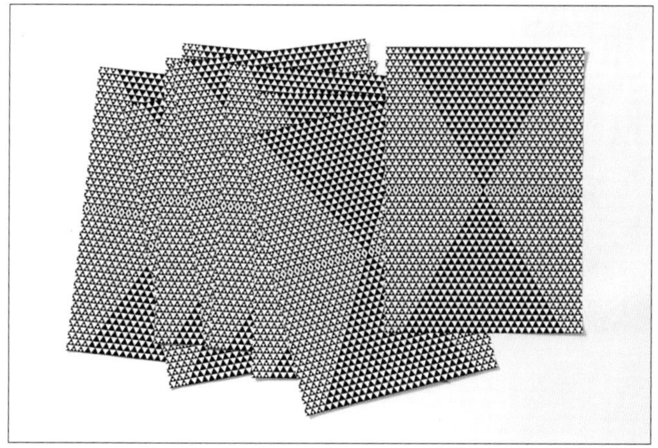

● 94.超大化

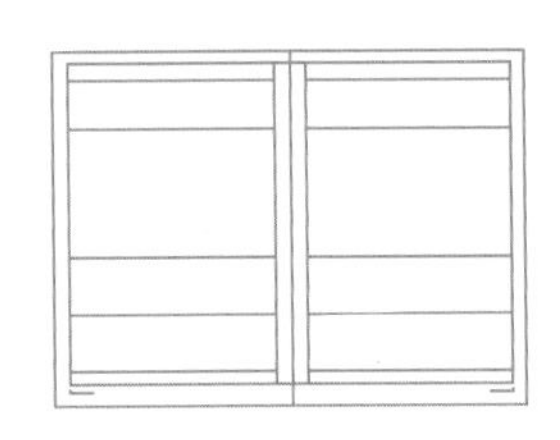

在大型的版式设计中可以使用超大图形（supergraphics），但要考虑如下问题：

◎ 改变元素的大小、粗细和色值，以产生动态的版式效果。

◎ 考虑字体的维度。

◎ 考虑动态效果。与页面上的文字相比，移动的文字需要更大的字间距才能保证清晰。

项目
彭博动态数字展示屏

客户
彭博资讯公司（Bloomberg LLP）

设计公司
Pentagram公司

艺术总监/设计师（环境制图）
保拉•舍尔（Paula Scher）

艺术总监/设计师（动态显示）
丽莎•斯特劳斯费尔德（Lisa Strausfeld）

设计师
伊•金（Jiae Kim），安德鲁•弗里曼（Andrew Freeman），里昂•伯德（Rion Byrd）

建筑设计
STUDIOS建筑公司

摄影师
彼得•毛斯/埃斯欧（Peter Mauss/Esto）

电子显示屏上的超大黑体数字图形不断滚动，将信息与品牌相结合

两页图：超大图形将内容、数字和设计风格融合在了一起

四个水平横版上的动态符号的颜色可以改变，字号的大小和字母的颜色可以随着信息的不同发生变化，在展示数据的同时，也传达出了一种观点

NIKKEI
VALUE 11276.59 % CHANGE +0.09 LAST UPDATED 4:29

Celsius
HIGH 29° LOW 16° HUMIDITY 81%
Clear

● 95.移动模块

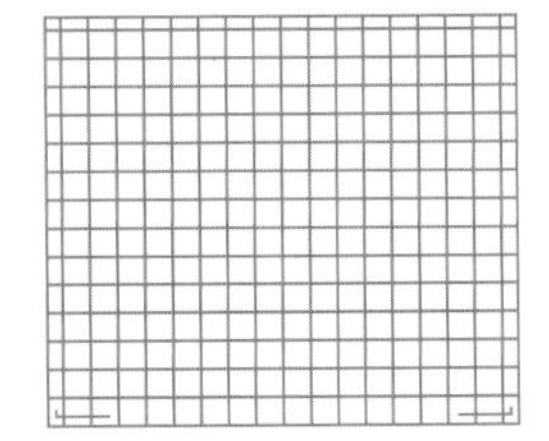

在网页上与在印刷品上一样，均等的模块可以提供多种划分内容的方式，包括能让网页更生动的视频区。

流动性

在交互设计这个新领域里，值得一提的就是流动网格和布局。当纸张尺寸已经不再重要时要怎么做？是否可以使用任意的尺寸，并将内容都集中在屏幕中心？还是根据屏幕大小重新创建流动的版式布局？网页设计专家可能更喜欢后一种做法，但这样的版式设计需要复杂的技术。

项目
哥伦比亚大学地球研究所
网页

客户
哥伦比亚大学地球研究所

创意总监
马克•英格利斯

设计
金圣熙，约翰•斯蒂斯洛

模块化的部分可以显示丰富多样的信息

本页图和上页图：
在主页上，模块设置在主导航栏下面，并且可以组合成各种配置：
• 所有横跨整个页面的模块都可以用作版头，包括链接
• 一个单独的模块表达一个主题
• 两个模块合在一起可以组成一个导航条
• 在页面一侧的模块可以组成一个长条的垂直分栏，作为发布新闻和活动的公告栏
• 模块可以包含视频

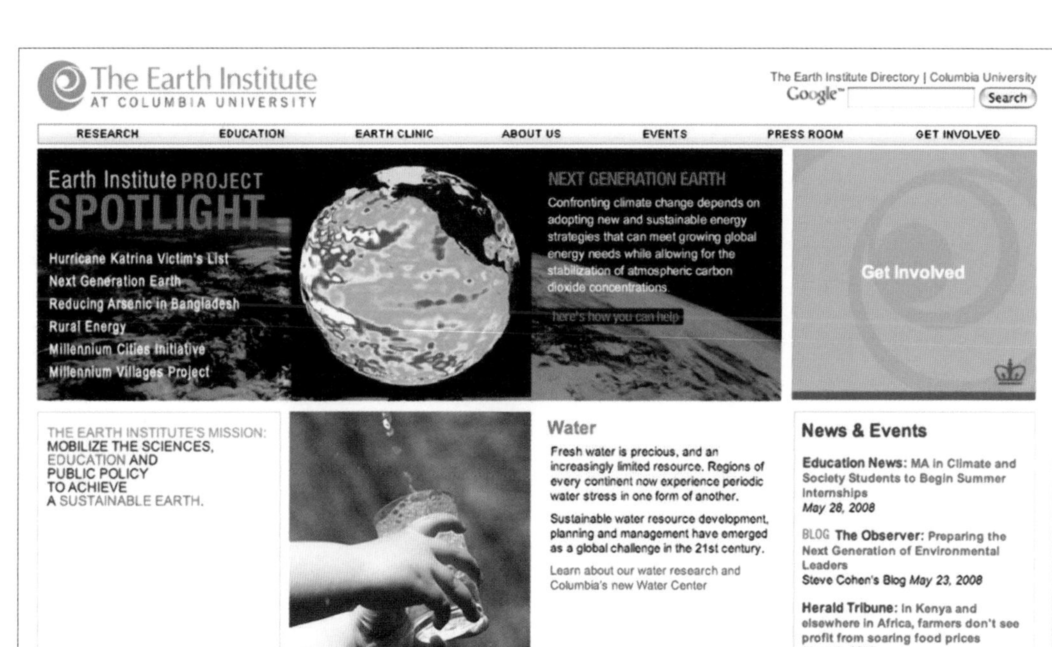

导航栏可以指引用户离开主页，给用户带来深层次的阅读体验

子页面使用模块化组织，并根据信息的需要趋向水平层级的结构

96.发挥优势

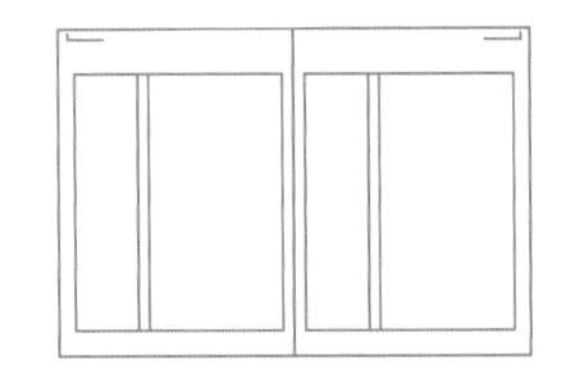

许多优秀的设计师声称他们不使用网格，但他们的设计依然具有良好的空间感、清晰的结构、明确的中心。大多数设计师是在无意识的情况下遵守了设计的基本原则，从而凸显素材并使之更清晰。

另见21页

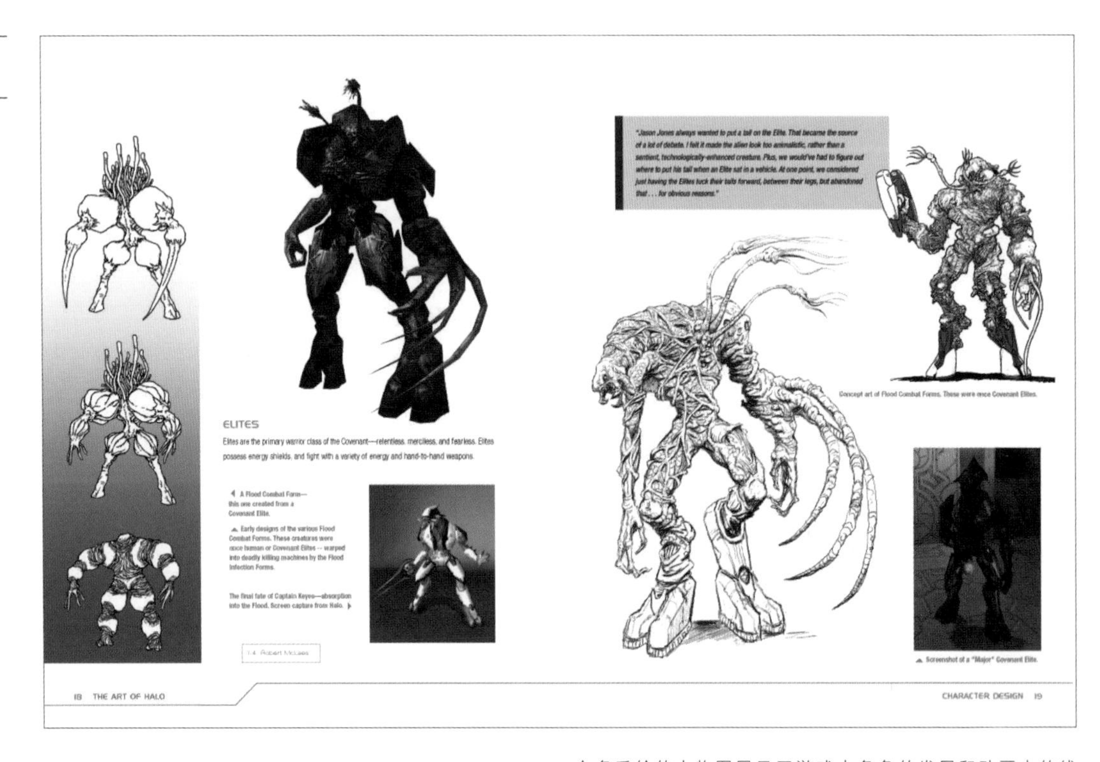

众多手绘的人物图展示了游戏中角色的发展和动画中的线索。水平方向的分隔线起到了基底线的作用，在分隔线的一端，线条的变化增添了页面的动感

项目
《光环的艺术》（*The Art of Halo*）

客户
兰登书屋（Random House）

设计师
李琳（Liney Li）

这是一本关于游戏“光环”的艺术设计书，图书的版式设计凸显了英雄角色的不死之身

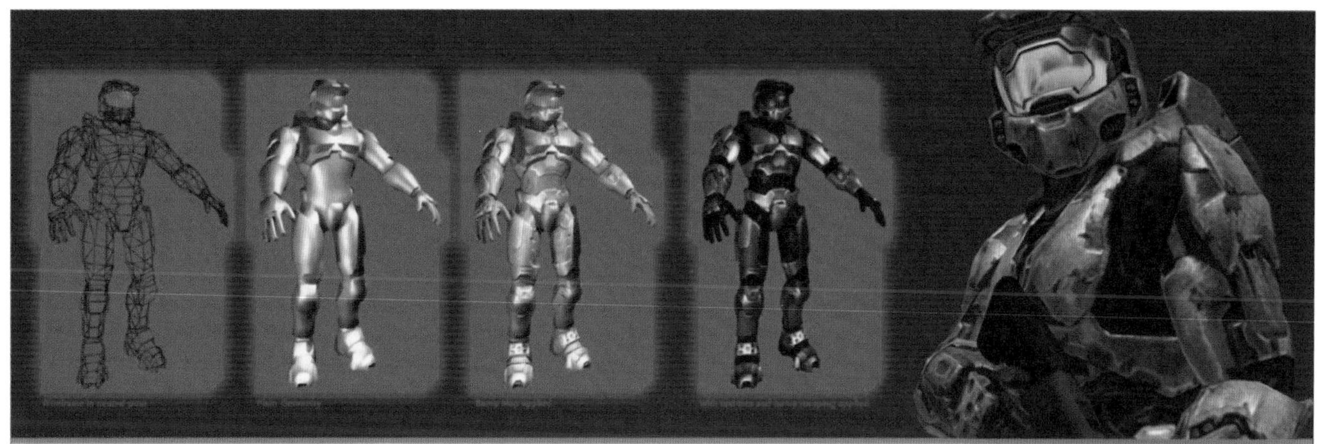

1

▲ The evolution of Master Chief from wire frame, rendering, and finally clad in his battle armor.

2

▲ One of the public's first looks at Halo came in the November, 1999 issue of Computer Gaming World.he evolution of Master Chief from wire frame, rendering, and finally clad in his battle armor.

The collaborative process at Bungie wasn't confined to the *Halo* team. There were several Bungie artists and programmers working on other titles during the various stages of *Halo*'s development. "I didn't do a lot on *Halo*—I was assigned to a team working on a different project," said character artist Juan Ramirez. "But most of us would weigh in on what we saw. I like monsters and animals and creatures—plus I'm a sculptor, so I did some sculpture designs of the early Elite.

"When I came on, I wasn't really a 'computer guy'—I was more into comics, film, that kind of thing. I try and apply that to my work here—to look at our games as more than just games. Better games equals better entertainment. A lot of that is sold through character design."

THE MASTER CHIEF

Seven feet tall, and clad in fearsome MJOLNIR Mark V battle armor, the warrior known as the Master Chief is a product of the SPARTAN Project. Trained in the art of war since childhood, he may well hold the fate of the human race in his hands.

MARCUS LEHTO, ART DIRECTOR: *"At first, Rob [artist Robt. McLees] and I were the only artists working on Halo. After that we hired Sheik [artist ShiKai Wang], who's just great from the conceptual standpoint. I'd do a preliminary version of something, then Sheik would work from that, and really enhance the concept.*

"The Master Chief design sketch that really took hold came after heavy collaboration with ShiKai. One of his sketches—this kind of manga-influenced piece, with ammo bandoliers across his chest, and a big bladed weapon on his back—really caught our imagination.

"Unfortunately, when we got that version into model form, he looked a little too slender, almost effeminate. So, I took the design and tried to make it look more like a modern tank. That's how we got to the Master Chief that appears in the game."

7

4 THE ART OF HALO

CHARACTER DESIGN 5

The Spartan was huge, easily seven feet tall. Encased in pearlescent green battle armor, the man looked like a figure from mythology—otherworldly and terrifying. Master Chief SPARTAN-117 stepped from the tube and surveyed the cryo bay. The mirrored visor on his helmet made him all the more fearsome, a faceless impassive soldier built for destruction and death.

The technician felt a pang of fear—and sorrow for the Covenant troops that would have to face this Spartan in combat.

—Excerpt from Halo: The Flood by William C. Dietz, the novelization of the game

1

2

3

4

An integral part of creating a good story is the creation of believable and interesting characters. Bungie's 3-D modelers craft designs of the various characters that appear in-game, which must then be "textured"—telling the game engine how light and shadow react with the model. From there, the models must be rigged so they can be animated. "Overlap is vital, particularly among modelers and animators," says animator William O'Brien. "We depend on each other for the final product to work—and none of us can settle. We always have to up it a notch."

"Our job is to bring the characters to life in the game," said Nathan Walpole, animation lead for *Halo 2*. "It's what we're best at. We don't use motion capture—most of us are traditional 2-D animators, so we prefer to hand-key animation. Motion capture just looks so bad when it's done poorly. We have more control over hand-keyed animation, and can produce results faster than by editing mocap."

Crafting the animations that bring life to the game characters is a painstaking process. "Usually, we start with a thumbnail sketch to build a look or feel," explained Walpole. "Then, you apply it to the 3-D model and work out the timing."

"Sometimes the timing's so off, it's hilarious," adds animator Mike Budd. "Everyone comes over and has a good laugh. Working together like we do keeps us fresh. There's such a variety of characters—human and alien. And you work on them in a matter of weeks. You're always working on something new and interesting."

▲ A pair of Grunts prepare to engage the enemt. Screen capture from Halo.

5

To design the characters' motions, the animators study virtually any source of movement for inspiration—though this can create some challenges for animator William O'Brien: "Just being surrounded by people with good senses of humor makes it easier to do your job. The drawback is, I've always had my own office. To animate a character, I often act out motions and movements; this gives you a sense of what muscle and bone actually do. But now, I have an audience. 'Hey, look at the crazy stuff Bill's doing now!' So now, I tend to do that kind of work on video, in private."

◀ Opposite page: Captions needed for illustrations 1, 2 ,3 , and 4.

2

CHARACTER DESIGN 3

这本书将古典风格与风格化的未来主义排版方式结合起来，蓝色将插图说明与正文区分开。分隔线和指向标志（箭头和诸如“左”“右”之类的单词）设置为橙色

沿着页面一侧的屏幕区域创建了侧边栏，并将各个角色分开

● 97.灵活设计

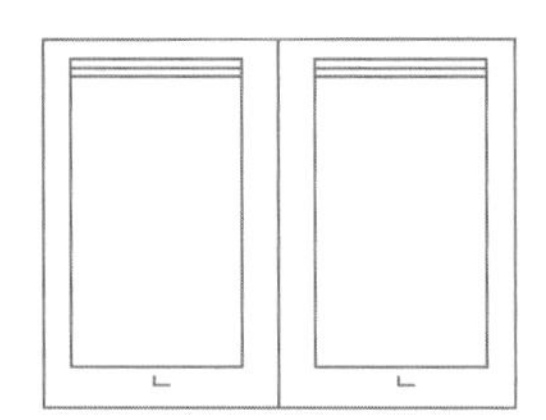

有时，设计中的一些形式，如充足的页边距、易读的字体和适当的斜体字需要先放在一边。在某些情况下，“错误”的设计可能是正确的。如果想要带给读者刺激和联想，那么最好的办法就是打破规则。

这个设计灵活，富有层次感，趣味十足

Foreword

With Design and the Elastic Mind, The Museum of Modern Art once again ventures into the field of experimental design, where innovation, functionality, aesthetics, and a deep knowledge of the human condition combine to create outstanding artifacts. MoMA has always been an advocate of design as the foremost example of modern art's ability to permeate everyday life, and several exhibitions in the history of the Museum have attempted to define major shifts in culture and behavior as represented by the objects that facilitate and signify them. Shows like Italy: The New Domestic Landscape (1972), Designs for Independent Living (1988), Mutant Materials in Contemporary Design (1995), and Workspheres (2001), to name just a few, highlighted one of design's most fundamental roles: the translation of scientific and technological revolutions into approachable objects that change people's lives and, as a consequence, the world. Design is a bridge between the abstraction of research and the tangible requirements of real life.

The state of design is strong. In this era of fast-paced innovation, designers are becoming more and more integral to the evolution of society, and design has become a paragon for a constructive and effective synthesis of thought and action. Indeed, in the past few decades, people have coped with dramatic changes in several long-standing relationships—for instance, with time, space, information, and individuality. We must contend with abrupt changes in scale, distance, and pace, and our minds and bodies need to adapt to acquire the elasticity necessary to synthesize such abundance. Designers have contributed thoughtful concepts that can provide guidance and ease as science and technology proceed in their evolution. Design not only greatly benefits business, by adding value to its products, but it also influences policy and research without ever reneging its poietic, nonideological nature—and without renouncing beauty, efficiency, vision, and sensibility, the traits that MoMA curators have privileged in selecting examples for exhibition and for the Museum's collection.

Design and the Elastic Mind celebrates creators from all over the globe—their visions, dreams, and admonitions. It comprises more than two hundred design objects and concepts that marry the most advanced scientific research with the most attentive consideration of human limitations, habits, and aspirations. The objects range from

狭窄的页边距、变异的字体、几乎消失的页码和脚注都是整体设计的一部分，其目的是在反映主题的同时吸引参观者的关注

项目
“设计和弹性思维”（*Design and the Elastic Mind*）展览

客户
现代艺术博物馆（Museum of Modern Art）

设计师
伊尔玛•布姆（Irma Boom）

封面字体设计师
丹尼尔•马列维尔德（Daniël Maarleveld）

在“设计和弹性思维”展览的目录中，设计师避开了传统形式的设计。最后页面的效果就像这个展览一样具有挑衅的感觉，甚至令人有点儿恼火

sometimes for hours, other times for minutes, using means of communication ranging from the most encrypted and syncopated to the most discursive and old-fashioned, such as talking face-to-face—or better, since even this could happen virtually, let's say nose-to-nose, at least until smells are translated into digital code and transferred to remote stations. We isolate ourselves in the middle of crowds within individual bubbles of technology, or sit alone at our computers to tune into communities of like-minded souls or to access information about esoteric topics.

Over the past twenty-five years, under the influence of such milestones as the introduction of the personal computer, the Internet, and wireless technology, we have experienced dramatic changes in several mainstays of our existence, especially our rapport with time, space, the physical nature of objects, and our own essence as individuals. In order to embrace these new degrees of freedom, whole categories of products and services have been born, from the first clocks with mechanical time-zone crowns to the most recent devices that use the Global Positioning System (GPS) to automatically update the time the moment you enter a new zone. Our options when it comes to the purchase of such products and services have multiplied, often with an emphasis on speed and automation (so much so that good old-fashioned cash and personalized transactions—the option of talking to a real person—now carry the cachet of luxury). Our mobility has increased along with our ability to communicate, and so has our capacity to influence the market with direct feedback, making us all into arbiters and opinion makers. Our idea of privacy and private property has evolved in unexpected ways, opening the door

top: James Powderly, Evan Roth, Theo Watson, and HELL. Graffiti Research Lab. **L.A.S.E.R. Tag.** Prototype. 2007. 60 mW green laser, digital projector, camera, and custom GNU software (L.A.S.E.R. Tag V1.0, using OpenFrameworks)

New forms of communication transcend scale and express a yearning to share opinions and information. This project simulates writing on a building. A camera tracks the beam painter of a laser pointer and software transmits the action to a very powerful projector.

16 Design and the Elastic Mind

17 Paola Antonelli Design and the Elastic Mind

bottom: James Powderly, Evan Roth, Theo Watson, DASK, FOXY LADY, and BENNETT4SENATE. Graffiti Research Lab. **L.A.S.E.R. Tag graffiti projection system.** Prototype. 2007. 60 mW green laser, digital projector, camera, custom GNU software (L.A.S.E.R. Tag V1.0, using OpenFrameworks), and mobile broadcast unit

for debates ranging from the value of copyright to the fear of ubiquitous surveillance.[2] Software glitches aside, we are free to journey through virtual-world platforms on the Internet. In fact, for the youngest users there is almost no difference between the world contained in the computer screen and real life, to the point that some digital metaphors, like video games, can travel backward into the physical world: At least one company, called area/code, stages "video" games on a large scale, in which real people in the roles of, say, Pac Man play out the games on city streets using mobile phones and other devices.

Design and the Elastic Mind considers these changes in behavior and need. It highlights current examples of successful design translations of disruptive scientific and technological innovations, and reflects on how the figure of the designer is changing from form giver to fundamental interpreter of an extraordinarily dynamic reality. Leading up to this volume and exhibition, in the fall of 2006 The Museum of Modern Art and the science publication Seed launched a monthly salon to bring together scientists, designers, and architects to present their work and ideas to each other. Among them were Benjamin Aranda and Chris Lasch, whose presentation immediately following such a giant of the history of science as Benoit Mandelbrot was nothing short of heroic, science photographer Felice Frankel, physicist Keith Schwab, and computational design innovator Ben Fry, to name just a few.[3] Indeed, many of the designers featured in this book are engaged in exchanges with scientists, including Michael Burton and Christopher Woebken, whose work is influenced by nanophysicist Richard A. L. Jones; Elio Caccavale, whose interlocutor is Armand Marie Leroi, a biologist from the Imperial

装订后，图片会被挡住，这在传统的版式设计中是不会出现的

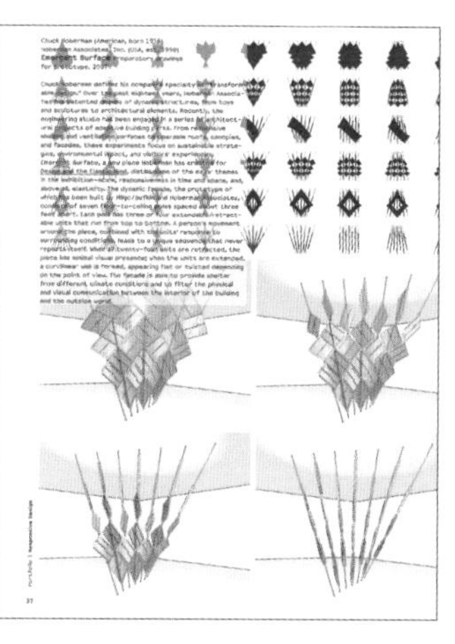

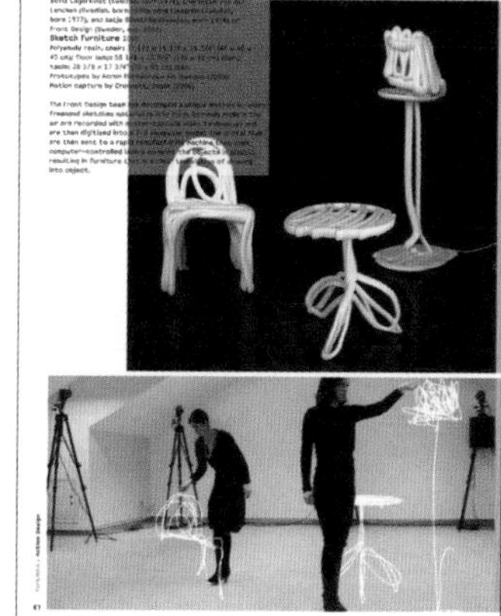

在这本展览目录中，设计师故意把文字加在了图案上

悬浮着的文本框叠加在图片上

98.追随自己的心

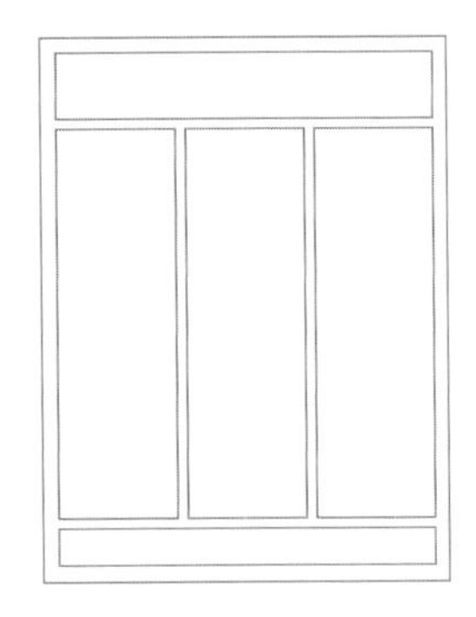

确定一个作品的网格就像解开一个谜题，有时，内容的主题和其传达的思想本身就是一个谜。有意义的设计可以激发设计师强大的潜力，让作品传达有价值的信息，滋养人们的心灵。在设计作品时，反映出对读者的热爱和对传播的理解是吸引读者的重要因素。

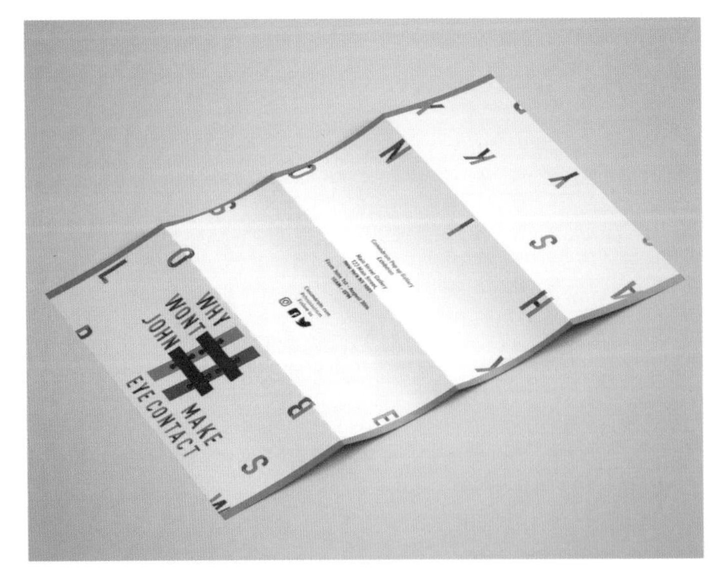

本页图： 为了平衡“谜语”品牌的标识，宣传册采用了清晰的双栏网格引导读者，蓝色的垂直分隔线在整个页面中反复出现

下页图： 网站顶端的分隔线与宣传册呼应，社交媒体平台上的内容被简化，地铁和公交车站的海报将品牌标识融入文字中

项目
“谜语”品牌系统设计

设计师
戴娜•伊普希尔（Dayna Iphill）

教授
道格拉斯•戴维斯（Douglas Davis）

“谜语”这个组织旨在提高人们对自闭症的认识。因为对那些不知道自闭症是什么的人来说，“发展性障碍”这个词可能会让他们产生困惑，所以，这个设计将一些观点以谜题的形式呈现出来。“谜语”这个组织的品牌系统包括标识、海报、宣传册、网站和社交媒体平台

C
CONUNDRUM AUTISM AWARENESS PROJECT
WHAT IS AUTISM?
GET INVOLVED
PERSONAL STORIES

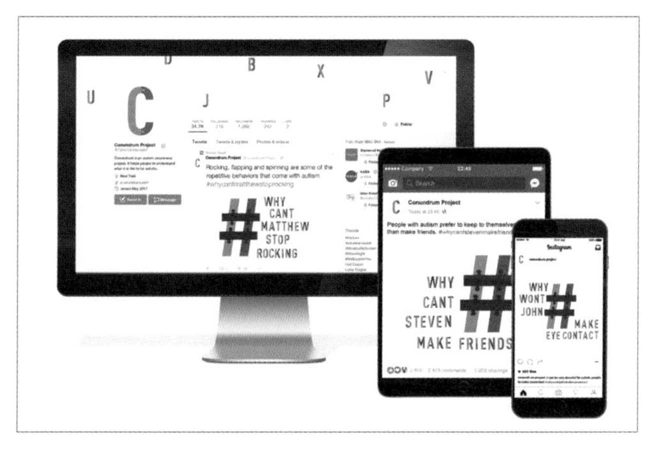

C
WHY CANT MATTHEW STOP ROCKING
WHY CANT STEVEN MAKE FRIENDS
WHY WONT JOHN MAKE EYE CONTACT

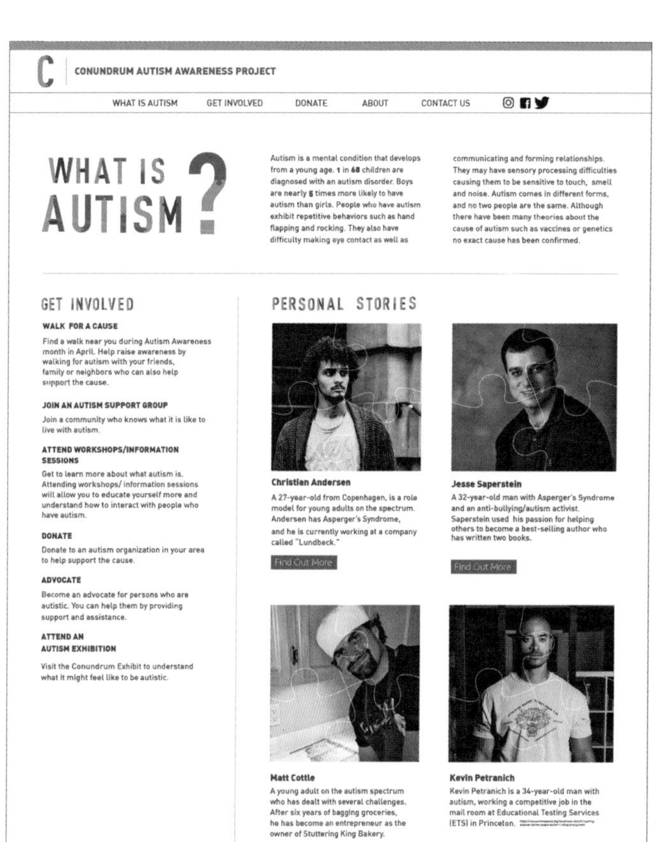

C
CONUNDRUM AUTISM AWARENESS PROJECT
WHAT IS AUTISM
GET INVOLVED
DONATE
ABOUT
CONTACT US
WHAT IS AUTISM?
Autism is a mental condition that develops from a young age. 1 in 68 children are diagnosed with an autism disorder. Boys are nearly 5 times more likely to have autism than girls. People who have autism exhibit repetitive behaviors such as hand flapping and rocking. They also have difficulty making eye contact as well as
communicating and forming relationships. They may have sensory processing difficulties causing them to be sensitive to touch, smell and noise. Autism comes in different forms, and no two people are the same. Although there have been many theories about the cause of autism such as vaccines or genetics no exact cause has been confirmed.
GET INVOLVED
WALK FOR A CAUSE
Find a walk near you during Autism Awareness month in April. Help raise awareness by walking for autism with your friends, family or neighbors who can also help support the cause.
JOIN AN AUTISM SUPPORT GROUP
Join a community who knows what it is like to live with autism.
ATTEND WORKSHOPS/INFORMATION SESSIONS
Get to learn more about what autism is. Attending workshops/ information sessions will allow you to educate yourself more and understand how to interact with people who have autism.
DONATE
Donate to an autism organization in your area to help support the cause.
ADVOCATE
Become an advocate for persons who are autistic. You can help them by providing support and assistance.
ATTEND AN AUTISM EXHIBITION
Visit the Conundrum Exhibit to understand what it might feel like to be autistic.
PERSONAL STORIES
Christian Andersen
A 27-year-old from Copenhagen, is a role model for young adults on the spectrum. Andersen has Asperger's Syndrome, and he is currently working at a company called "Lundbeck."
Find Out More
Jesse Saperstein
A 32-year-old man with Asperger's Syndrome and an anti-bullying/autism activist. Saperstein used his passion for helping others to become a best-selling author who has written two books.
Find Out More
Matt Cottle
A young adult on the autism spectrum who has dealt with several challenges. After six years of bagging groceries, he has become an entrepreneur as the owner of Stuttering King Bakery.
Find Out More
Kevin Petranich
Kevin Petranich is a 34-year-old man with autism, working a competitive job in the mail room at Educational Testing Services (ETS) in Princeton.
Find Out More
WHAT IS AUTISM
GET INVOLVED
DONATE
ABOUT
CONTACT US

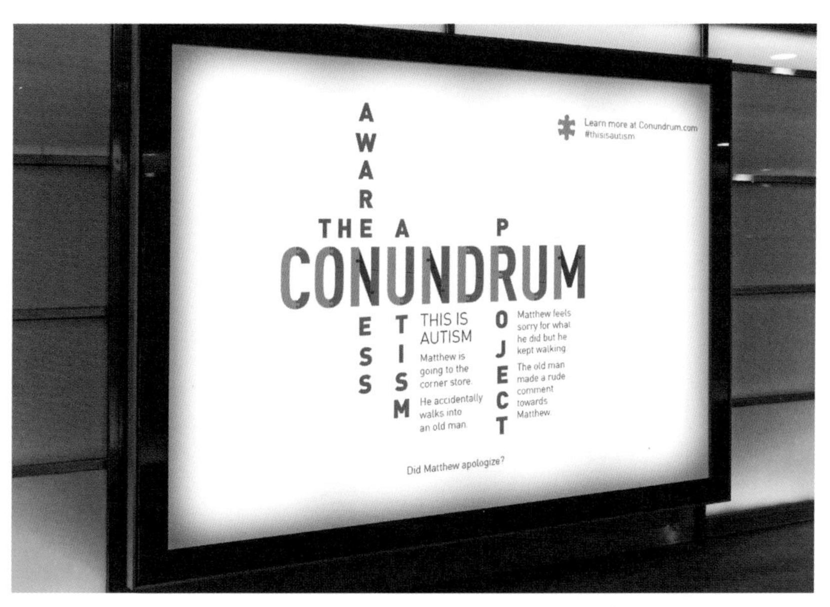

Learn more at Conundrum.com
#thisisautism
AWARENESS
THE CONUNDRUM
AUTISM
PROJECT
THIS IS AUTISM
Matthew is going to the corner store.
He accidentally walks into an old man.
Matthew feels sorry for what he did but he kept walking
The old man made a rude comment towards Matthew
Did Matthew apologize?

THE
CONUNDRUM
AUTISM
AWARENESS
PROJECT
THIS IS AUTISM

● 99.忘记规则

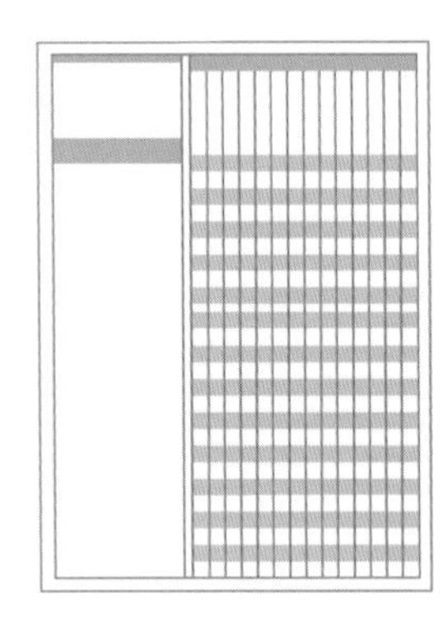

本书展示了使用网格系统的一系列规则，关于排版、空间和色彩等。如前所述，版式设计的主要规则是确保网格系统能将设计与素材联系起来，保证信息的层次结构清晰，并要注意排版风格，无论是经典、清晰的风格，还是活泼的混搭风格。在版式设计上，匠人精神很重要，而且，设计要保持平衡和一致。

在学习这本书中的规则的同时，也要有自己的思考。虽然了解规则很重要，但偶尔也要打破规则。没有哪一本书或哪一个网站可以教你所有知识。你应该认真观察、勤于提问，善于向他人学习，必要时请求帮助，并保持幽默感。学习的方式要灵活，并持之以恒，不断练习。一个设计的成功依赖于反复的修改，要享受修改的过程。欢迎向我提问。

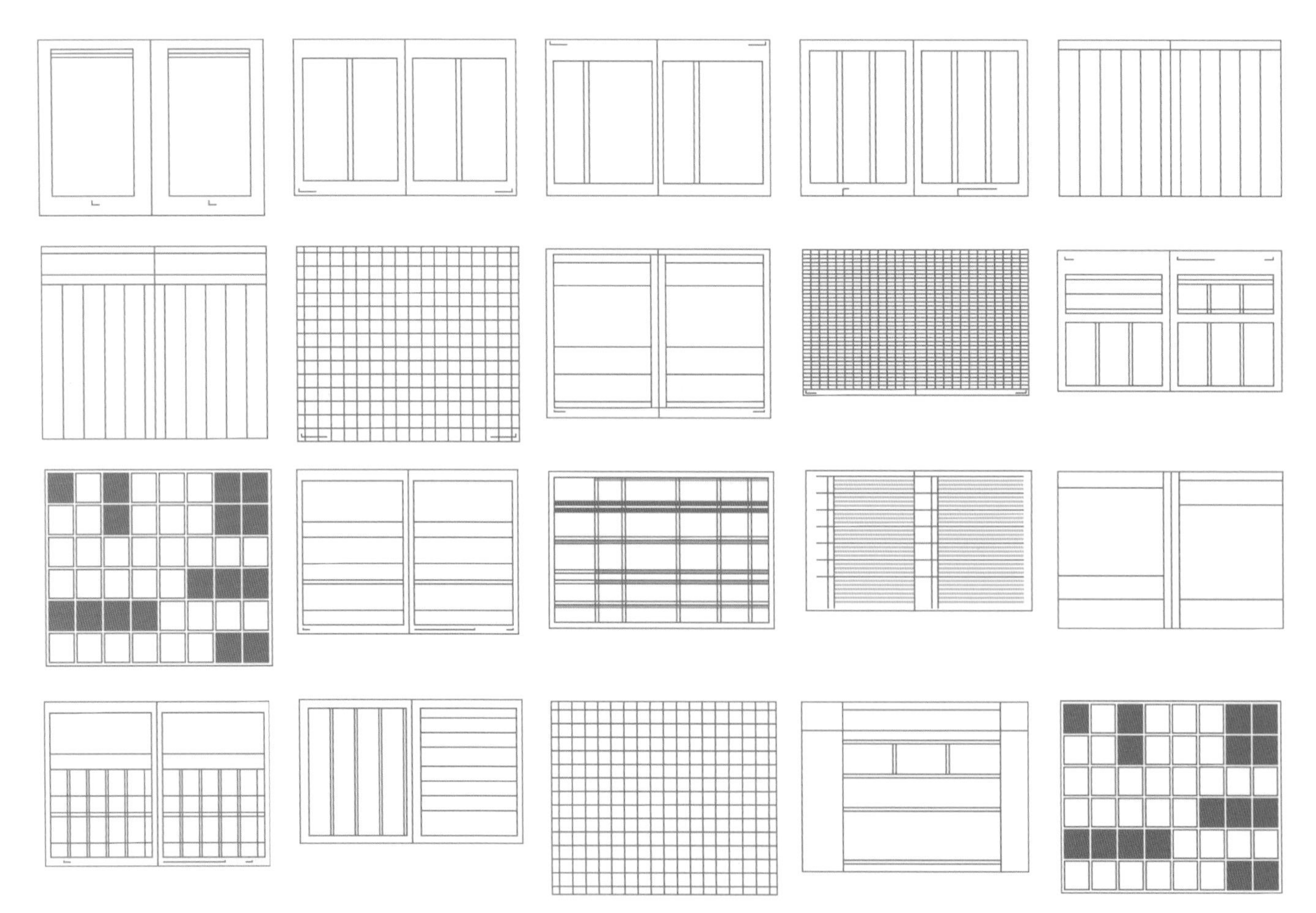

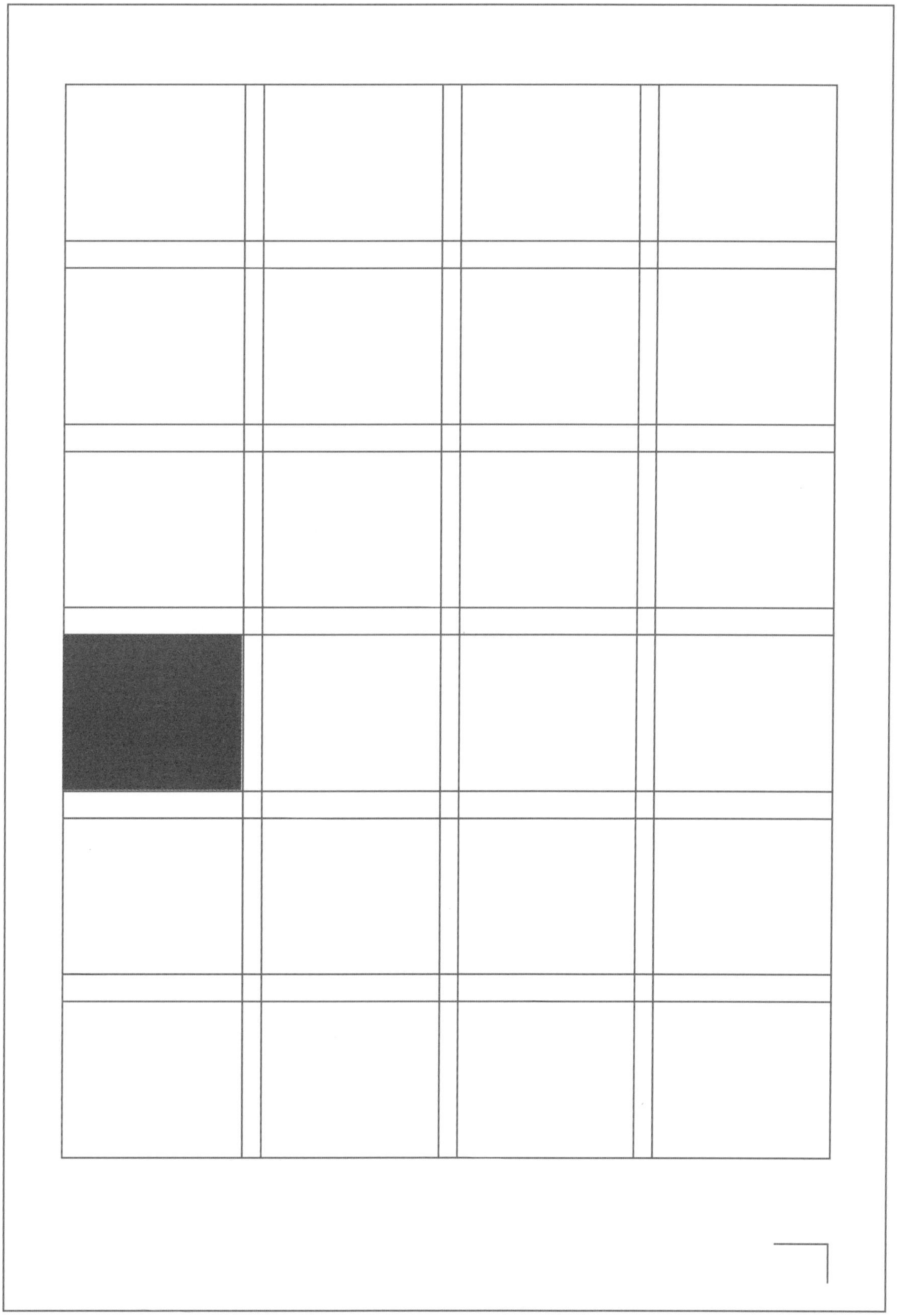

术语表

版式设计中的术语精选

附加材料（Back Matter）：支持性材料，不属于正文的内容，包括附录、注释、书目、术语表和索引等。

CMYK：青色（cyan）、品红（magenta）、黄色（yellow）和黑色（black），这四种颜色在全彩印刷中使用。

栏（Column）：垂直的框架，用于放置文字或图片。栏中的文本通常是水平的。

标题组（Deck）：类似于标语。

左对齐（Flush Left）：与左侧边距垂直对齐的文本，其右侧边距的宽度不等，但变化幅度不大。不整齐的边距也被称为"行尾不齐（ragged）"。

右对齐（Flush Right）：与右侧边距垂直对齐的文本，其左侧边距的宽度不等。

字型（Font）：在数字印刷中，字型是一种字体的风格，用于排版。字型经常与字体（typeface）互换使用(但经常用错)。把字型想象成产品，它是一种形式与大小相同的文字的完整组合。

前辅文（Front Matter）：一本书中正文之前的内容，如扉页、版权页、目录等。

JPEG：联合摄影专家组（Joint Photographic Experts Group）的简称。一种压缩格式，用于网页上使用的图片，不适合传统印刷。

两端对齐（Justify）：使栏中的文本左右对齐。

布局（Layout）：印刷品或屏幕上的元素（如字体和视觉元素）的排列。

版头（Masthead）：载有与出版物有关的人员名单及其职务，还包含出版物的信息。

负空间（Negative Space）：图形与实物之间的空间，主要出现在美术、雕塑或音乐相关的内容中。

孤行（Orphan）：段落的第一行或最后一行与该段的其余部分分开，单独位于页面或栏的底部或顶部。

胶装（Perfect Binding）：一种使用胶水黏合的装订技术。印刷帖码的边缘被粘上，加上书封，然后将封装好的书其余三面修剪干净。

派卡（Pica）：文字大小的计量单位。1派卡等于12点。在印刷中，1派卡等于1/6英寸（1英寸相当于2.54厘米）。

像素（Pixel）：一个正方形的点，表示计算机屏幕上显示的最小单位(代表图片的基本元素)。

点（Point）：排版中的一种计量单位。1派卡中有12个点，1英寸中大约有72个点。

RGB：红色（red）、绿色（green）、蓝色（blue），电脑显示器上的颜色。Photoshop软件在扫描时以RGB的形式表现图像。对于大多数的卷筒化印刷来说，图片必须是CMYK的TIFF格式。

页首标题（Running Head）：在页面顶部的标题，表示材料的内容和位置。页首标题可以包含页码，栏外脚注与在页脚处放置的内容是相同的。

正文（Running Text）：结构紧凑的文字，通常不被标题、表格、插图等打断。

骑马订装订（Saddle Stitched）：用金属钉（类似于订书钉）装订。

饱和色（Saturated）：含有极少量灰色的颜色，是一种强烈的颜色。随着饱和度的提高，灰度值逐渐降低。

轮廓图（Silhouette）：一种背景被清除，只留下一个图形或物体的图像。

下沉（Sink）：也被称作章头空白，指页面最上面的空白。

准则（Spec）：全称是specification，指排版的说明，现在通常表示页面版式的风格和功能。

叠印（Surprint）：一层墨水盖在另一层墨水上。

标语（Tagline）：从文本中摘录的语录或几行文字。

TIFF：标记图片文件格式（Tagged Image File Format）的英文缩写。一种用于电子存储和传输的位图，包含灰度和彩色图片的格式。TIFF格式是传统印刷需要的图片格式。

字体（Typeface）：具有特定特征的一种文字体式。字体有共同的特征，一种字体可以有斜体、大写和不同粗细的设计。

排版（Typography）：版面的样式、排列或者说外观，是一种选择和设计样式的艺术。

卷筒纸胶印（Web Offset）：使用卷筒纸在印刷机上印刷（印刷厂常使用“web”来指卷筒纸），将承载物上的图片印到纸上。

留白（White Space）：页面或屏幕上不包含文本或插图的空白区域。

寡行（Widow）：一行短句、一个单词或单词的一部分在段尾被分开。寡行和孤行经常被混淆。

本术语表的定义源于《芝加哥格式手册》（*The Chicago Manual of Style*）。

快速入门指南

1
评估素材

◎ 素材主题是什么？
◎ 正文内容很多吗？
◎ 有很多元素吗？有子标题吗？有页首标题吗？有图表、表格和图片吗？
◎ 编辑人员是否已经确定并标记了信息的层次结构，还是需要自己解决这个问题？
◎ 素材是否需要再创作或拍摄？
◎ 这个作品是印刷出来，还是在网上发布？

2
提前规划，了解项目规范

◎ 产品将如何印刷？
◎ 是单色、双色，还是四色？
◎ 如果材料将以传统方式印刷，最好使用300 dpi的TIFF格式。
◎ 72 dpi的JPEG格式不适合印刷，只适用于网页。
◎ 页面的实际尺寸是多少？
◎ 项目是否有指定的页数？还有增加的余地吗？
◎ 客户或印刷厂有最小页边距的要求吗？

3
选择模板、边距和字体

◎ 根据你所需的页面数量确定一个最佳模板。
◎ 如果是技术性的素材，或者页面较大，可以分成两栏或多栏。
◎ 确定间距。对初学者来说，这是最棘手的一步。给自己一些时间去尝试和犯错。记住，任何设计都需要留白，即使页面中的素材很多。
◎ 根据步骤1中评估的主题内容确定字体。只需要一种字体（可以变化字体的粗细）还是需要多种字体？
◎ 大多数计算机里都有很多默认字体，要熟悉字体和字体系列，不要总是使用流行的字体。
◎ 考虑字体大小和行间距。
◎ 经过可视化处理或者绘制草稿之后，直接将文本置于文档中，看看是否合适。

4

了解版式设计和字体排版的规则

◎ 在英文排版中，句号后只有一个空格。排版的程序不同于文字处理，最初用来模仿打字机的双空格已经是历史。

◎ 在英文段落中，如果需要中断文本以防有过多的连字符或不合适的断句，可以使用软回车。

◎ 使用拼写检查。

◎ 确保斜体和粗体是所用字体自带的斜体和粗体。即使排版程序允许你手动倾斜或加粗文字，也不要随意使用。

◎ 英文排版注意换行时出现的问题，如一个名称被分隔开，一行中有两个以上连字符，或是一行末尾的连字符后面紧跟一个长破折号。

连接号也有大作用

◎ 破折号：用于语法或叙述停顿。

◎ 一倍线：用于表示时间段或连接数字，它是半个破折号。

◎ 连字符：连接单词和短语，在行末连接单词。

5

分页的规则

分页

◎ 分页时，注意避免寡行和孤行（参见术语表）。

◎ 请参阅本书前几页中的案例，但请注意不要照搬。

◎ 印刷时，需要把字体、文档和图片一起打包。

本书参与者

黄色数字为本书中各原则的编号
本名单包括书中提到的设计机构、设计师，及未被提及但有贡献的参与机构及个人

3，23
Open，司各特•斯托厄尔

4，5，11，12，16
BTDnyc

5，25
玛莎•斯图尔特生活全媒体公司

6，61，73，77，86
安娜•图尼克

金字塔/《阶段》杂志

7，20
亚当斯•莫瑞卡公司

肖恩•亚当斯，克里斯•泰伦，诺林•莫瑞卡，莫妮卡•施劳格

8
肖恩•亚当斯

9，70，87
马克•迈尔尼克

10，12，42，50

OCD | 原创设计冠军

博比•马丁

13
弗里茨•梅兹

14，51
RADIA工作室

15，17，45
曾•西摩设计工作室

帕特里克•西摩，劳拉•霍威尔，苏珊•布拉佐夫斯基

17，65
圣约翰大教堂

18，46，75
《可颂》杂志

马场诚子

19
希维•梅塔（Heavy Meta），芭芭拉•格劳伯，希拉里•格林鲍姆

21
凯蒂•霍曼斯

22，25，35，43
芭芭拉•德怀尔德

24
皮尤慈善信托基金会

Iridium集团

26，59，78
博比•马丁

27
《大都会》杂志

克里斯威尔•拉宾

28
纽约大学医学中心

28，40，64
卡拉佩鲁齐设计公司

28，70，90
苏珊娜•戴尔•奥尔托

29，31，39，53
《每日笔记》杂志社

林修三，柳政明

30
《纽约时报》

设计总监：Tk

32，54，72，89
玛丽安•班杰斯，罗斯•米尔斯，理查德•图利

33
华伦天奴集团

罗伯特•瓦伦丁（Robert Valentine）

34
明尼苏达大学设计学院

珍妮特•艾布拉姆斯，西尔维亚•哈里斯

36
212联合公司

负责人：戴维•吉布森（David Gibson），安•川原（Ann Harakawa）

项目经理：布莱恩•斯科（Brian Sisco）

设计师：劳拉•瓦拉纳西（Laura Varacchi），朱莉•帕克（Julie Park）

插图：克里斯•格里格斯（Chris Griggs）

撰稿人：莱尔•雷克斯（Lyle Rexer）

37，38，56，58，93
马里奥•埃斯肯纳兹工作室

马里奥•埃斯肯纳兹，杰玛•维尔加斯，马克•费雷尔•维维斯，达尼•鲁比奥

41，82
Pentagram公司，艾米莉•欧本曼

合伙人

艾米莉•欧本曼，克里斯汀娜•霍根，伊丽莎白•古德斯皮德，乔伊•崔罗，安娜•梅克斯勒

42，48，62，91，95
哥伦比亚大学地球研究所

马克•英格利斯，金圣熙

44，58
纽约麦莫制作公司

道格拉斯•里卡迪

47
纽约城市中心

安德鲁•杰拉贝克，戴维•萨克斯（David Saks）

49，74
德鲁•霍奇斯，内奥米•米祖萨基

50
ThinkFilm公司

52，76
螺纹工作室

凯拉•克拉克，菲奥娜•格里夫，雷根•安德森

菲尔•凯利，德斯纳•瓦安加-肖卢姆，卡琳•吉本斯，特拉亚•内瓦

55
诺姆工作室（Noom Studio）

庞亚普•诺姆•其塔亚拉克

57
SpotCo

盖尔•安德森，弗兰克•加尔久洛，埃德尔•罗德里格斯

60
科林斯公司

布莱恩•科林斯，约翰•穆恩，迈克尔•潘吉里南

马丁公司

迈克•休斯，肖恩•莱利，雷蒙德•麦金尼,泰•哈珀

63，65
理查德•图利

65，94
Pentagram公司

保拉•舍尔，丽莎•斯特劳斯费尔德，伊•金，安德鲁•弗里曼，里昂•伯德，彼得•毛斯/埃斯欧

66
丽贝卡•罗斯

67
杰奎琳•陶设计公司

杰奎琳•陶

68
赛马•扎伊迪

69
安德里亚•德佐

71
西蒙和舒斯特公司

Picador公司

亨利•塞内•绮（Henry Sene Yee），亚当•奥尔巴赫（Adam Auerbach），朱利安•永（Julyanne Young）

79
威曼•德鲁克斯，康斯莱设计工作室

贝丝•托恩德罗

80
维格尼利联合公司

马西莫•维格尼利，达尼•皮德曼

81
INDUSTRIES文具公司

德鲁•苏扎

83
触手可及设计公司

詹妮弗•莫拉，迈克尔•塞纳托，蒂姆•袁，格温多林•霍顿

84
辛迪•海勒

85
日经出版社

88
实践工作室/泰国

桑提•拉拉查韦，埃卡拉克•潘帕纳瓦特，蒙柴•桑提维

92
设计出租车

96
李琳

97
现代艺术博物馆

伊尔玛•布姆

98
戴娜•伊普希尔

第5页的图片：
上图来自Astronomy 365，由哈里•N.艾布拉姆斯公司出版。©2006 Jerry T. Bonnel and Robert J. Nemiroff.，经授权使用。

下图来自Symbols of Power，由哈里•N.艾布拉姆斯公司出版。©2007 American Federation of Arts and Les Arts Décoratifs，经授权使用。

第6页的图片：
左图来自Symbols of Power，由哈里•N.艾布拉姆斯公司出版。©2007 American Federation of Arts and Les Arts Décoratifs，经授权使用。

第15页图片：
待补充。

第16页的图片：
图片来自Sauces，由约翰•威利父子出版社出版。©2008 James Peterson，经由约翰•威利父子出版社授权再版。

致谢

策划这样一本书是一次冒险。感谢史蒂文•海勒（Steven Heller）建议我做这项工作，还要感谢朱迪思•克西（Judith Cressy）的耐心和戴维•马丁内尔（David Martinell）所做的沟通工作。

书中介绍的许多专业人士花了很长时间收集材料，回答问题，并慷慨地允许我使用他们的项目。在此向他们致以感谢和敬意，我从他们的作品中学到了很多。

感谢唐娜•戴维（Donna David）多年前给我教学的机会，本书中使用的术语表也由唐娜提供。这是一次令人紧张的经历，让我受益良多。自始至终我都强调，平面设计是一种协作。贾尼斯•卡拉佩鲁齐为本书第一版和新版本提供了支持。小乔治•加拉斯蒂吉（George Garrastegui, Jr.）在阅读了第一版后，提供了富有洞察力的评论和建议，让新版更适合学生和相关从业者。还要感谢庞亚普•诺姆•其塔亚拉克、苏珊娜•戴尔•奥尔托，尤其要感谢帕特里夏•张（Patricia Chang），她目光敏锐、才华横溢、诚实可靠，是完美的合伙人。

还有我最爱的生活上的搭档——帕特里克•詹姆斯•奥尼尔，再次强调一遍，他是一个善良又有趣的人，对这本书来说，他是不可缺少的存在。